Lecture Notes in Computer Science 5411

Commenced Publication in 1973
Founding and Former Series Editors:
Gerhard Goos, Juris Hartmanis, and Jan van Leeuwen

Editorial Board

D0813642

Robert O. Briggs   Pedro Antunes
Gert-Jan de Vreede   Aaron S. Read (Eds.)

Groupware: Design, Implementation, and Use

14th International Workshop, CRIWG 2008
Omaha, NE, USA, September 14-18, 2008
Revised Selected Papers

 Springer

Volume Editors

Robert O. Briggs
Gert-Jan de Vreede
Aaron S. Read
University of Nebraska at Omaha, Omaha, NE 68135, USA
E-mail: {rbriggs,gdevreede}@mail.unomaha.edu;aarons.read@gmail.com

Pedro Antunes
University of Lisbon, 1749–016 Lisbon, Portugal
E-mail: paa@di.fc.ul.pt

Library of Congress Control Number: 2008942102

CR Subject Classification (1998): H.5.2, H.5.3, H.5, K.3.1, K.4.3, C.2.4

LNCS Sublibrary: SL 3 – Information Systems and Application, incl. Internet/Web
and HCI

ISSN   0302-9743
ISBN-10   3-540-92830-8 Springer Berlin Heidelberg New York
ISBN-13   978-3-540-92830-0 Springer Berlin Heidelberg New York

springer.com

© Springer-Verlag Berlin Heidelberg 2008
Printed in Germany

Typesetting: Camera-ready by author, data conversion by Scientific Publishing Services, Chennai, India
Printed on acid-free paper   SPIN: 12592296   06/3180   5 4 3 2 1 0

Preface

This volume presents the proceedings of the 14th International Workshop of Groupware (CRIWG 2008). The conference was held in Omaha, Nebraska, USA during September 14–18, 2008. Previous conferences were held in Argentina (Bariloche) in 2007, Spain (Medina del Campo) in 2006, Brazil (Porto de Galinhas) in 2005, Costa Rica (San Carlos) in 2004, France (Autrans) in 2003, Chile (La Serena) in 2002, Germany (Darmstadt) in 2001, Portugal (Madeira) in 2000, Mexico (Cancun) in 1999, Brazil (Buzios) in 1998, Spain (El Escorial) in 1997, Chile (Puerto Varas) in 1996, and Portugal (Lisbon) in 1995.

The CRIWG workshops seek to advance theoretical, experimental, and applied technical knowledge of computer supported collaboration. In the CRIWG workshops, researchers and professionals report findings, exchange experiences, and explore concepts for improving the success of people making a joint effort toward a group goal. Topics of discussion are wide ranging, encompassing all aspects of design development, deployment, and use of groupware.

CRIWG embraces both mature works that are nearly ready for publication in peer-review journals, and new, cutting-edge works in progress. A total of 30 papers were accepted for presentation this year – 20 full papers and 10 works in progress. Papers were subjected to double-blind review by at least three members of the Program Committee.

The papers are organized into 12 sessions, each on a different theme: Groupware Solutions, Co-located Groups, Groupware for Health Care, Collaborative Systems Development, Collaborative Emergency Response, Groupware Approaches, Patterns of Collaboration, ThinkLets-Based Process Design, Mobile Applications, Knowledge and Learning, Groupware Technologies, and Collaborative Modeling. Our keynote speaker for the event was Alexander Verbraeck, Chair of the Systems Engineering Department in the Faculty of Technology, Policy, and Management at Delft University of Technology in The Netherlands.

CRIWG 2008 would not have been possible without the work and support of a great number of people. We thank the members of the Program Committee for their valuable reviews and the CRIWG Steering Committee for its timely and sagacious advice and support. We extend a special acknowledgement to our sponsor organizations: The Institute for Collaboration Science, the College of Business Administration, and the College of Information Sciences and Technology at the University of Nebraska at Omaha. We owe a special debt of gratitude to our local Organizing Committee, who worked long hours to produce a fine workshop. Finally, we honor the authors and attendees for their substantial contributions that made CRIWG 2008 a valuable experience for all involved.

September 2008

Robert O. Briggs
Pedro Antunes

Organization

Conference Chair

Gert-Jan de Vreede   University of Nebraska at Omaha, USA

Program Committee Chairs

Robert O. Briggs   University of Nebraska at Omaha, USA
Pedro Antunes   University of Lisbon, Portugal

Steering Committee

Pedro Antunes   University of Lisbon, Portugal
Marcos Borges   Federal University of Rio de Janeiro, Brazil
Gert-Jan de Vreede   University of Nebraska at Omaha, USA
Jesus Favela   CICESE, Mexico
Jörg M. Haake   FernUniversität in Hagen, Germany
José A. Pino   Universidad de Chile, Chile
Carolina Salgado   Universidade Federal de Pernambuco, Brazil

Program Committee

Mark S. Ackerman   University of Michigan, USA
Marcos Borges   Universidade Federal do Rio de Janeiro, Brazil
Fabio Calefato   University of Bari, Italy
Traci Carte   University of Oklahoma, USA
César Collazos   Universidad del Cauca, Colombia
Atanasi Daradoumis   Open University of Catalonia, Spain
Bertrand David   Ecole Centrale de Lyon, France
Gert-Jan de Vreede   University of Nebraska at Omaha, USA
Dominique Decouchant   Laboratoire LIG, Grenoble, France
Alicia Díaz   Universidad Nacional de La Plata, Argentina
Yannis Dimitriadis   Universidad de Valladolid, Spain
Jesus Favela   CICESE, Mexico
Alejandro Fernández   Universidad Nacional de La Plata, Argentina
Christine Ferraris   Université de Savoie, France
Hugo Fuks   Pontifícia Universidade Católica do Rio de Janeiro,
                         Brazil
Werner Geyer   IBM T. J. Watson Research Center, USA
Eduardo Gómez-Sánchez   Universidad de Valladolid, Spain

Table of Contents

Collaborative Emergency Response

Groupware Approaches

Patterns of Collaboration

ThinkLets-Based Process Design

Mobile Applications

Knowledge and Learning

Groupware Technologies

Collaborative Modeling

Attention-Based Management of Information Flows in Synchronous Electronic Brainstorming

Antonio Ferreira[1], Valeria Herskovic[2], and Pedro Antunes[1]

[1] Department of Informatics, University of Lisbon, Portugal
{asfe,paa}@di.fc.ul.pt
[2] Department of Computer Science, Universidad de Chile, Chile
vherskov@dcc.uchile.cl

Abstract. In this paper we argue for buffering group awareness information to mitigate information overload and help users keep up with the group. We propose an attentive groupware device, called the opportunity seeker, that leverages the natural alternation between a user doing individual work and attending to the group to automatically manage the timing and quantity of information to be delivered based upon each user's state of attention. We explain how this device can be applied to synchronous electronic brainstorming and present results from a laboratory experiment, which indicate that groups produced 9.6% more ideas when compared to the immediate broadcast of ideas. In addition, a user-level post-hoc analysis suggests that information overload was attenuated with the opportunity seeker as users had 7.5 seconds of extra uninterrupted time to think about and type an idea, which they began to write 6.4 seconds sooner, and completed in 4.2 seconds less time.

1 Introduction

Attention management is an important topic in our information-rich world and is gaining momentum in the Human-Computer Interaction (HCI) field as evidenced by recent research on Attentive User Interfaces (AUI) [1,2]. The main motivation for AUI is the recognition that as the needs for information and communication rise so do the costs of not paying attention and being interrupted. So, instead of assuming the user is always focused on the entire computer screen, AUI negotiate the users' attention by establishing priorities for presenting information.

Most research on AUI is directed towards single-user work and assumes user performance degrades with the number of simultaneous requests for attention. Therefore, researchers are enhancing input/output devices so that the user remains focused on a primary task without getting too much distracted by secondary—typically unrelated and unexpected—tasks, e.g., by using eye-gaze and body orientation sensors [3], statistical models of interruptibility [4], and displays capable of showing information at various levels of detail [5].

Regarding multi-user work, the research is situated in video conferencing [6,7], making the study of AUI for groupware systems a largely unexplored area. We present three arguments to promote further investigations on this matter.

R.O. Briggs et al. (Eds.): CRIWG 2008, LNCS 5411, pp. 1–16, 2008.

Firstly, the convergence of AUI and groupware systems poses new challenges to researchers due to differences in individual and group work: a) people working in a group are more occupied with requests for attention because they have to manage more information flows; b) instead of doing a single extensive task, group members usually execute a series of intertwined tasks; c) group members have to explicitly manage the trade-offs of attending to the group and doing individual work; and d) in group work the primary and secondary tasks are typically related and may both contribute to the shared goal.

Secondly, the current emphasis of AUI applied to groupware systems is still, to the best of our knowledge, on evaluating the enhanced devices *per se* (for example, the perception of movement or sudden brightness changes [6]), in contrast with determining the outcomes of using these devices in work settings.

Thirdly, groupware researchers are designing systems that provide ever greater awareness information about the presence and actions performed by users on a group using devices such as radar views, multi-user scrollbars, and telepointers [8]. However, a problem with this trend is that it fails to recognise that sometimes more is less due to the limitations of the human attentive capacity.

Given this situation, we must consider the *group attention problem*: as the needs for collaboration rise so do the costs of not attending to the group and becoming overloaded with information.

We argue that this problem is poorly addressed by existing group awareness devices due to the lack of assumptions regarding human attention and because these devices require manual control of the type and quantity of information to be displayed, e.g., via filters, thus penalising individual performance.

This trade-off between limiting group awareness information and manual intervention by the users sets the stage for introducing an attentive device that automatically adjusts the delivery of group awareness information using a buffering technique grounded on each user's predicted state of attention. We explain how the device can address information overload in synchronous electronic brainstorming sessions and report the results of a laboratory experiment to evaluate group performance with and without the attentive device. Next, we discuss the validity of the model of user behaviour that we used for the brainstorming context and identify some limitations of this study. We conclude the paper with a summary of contributions and paths for future work.

2   Related Work

The study of AUI for groupware systems is, for the most part, an unexplored research area, with the exception of video conferencing. GAZE-2 is a system developed to facilitate the detection of who is talking to whom in remote meetings [6]. It shows video images of the users' faces on the computer screen, which can be automatically rotated by intervention of eye-trackers placed in front of each user, so that the faces appear to be staring at the user who is speaking. In this way, group turn taking may be more natural and require fewer interruptions to determine who will speak next.

Another feature of GAZE-2 is the automatic filtering of voices when multiple conversations are being held at the same time. Depending upon the user in focus, the respective audio stream is amplified, and the other streams are attenuated. If the focus of interest suddenly changes, as sensed by the eye-tracker, the audio is again adjusted. To save network bandwidth, filters are also applied to the video images by decreasing their quality as the angle of rotation increases.

The eyeView system explores the GAZE-2 ideas in the context of large meetings. It controls the size of video windows, arranged side-by-side, as well as the users' voice volumes as a function of the user's current focus of attention [7].

GAZE-2 and eyeView utilise audio and video filters to manipulate the amount of group awareness information that users are exposed to during electronic meetings. However, we found no evidence that group work benefited. Instead, the literature describes technological evaluations via user questionnaires concerning the self-subjective perception of eye-contact and distraction, as well as changes in colour and brightness during camera shifts [6].

Some studies do address the evaluation of AUI from the perspective of task execution, but are restricted to single-user activity. One study measured the effects of interruptions on task completion time, error rate, annoyance, and anxiety, and suggests that AUI should defer the presentation of peripheral information until task boundaries are reached [9]. In another study, the effectiveness and efficiency of users were evaluated as they performed two types of tasks under the exposure of four methods for coordinating interruption, and the authors recommend that AUI should let users manually negotiate their own state of availability, except when response time for handling the interruptions is critical [10].

However, as we mentioned earlier, there are numerous differences in individual and group work, which opens an opportunity for doing research on AUI for groupware systems.

3 Addressing the Group Attention Problem

To deal with the group attention problem—highlighting the need to keep users mindful of the group and mitigate information overload—we developed an attentive device for synchronous groupware systems, called the opportunity seeker, which collects group awareness information in a buffer and automatically manages the timing and quantity of information to be delivered to each user based upon his or her state of attention.

There is a trade-off in managing the delivery timing and quantity of group awareness information, in that too few updates may give the wrong impression about what the group is doing, while too many may provide up-to-date awareness information but be too distracting. We address this trade-off by leveraging the typical alternation between primary and secondary tasks in group work to find natural opportunities to interrupt the user. According to Bailey and Konstan [9] these opportunities should occur at the boundaries between consecutive tasks, i.e., for group work, at the transitions between the user doing individual work and paying attention to the group (see Fig. 1).

Fig. 1. Natural task switching during group work

Thus, regarding the delivery timing, the **opportunity seeker** only displays group awareness information to the user when s/he is likely *not* doing individual work. Concerning the limit on the quantity of information to deliver at once, the purpose is to avoid overloading the user if his or her work pace differs too greatly from the rhythm of the group.

3.1   Tackling Information Overload in Electronic Brainstorming

The rules of brainstorming [11] encourage users to do two cognitive tasks: the first is to produce as many ideas as possible because quantity is wanted; and the second is to read, or at least look at, the other users' ideas because combination and improvement of ideas is sought (cf. tasks in Fig. 1). In electronic brainstorming users can submit ideas in parallel, which puts more effort in the second cognitive task. As the number of ideas increases, e.g., because the group is inspired or group size is large, users may no longer be able to process the ideas, and may even become distracted by them, thus causing information overload.

It was for this work context that we created the first implementation of the **opportunity seeker**. The result is ABTool, or Attentive Brainstorming Tool, a

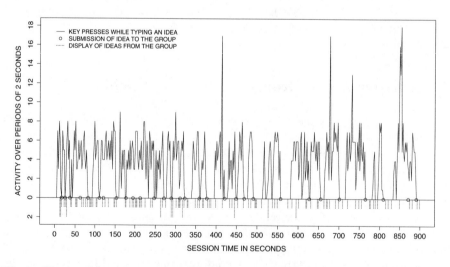

Fig. 2. User and group activity during a brainstorming session with ABTool, with immediate broadcast of ideas to everyone on the group (i.e., with the **opportunity seeker** disabled). Above the X-axis are aggregated counts of user key presses. The spikes occurred when the user pressed the delete or cursor keys. The circles on the X-axis show when the user submitted the idea s/he was typing to the group. Below the X-axis are the instants in time when the user received ideas from the other users.

custom-made tool for synchronous electronic brainstorming with built-in sensors of user performance that automatically manages the timing and quantity of ideas to be delivered to each user over a brainstorming session.

Two major challenges in applying the opportunity seeker to ABTool were to characterise how users work in a scenario with immediate broadcast of ideas to the group, and to detect task switching during electronic brainstorming activity. To this end we asked groups of five volunteers to simulate a distributed work setting by only using the tool to communicate, i.e., no face-to-face interaction was allowed. We recorded three types of events: a) user key presses while typing ideas; b) the moments when the user submitted an idea to the group; and c) the instants when group ideas were delivered to the user's computer screen.

Figure 2 shows a sample of the data we obtained and illustrates the results for an entire fifteen minute session, in which 152 ideas were produced.

From the evidence we collected three patterns of user activity emerged: a) users usually did not stop typing when they received ideas from the other users, thus, we assume they continued focused on the individual task of generating ideas; b) users typically paused after putting forward an idea, presumably to keep up with the group; and c) there were numerous periods of time with no typing activity (not shown in Fig. 2).

Based upon these three patterns, we hypothesise that a task boundary, i.e., an opportunity to display ideas from others, occurs when the user submits an idea to the group. In addition, new ideas should be delivered after a period of inactivity (currently, ten seconds) so that the user does not get the impression that the group is not producing ideas too.

Figure 3 shows the state transition diagram that models the behaviour of the user as assumed by the opportunity seeker on ABTool (also cf. Fig. 1).

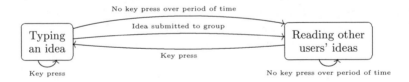

Fig. 3. Model of user behaviour assumed by the opportunity seeker on ABTool

Another feature of the opportunity seeker is that it imposes a limit on the number of ideas from others that can be displayed at once (currently, ten). This is to avoid overloading the user, e.g., by filling up the entire computer screen with new ideas, when the user is working at a slower pace than the other group members. Figure 4 shows a simulation that exemplifies the delivery of ideas with the opportunity seeker compared to the immediate broadcast of ideas.

3.2   Software Architecture and Design

Technically, ABTool is characterised by a client-server architecture, in which the server mediates the group information flows. The server also collects performance

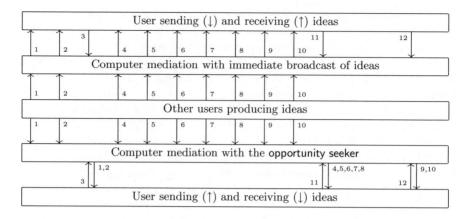

Fig. 4. Simulation of group and user activity during a brainstorming session with immediate broadcast of ideas (*upper region*) and with the **opportunity seeker** (*lower region*). In both cases the user produces three ideas (numbered 3, 11, and 12) but the exposure to the nine ideas s/he received from the other users is different. For illustration purposes, we do not show the propagation of ideas 3, 11, and 12 to the group, and limit the number of ideas delivered at once to five.

data, which are stored in an XML log. The purpose of the clients, one per user, is to receive input from the users and pass it on to the server, and to display new ideas as they become available from the server.

ABTool is written in C# and is based upon the Microsoft .NET Framework 2.0. Communication between the clients and the server is done via TCP/IP sockets and all messages (ideas, key presses, users joining or retiring the group, sessions starting or ending) are automatically serialised and deserialised using BinaryFormatter objects attached to NetworkStream instances.

Within the client and server applications, messages are propagated using events, to which consumer objects can subscribe themselves. Given that almost all classes on ABTool handle message events, namely the user interfaces, the opportunity seeker, and the classes responsible for receiving and sending messages from/to the network, we defined an IHandlesMessages interface and a default implementation for it, DefaultHandlesMessages, which relies on reflection to allow those classes to delegate the determination of the method to run as a function of the type of message associated with the event.

Figure 5 shows that the opportunity seeker derives from the AttentiveDevice generalisation, which actually implements immediate delivery of ideas from the users to the group. The OpportunitySeeker class alters this default behaviour by maintaining separate buffers, one per user, containing ideas that have been put forward by the other users. The buffer is stored in the UserNode, which also keeps a Timer object that every verificationPeriod milliseconds verifies the time of the most recent key press by the user, and if it was more than activationTimeSpan milliseconds ago, then it delivers up to ideasAtOnce ideas to the user.

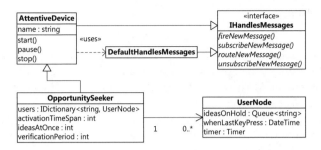

Fig. 5. Class diagram showing details of the opportunity seeker on ABTool

The AttentiveDevice and OpportunitySeeker classes implement three methods: start() is run when a session starts or resumes; pause() is executed when, for some reason, the session needs to be paused; and stop() is run at the end of a session. Other methods handle the reception and forwarding of messages, but we omitted those for brevity.

To conclude the presentation of ABTool, we show in Fig. 6 two screen shots of the client application with the opportunity seeker running.

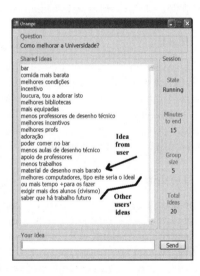

Fig. 6. Opportunity seeker managing the delivery of ideas on ABTool. *Left:* while typing an idea, the user receives no new ideas from the group. *Right:* when the user submits an idea to the group, new ideas from others are displayed.

4 Laboratory Experiment

We now describe a laboratory experiment that we set up using ABTool to test the hypothesis that group performance, measured by the number of ideas produced, improves when groups are exposed to the opportunity seeker device.

4.1 Participants

A total of 11 groups of 5 people, for a total of 55 volunteers (44 men and 11 women) participated in the experiment. The median age was 23 years (min. 20 and max. 29). 51 participants were students (40 undergraduate, 10 MSc, 1 PhD), and the remaining 4 comprised researchers, a software developer, and a translator. A convenience sampling was used to select participants, who were recruited from social contacts and posters on corridors at the University of Lisbon. No monetary reward was offered and the only information available was that the experiment would concern brainstorming.

4.2 Apparatus

The experiment was conducted in a laboratory room having five laptops with identical hardware (Intel Pentium M at 1.2 GHz, 1 GByte of RAM) and software specifications (Microsoft Windows XP SP2, .NET Framework 2.0), interconnected by a dedicated 100 Mbit/s Ethernet network. Keyboard sensitivity, desktop contents, display resolution, and brightness were controlled. Each computer had screen-recording software (ZD Soft Screen Recorder 1.4.3), and a webcamera (Creative WebCam Live!) affixed to the top of the screen. The client application of ABTool was installed on all five laptops.

4.3 Task

Participants completed practice and test tasks, both related to brainstorming. The practice task allowed participants to get familiar with ABTool. In the test task, participants were given a question and then asked to generate as many ideas as possible, by typing on the keyboard and by looking at the computer display. Speech and other forms of communication were disallowed to simulate a distributed work environment and to mitigate extraneous influences.

4.4 Design

A repeated measures design was chosen for the experiment. The independent variable was *device type* and every group of participants was under the influence of a control treatment, with immediate broadcast of ideas to the group, and an experimental treatment, with the opportunity seeker. The dependent variable, *group performance*, was calculated from the sum of the number of ideas produced by each user on the group per brainstorming session.

The order of exposure to the treatments and the brainstorming questions are depicted in Table 1. We note that, sometimes, session order is greater than two and that four questions were used, because we are reporting here a part of a larger experiment with two additional treatments, involving similar brainstorming tasks.

4.5 Procedure

A trial started when a group of participants arrived at the laboratory room. An introduction to this research was given and participants were informed on their

Table 1. Session order/brainstorming question per group and treatment. The questions were: A, how to preserve the environment; B, how to promote tourism; C, how to improve the university; and D, how to stimulate sports practice.

						Groups					
	1	2	3	4	5	6	7	8	9	10	11
Control	1/C	2/D	4/C	3/B	1/B	1/A	2/C	3/B	2/B	3/C	1/A
Experimental	3/B	1/A	2/B	4/C	3/C	2/B	3/A	1/C	1/C	2/A	3/B

privacy rights and asked to sign a consent form. Next, participants filled in an entrance questionnaire about gender, age, and occupation. Written instructions on the rules of brainstorming and on the ABTool application were then handed in to all participants and read out loud by the experimenter.

Participants were asked to carry out the practice task for 5 minutes, after which questions about ABTool were answered. The group then performed the test tasks in succession, each lasting for 15 minutes, with a brief rest period in between. At the end of the trial, answers were given to the questions participants had about this research, comments were annotated, and the experimenter gave thanks in acknowledgement of their participation in the experiment.

5 Results

Results are organised in three parts: firstly, an analysis of overall group performance, which is central to our research hypothesis; secondly, a decomposition of group performance into consecutive periods over a brainstorming session; finally, results from a post-hoc analysis based upon more fine-grained data, collected at the user level.

5.1 Group Performance

Groups produced an average of 9.6% extra ideas per session when under the exposure of the opportunity seeker than under the control treatment, totalling 1251 vs. 1141 ideas for the 11 sessions (see Table 2).

The Shapiro-Wilk normality test indicated that the normality assumption could not be accepted for both the control and experimental data distributions

Table 2. Number of ideas produced by groups under the two treatments

						Groups								
	1	2	3	4	5	6	7	8	9	10	11	Total	M	SD
Control	152	83	133	91	264	77	48	53	66	104	70	1141	103.7	62.0
Experimental	192	108	113	117	258	77	68	61	76	116	65	1251	113.7	60.8
Difference	40	25	−20	26	−6	0	20	8	10	12	−5	110	10.0	17.2

($W = 0.795$, $p = 0.008$; and $W = 0.797$, $p = 0.009$, respectively). Therefore, we applied the non-parametric Wilcoxon signed-ranks test, which revealed a 3.7% probability of chance explaining the difference in group performance, $W_+ = 45.5$, $W_- = 9.5$.

We also analysed possible confounding influences from the questions or session order on group performance to see if there was a bias introduced by popular questions or a learning effect due to the nature of the repeated measures design. We applied the Wilcoxon signed-ranks test to both scenarios, which found no significant influences: $p > 0.205$ and $p > 0.343$, respectively.

Given this evidence, we can accept the hypothesis that group performance improved when groups were exposed to the **opportunity seeker** device in electronic brainstorming tasks with **ABTool**.

5.2   Group Performance over Time

Concerning the analysis of group performance through the duration of the brainstorming sessions, we broke down the 900 seconds that each session lasted into consecutive periods of 300, 150, and 30 seconds and counted the number of ideas put forward during each period. By following this approach we intend to highlight specific periods when one of the devices would enable better group performance. For example, a brainstorming session may be decomposed into at the beginning, when users usually have plenty of ideas, at the middle, and at the end, when users are typically more passive.

This decomposition is depicted in the top region in Fig. 7, which shows that in all three periods of 300 seconds groups produced more ideas with the **opportunity seeker** than with the control device. We obtained similar results at the

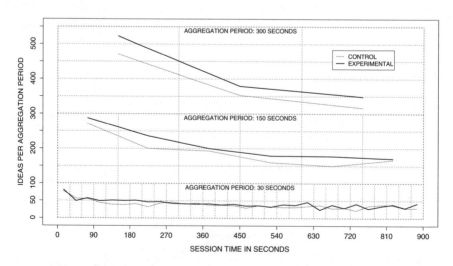

Fig. 7. Group performance through the duration of the brainstorming sessions under the control and experimental treatments. *Top:* number of ideas per period of 300 seconds. *Middle and bottom:* same, considering periods of 150 and 30 seconds, respectively.

150 seconds level of aggregation (see middle region in Fig. 7). Finally, if we consider periods of 30 seconds (see bottom region in Fig. 7) then groups performed better with the **opportunity seeker** in 21 out of 30 cases. From the evidence collected, there seems to be no particular phase when group performance with the **opportunity seeker** could be considered worse than with the control device.

These results encouraged us to extend the analysis of group performance over time to other aggregation periods. We looked at the 26 divisors of session duration (in seconds), from counts of ideas generated in the two halves of a session (each lasting 450 seconds) down to the 900 aggregation periods of one second each. Then, for all 26 aggregation periods we measured the percentage of cases over the duration of a session in which group performance was better, worse, and equal with the **opportunity seeker** compared to the control device.

In these circumstances, group performance with the **opportunity seeker** was better than with the control device in *at least* 40% of the cases, with an average of 68.3% ($SD = 18.9$), which contrasts with the percentage of cases in which it was worse: *at most* 40%, with a mean value of 24.4% ($SD = 14.0$). In other words, for all 26 aggregation periods considered, the **opportunity seeker** always had a higher proportion of cases over the session duration in which group performance was better than with the control device.

5.3 *Post-hoc* Analysis

We also performed a post-hoc analysis based upon fine-grained data collected with ABTool to characterise the actual delivery of ideas and the performance of the users during the brainstorming sessions.

We considered the following variables: DLVR, deliveries of ideas per session; TBDL, seconds between consecutive deliveries; TIDEA, seconds to write an idea; PAUSE, seconds between a user submitting an idea to the group and restart typing; CIDEA, characters per idea; CHARS, total number of characters typed per user in a session; and DCHARS, total characters deleted per user per session.

Table 3 shows a summary of the results we obtained for all users that participated in the experiment, including descriptive statistics and the output of the Wilcoxon signed-ranks test, which we use here to prioritise the presentation of

Table 3. Results of *post-hoc* analysis at the user level, ordered by p-value

Variable	Control		Experimental		Difference		Wilcoxon test		
	M	SD	M	SD	M	SD	W_+	W_-	p
DLVR	82.7	48.1	46.2	4.6	−36.5	37.4	0.0	1540.0	0.000
TBDL	13.7	5.9	21.2	6.1	7.5	3.2	1540.0	0.0	0.000
TIDEA	25.7	17.3	21.5	11.8	−4.2	12.9	422.0	1118.0	0.004
PAUSE	34.1	34.3	27.7	19.2	−6.4	21.7	469.0	1071.0	0.012
CHARS	1044.8	511.2	1110.4	529.8	65.6	321.4	936.5	603.5	0.164
CIDEA	45.6	12.7	43.9	12.9	−1.7	9.5	613.0	872.0	0.266
DCHARS	206.7	163.0	199.3	133.3	−7.4	121.9	724.0	816.0	0.703

further details rather than to do null hypotheses significance testing. Thus, no family-wise corrections were made.

Starting with the DLVR variable, the experimental device reduced by an average of 44.1% the number of deliveries of group ideas that reached a user per session. This difference, from a mean value of 82.7 deliveries per session to 46.2, was due to each delivery having comprised a batch of 1.9 ideas on average ($SD = 1.2$), with up to 5 ideas in 99% of the cases and a maximum batch size of 9 ideas (happening only once), unlike when under the control treatment, in which new ideas were immediately broadcasted, one by one, to the group.

Another consequence of the opportunity seeker device, captured in variable TBDL, is that users had 54.7% more time, on average, to think about and type ideas without receiving new ideas from others, corresponding to uninterrupted periods with a mean duration of 21.2 seconds instead of 13.7 seconds with the control device.

The opportunity seeker trades up-to-date broadcasts of new ideas for less frequent deliveries of batches of ideas. This could have aggravated the alternation between doing individual work and attending to the group if, for instance, users had slowed down because of the apparent delays in group awareness updates or had become overloaded by the quantity of information in the batches.

In fact, variable TIDEA reveals that users needed a mean value of -16.3% of time to write an idea under the experimental treatment, corresponding to an average cut down of 4.2 seconds per idea when users typed their ideas without being interrupted with ideas from the other users. We also found, through variable PAUSE, that users switched 18.8% more rapidly, or 6.4 seconds faster, on average, from submitting an idea to the group to start typing the next idea, presumably reading ideas from others in between (see motivation near Fig. 2).

For the remaining variables in Table 3, results revealed small differences between the control and experimental treatments, thus likely explained by chance. The number of characters typed per user in a session, CHARS, was 6.3% higher, on average, with the opportunity seeker, influenced by the higher number of ideas produced (see Table 2), but balanced by slightly fewer characters per idea (CIDEA had a mean difference of -3.7%). Finally, the number of deleted characters, DCHARS, was 3.6% lower under the experimental treatment, on average.

6   Discussion

In this section, we elaborate on how users act when they receive new ideas from others and submit their ideas to the group, then we analyse the potential problem of some of the ideas not being delivered because of the buffering technique employed by the opportunity seeker, and, finally, we discuss the limitations of this study, in particular concerning the lack of a qualitative evaluation.

6.1   Validation of Patterns of User Activity

Earlier, we identified three patterns of user activity in brainstorming sessions with immediate broadcast of ideas, from the visual analysis of plots such as the one shown in Fig. 2. These patterns are important because they supply the basis for the model of user behaviour depicted in Fig. 3.

We now provide evidence for the first two patterns—that users typically do not stop typing when they receive new ideas from the other users and that they usually pause after putting forward an idea—based upon fine-grained data collected during the laboratory experiment.

On the one hand, in the first five seconds after the reception of new ideas from others, users continued typing their idea at a mean rate between 1.4 and 1.6 key presses per second (SD between 0.7 and 0.8). On the other hand, after submitting an idea to the group, users almost stopped typing for at least five seconds, with a mean rate between 0.1 and 0.2 key presses per second (SD between 0.2 and 0.3). This provides evidence to validate the two patterns mentioned above.

6.2   Undelivered Ideas

One of the concerns of buffering ideas during brainstorming sessions, instead of immediately broadcasting them, is that the ideas submitted near the end of the session may not be delivered to some of the users. This can happen when a user is less productive than the others, either because s/he types very slowly or does not type at all due to lack of inspiration. As explained earlier, in these circumstances the **opportunity seeker** delays the delivery of new ideas from others, limited to a predefined quantity, until the user finally submits the idea to the group or until a timeout occurs, respectively.

Since it is undesirable to have undelivered ideas, we measured group production in each session with the **opportunity seeker** and subtracted from it the number of ideas from others actually received by each user. Table 4 shows that in 72.7% of the cases (or 40 sessions out of a total of 55) all ideas were delivered to the users and that in 20.0% of the times one or two ideas were not delivered; the remaining 7.3% were for cases with between three and seven undelivered ideas, each occurring only once.

In other words, these data reveals that the users' natural work rhythm was rapid enough so that less than one idea ($M = 0.6$, $SD = 1.4$) was not delivered at the end of a session with the **opportunity seeker**, which seems reasonable.

6.3   Limitations

We had to accept several compromises for this study, most of them related to the absence of a qualitative analysis of both the users' ideas and the videos that were captured during the brainstorming sessions.

Table 4. Sessions with undelivered ideas. Column 0 represents the special case in which all ideas were delivered to the users. No more than seven ideas remained to be delivered at the end of a session, and this happened only once.

Undelivered ideas	0	1	2	3	4	5	6	7
Number of sessions	40	7	4	1	1	1	0	1

Firstly, we did not evaluate nor compare the quality of the ideas due to the subjective nature of this task and also because it would have required several evaluators, which have not been available so far. Then again, quantity is one the goals of brainstorming [11] and there is evidence that quality is positively linked to quantity [12].

Secondly, we did not investigate duplicate ideas, something that could be explicitly addressed in a qualitative analysis. The interest here would be to know if the buffering mechanism on the opportunity seeker artificially inflated the number of generated ideas by causing users to unknowingly submit ideas equivalent to those stored in the buffer but not yet displayed. However, with immediate broadcast of ideas users may not be able to keep up with the others, which might also lead to duplicate ideas. Thus, a comparison between the two conditions on this topic is appealing and its results could eventually help fine-tune the opportunity seeker.

Thirdly, we always used the same values for the parameters of the opportunity seeker: no more than ten ideas were presented at once and the inactivity period after which ideas would be delivered to the user was ten seconds. We could have experimented with other values (keeping in mind the objectives explained earlier, e.g., not filling up the computer screen with new ideas) but that would have increased the complexity of the experimental design beyond our current logistic capacity.

Fourthly, we faced many difficulties while examining the video feeds of the computer screen and the user's face. The purpose was to make observations related to the three patterns of user activity identified earlier: a) if users are able to attend to other users' ideas and write an idea simultaneously; b) if the pause in typing activity after the submission of an idea coincides with the user looking at others' ideas; and c) if periods of inactivity correspond to lack of imagination, distraction, or to engaged reading. However, the videos showed users who appear to be focused on the task and computer screen most of the time. Very occasionally, there was an outward reaction to reading an idea, e.g., a frown or smile. It was also infrequent to observe users acting distracted, for instance, staring somewhere else than the computer screen. Given this data, it was impossible to accurately distinguish when a user was reading ideas, pausing, or distracted, so we had to discard these data.

Finally, we did not assess the degree to which users experienced information overload, if any. There exist several techniques that could provide insight into this, such as physiological measures and self-assessments of mental workload [13], which could be applied in future experiments.

7   Conclusions and Future Work

We highlighted the need to apply Attentive User Interfaces to groupware systems and made contributions to address the group attention problem in synchronous electronic brainstorming settings.

Firstly, we presented an attentive device, the opportunity seeker, which applies buffering of group awareness information to mitigate information overload. The opportunity seeker considers the natural rhythms of group work to time the delivery of ideas with the situations in which users are most likely to benefit from them. Secondly, we showed how this device can be implemented on an electronic brainstorming tool and how task boundaries can be detected via keyboard activity. Thirdly, we provided evidence that the opportunity seeker can increase the work done by groups, and that the improvement amounts to 9.6% in the number of ideas produced in electronic brainstorming tasks.

In addition, results from a post-hoc analysis show that the opportunity seeker reduced the number of deliveries of ideas by 44.1% by combining ideas in small batches and that this translated into 54.7% more time for users to think about and type ideas without receiving new ideas from others. In these conditions, users were 18.8% faster in alternating between generating an idea, which they did in 16.3% less time, and reading other users' ideas.

We believe that the attentive device we propose in this paper provides benefits for today's and tomorrow's demands: on the one hand, even if the users in our experiment were not overloaded with information, the number of ideas produced was, nonetheless, higher; on the other hand, the opportunity seeker facilitates the creation of electronic brainstorming sessions with larger group sizes because it ensures that each user will be exposed to new ideas from others at his or hers own natural rhythm, thus automatically mitigating information overload.

As for future work, we are considering several research paths: one is to address the limitations presented earlier; another is to experiment with the opportunity seeker in other types of computer-mediated group tasks, especially in convergence tasks, such as negotiation; finally, we have plans to introduce an eye-tracker in future experiments.

Acknowledgments

This work was supported by the Portuguese Foundation for Science and Technology, through project PTDC/EIA/67589/2006 and the Multiannual Funding Programme, and by the Departamento de Postgrado y Postítulo of the Vicerrectoría de Asuntos Académicos of the Universidad de Chile.

References

1. Vertegaal, R.: Attentive user interfaces: Introduction. Communications of the ACM 46(3), 30–33 (2003)
2. Roda, C., Thomas, J.: Attention aware systems: Introduction to special issue. Computers in Human Behavior 22(4), 555–556 (2006)
3. Vertegaal, R., Shell, J.S., Chen, D., Mamuji, A.: Designing for augmented attention: Towards a framework for attentive user interfaces. Computers in Human Behavior 22(4), 771–789 (2006)

4. Fogarty, J., Ko, A.J., Aung, H.H., Golden, E., Tang, K.P., Hudson, S.E.: Examining task engagement in sensor-based statistical models of human interruptibility. In: CHI 2005: Proceedings of the SIGCHI conference on Human factors in computing systems, April 2005, pp. 331–340. ACM Press, New York (2005)
5. Baudisch, P., DeCarlo, D., Duchowski, A.T., Geisler, W.S.: Focusing on the essential: Considering attention in display design. Communications of the ACM 46(3), 60–66 (2003)
6. Vertegaal, R., Weevers, I., Sohn, C., Cheung, C.: GAZE-2: Conveying eye contact in group video conferencing using eye-controlled camera direction. In: CHI 2003: Proceedings of the SIGCHI conference on Human factors in computing systems, pp. 521–528. ACM Press, New York (2003)
7. Jenkin, T., McGeachie, J., Fono, D., Vertegaal, R.: eye View: Focus+context views for large group video conferences. In: CHI 2005: Extended abstracts on Human factors in computing systems, pp. 1497–1500. ACM Press, New York (2005)
8. Pinelle, D., Gutwin, C.: Groupware walkthrough: Adding context to groupware usability evaluation. In: CHI 2002: Proceedings of the SIGCHI conference on Human factors in computing systems, pp. 455–462. ACM Press, New York (2002)
9. Bailey, B.P., Konstan, J.A.: On the need for attention-aware systems: Measuring effects of interruption on task performance, error rate, and affective state. Computers in Human Behavior 22(4), 685–708 (2006)
10. McFarlane, D.C.: Comparison of four primary methods for coordinating the interruption of people in human-computer interaction. Human-Computer Interaction 17(1), 63–139 (2002)
11. Osborn, A.F.: Applied imagination: Principles and procedures of creative problem-solving, 3rd edn., Scribner, New York, NY, USA (1963)
12. Briggs, R.O., Reinig, B.A., Shepherd, M.M., Yen, J., Nunamaker, J.F.: Quality as a function of quantity in electronic brainstorming. In: HICSS 1997: Proceedings of the thirtieth Hawaii international conference on System sciences, Washington, DC, USA, pp. 94–103. IEEE Press, Los Alamitos (1997)
13. Wickens, C.D., McCarley, J.S.: Applied Attention Theory. CRC Press, Boca Raton (2008)

Implementing a System for Collaborative Search of Local Services

Tiago Conde[1], Luís Marcelino[2], and Benjamim Fonseca[3]

[1] UTAD, Quinta de Prados, Apartado 1013, 5001-801 Vila Real, Portugal
tiagolrconde@gmail.com
[2] Instituto Politécnico de Leiria – ESTG, Apartado 4163, 2411-901 Leiria
luis.marcelino@estg.ipleiria.pt
[3] UTAD/CITAB, Quinta de Prados, Apartado 1013, 5001-801 Vila Real, Portugal
benjaf@utad.pt

Abstract. The internet in the last few years has changed the way people interact with each other. In the past, users were just passive actors, consuming the information available on the web. Nowadays, their behavior is the opposite. With the so-called web 2.0, internet users became active agents and are now responsible for the creation of the content in web sites like MySpace, Wikipedia, YouTube, Yahoo! Answers and many more. Likewise, the way people buy a product or service has changed considerably. Thousands of online communities have been created on the internet, where users can share opinions and ideas about an electronic device, a medical service or a restaurant. An increasing number of consumers use this kind of online communities as information source before buying a product or service. This article describes a web system with the goal of creating an online community, where users could share their knowledge about local services, writing reviews and answering questions made by other members of the community regarding those services. The system will provide means for synchronous and asynchronous communication between users so that they can share their knowledge more easily.

Keywords: groupware, web 2.0, local services, social communities.

1 Introduction

The way information is available on the web has been changing in the past few years. Although the majority of information presented online comes from corporate websites, online journals and magazines, ordinary internet users started to join forces. The so-called Web 2.0 or Social Web is taking a major impact in peoples' lives and the way they use the internet. Websites with user created contents, like Wikipedia, MySpace, YouTube or Yahoo Answers, are growing exponentially [1].

Instead of simply being a way to distribute information, the internet is now an interactive world, where we can obtain the wisdom of its users in a collaborative way through wikis, blogs and virtual communities. Organizations don't use the web anymore just as a mechanism for spreading information and marketing but also as a way for obtaining feedback from their customers [2].

R.O. Briggs et al. (Eds.): CRIWG 2008, LNCS 5411, pp. 17–24, 2008.

Users contribute freely with their knowledge without having any monetary compensation. In most cases, their motivation is the acceptance and recognition by its online peers. Like in real life, reputation in virtual communities is very important. Many users identify themselves with other users because of common ideas, the attitude of that user to the community or just by the way he or she participates in that community.

In a study published by comScore [3], MySpace has grown 72% in the number of unique views between June 2006 (66 millions) and June 2007 (114 millions). Facebook has grown 270% in the same period from 14 million to 52 million unique views. The same has happen to other community websites like Hi5, Friendster or Orkut.

1.1 Local Services

Although we now live in the digital era where we can buy pretty much everything without getting outside our home, we still make our lives in the physical and real world. And in this world we have to get outside. We need to go to the dentist, put the car in the mechanic, leave the children at school, go to a restaurant or a bar at night. The information we own about the places that we go and the services that we use in our daily life is not only important for us but may well be important for other people living next to us. How many times do we ask a friend or neighbor if he knows where is the store that may have the product that we need, the best prices or best customer support. Everyday millions of people use their local journals, yellow pages books, magazines and the internet in a way to find information about everyday products and services. This kind of information is often available on the internet, but poorly organized and not in an intuitive form [4]. The internet can provide to its users new tools to enrich the level of information available for local services.

According to a survey carried out by BIGresearch, 92.5% of adults said they regularly or occasionally research products online before buying them in a store. Users also said they are most likely to communicate with others through face-to-face discussion (68.9%), though email (53.1%), telephone (50.9%), and cell phone (30%). Young adults 18-24 communicate about products and services by instant messaging (37.5%), text messaging (23.7%) and through online communities (20.6%) [5].

The emergence of the Social Web changed the way users search for local services online. People don't just want to know the contact of a service and its physical location. They want to know more about a local service before they actually go there. For instance we may like to know about service price policies, customer support or user opinions that already have used such service. This is valuable information that may affect the decision of buying some product or service. This way, users can rely on other users' experiences to make their own opinion and judgment about an unknown service. The owners of the information of local social web are no longer the traditional yellow pages type services, but the internet users. Therefore, local communities are created on the internet, grouped by region or city, where everyone can share its opinion and everyday knowledge to help a neighbor looking for a product or service. This way, local services can create and maintain an image and reputation in the online community that is impossible with the traditional local information services.

We analyzed five web sites of local services: Yelp, Yahoo! Local, welovelocal, MojoPages and Tupalo. All analyzed services have common features such as having the local services repository organized by location, category or tags. They also use geographic APIs like Google Maps or Yahoo! Maps for displaying the location of the local services in a map, allowing users to locate the service more easily. It is also possible not only to add textual information about a service, but also multimedia information like photos or video. These web sites have multiple services for asynchronous communication, like private messages, discussion groups or e-mail. Yahoo! Local gives little attention to the social and community aspect compared to the others, having only a repository of local services categorized by city, where is possible to read the reviews posted by the registered users. MojoPages distinguishes itself by how it presents its review form. Instead of a typical text box, it presents a form where the users can specify the service characteristics in a more organized way. However this approach can run against the service, because one of the flags of Web 2.0 is its simplicity, where the user doesn't have to fill complex forms. Concerning Yelp, it has a simple and well structured interface with very complete information and very accessible at the same time, with rating mechanisms of the users so that we can have a social hierarchy inside the community.

Table 1. Comparison table of the main features of the local service web sites analysed

	yelp	Yahoo! Local	weloveloca l	MojoPages	Tupalo
Organization	Location and category	Location and category	Location	Location and category	Location and tags
Map location	Google Maps	Yahoo! Maps	Google Maps	Google Maps	Google Maps
Multimedia	Images	Images	Images	Images and videos	Images (Flickr)
Services comparison	No	No	No	Yes	No
Reviews format	Text box	Text box	Text box	Form	Text box
Allow review comments	Yes (private)	Yes (public)	No	Yes (public)	No
Rate services/ reviews	Yes/Yes	Yes/Yes	Yes/Yes	Yes/Yes	Yes/Yes
Chat messages	No	No	No	Yes	No
Friends list	Yes	No	Yes	Yes	Yes
Discussion groups	No	Yes	Yes	Yes	Yes

1.2 Q&A Services

With the rapid growth of internet connections and personal computers, more people use the internet to obtain information and get the answers to daily questions. That way, some communities of Q&A were created, as is the case of Yahoo! Answers, Askville or AnswerBag [6]. In this type of communities' users ask questions that are answered by the community in a collaborative way. Every user can answer a question whatever the knowledge it may have, without needing to be an expert in the field. Through the exchange of questions and answers, users not only search for information, but they also share their experiences, opinions and advises. Of course the information that is generated is not always accurate and trustworthy and it's the user responsibility to judge and evaluate the quality of the answers that he obtains and get only the information that may fit his needs.

We analyzed five web sites of Q&A services, namely AnswerBag, Askville, Live QnA, Yahoo! Answers and Yedda. The majority of the web sites have their questions organized by category, but there are others, like Yedda and Live QnA, that have their questions organized by tags. All sites have mechanisms of notification for questions or categories that users might like to follow. We can highlight the Yahoo! Answers service not only by its large community but also by how it presents the information in its web site, in a well structured and organized way. Furthermore, it has a complex ranking system to promote the participation of registered users and prevent span. The Askville works like a computer video game where the user is rewarded or penalized by the questions he makes, earning experience points. This kind of mechanisms works positively because it stimulates users to participate and to obtain better results; at the same time it requires their constant availability and daily attention that many don't have the possibility or are not interest in. In Live QnA and Yahoo! Answers, answers are valid for a limited period of time. This approach will not be followed in this project because it can always arise something new that might not help the creator of the question but might help other users of the community that have a similar doubt.

Table 2. Comparison table of the main features of the q&a web sites analysed

	AnswerBag	**Askville**	**Live QnA**	**Yahoo! Answers**	**Yedda**
Organization	Category	Category	Tags	Category	Tags
Questions format	Text	Text	Text	Text	Text, Image, Video
Answers format	Text, Image, Video	Text	Text	Text	Text, Image, Video
Allow answer comments	Yes	Yes	Yes	No	No
Rate questions/ answers	Yes/Yes	No/Yes	No/Yes	No/Yes	No/Yes
Chat messages	No	No	No	Yahoo! Messenger	No
Friends list	Yes	Yes	No	Yes	Yes
Notification service	Categories and questions	Questions	Categories and users	Questions and users	Questions
Time to answer	Unlimited	Unlimited	Four days	One week	Unlimited

2 The System

The aim of this work is to create a web system in which we join the concepts of online local search, Q&A communities and social web communities like MySpace. The system will allow registered users to add a new local service in the system and/or review an existing one. The users can place questions about some issues that they might have, concerning a product or service near their location. Likewise, they share their knowledge answering the questions put by other community members, helping each other's out in a collaborative way. The system will allow every user, registered or anonymous, to search for local services or questions already put by registered users. Instead they might opt to navigate through predefined categories or by tags associated to a service. In order for a user to be up to date, the system will also provide a notification service that will alert the users about changes in services or questions that they choose to monitor. The system will also provide means to communicate synchronously and asynchronously using features such as web chat or private messages. Users will be able to collaborate by adding their knowledge into an existing local service that they might have just visited in their city simply by adding new information, by correcting the existing one, by adding a photo or simply by writing a review about that service, so that other users can form an opinion before they use the service.

2.1 Architecture

The internet is living a period of great growth where new applications and services appear every day. Many often it's hard to know which ones are going to grow and at what speed. One day a system may only have a community of a few thousands and after a short period it can have some millions. Therefore it must be taken in consideration questions like the availability, scalability and manageability of the system when we think of developing a specific architecture, so that we can be better prepared to the changes imposed by the internet users. Taking this into consideration, we chose a multi-tier architecture [7] for implementing the system that uses the client/server model. This allows us to implement the different services of the system in separate modules and libraries that are loosely coupled together. Each tier can be developed concurrently by different programmers and in different computer languages. The programming of a tier can be changed or relocated without affecting the other tiers, making it easy to continuously evolve the system as new opportunities arise and address issues of scalability and availability in a much simpler way.

The main functionalities of the system will be available through a public web service. This way the users can access the system services not only by using our client application, but also through external applications built by any user on the community. However the back office created to manage the services will access directly the system without passing through the web service. By using an n-tier architecture to expose the system functionalities through web services we ensure not only that we have a well defined interface, which we must follow and is interoperable between various known systems, but also that adding a new functionality or changing an existing one will not affect or compromise the other functionalities of the system.

Figure 1 shows the logical architecture of the system, where we can observe the 3-tier architecture, defined by a presentation tier for presenting the information, a logic tier divided into two distinct layers that provide the core operations of the system, and a data tier for storing the information generated by the users system.

2.2 Implementation

The system will be implemented using the Microsoft .NET Framework. This framework has a set of API libraries that cover a wide range of programming needs including web applications development, data access, database connectivity, file and xml manipulation, cryptography, security, etc. With this set of tools we can put our efforts in the developing of what really matters, that is, the system itself.

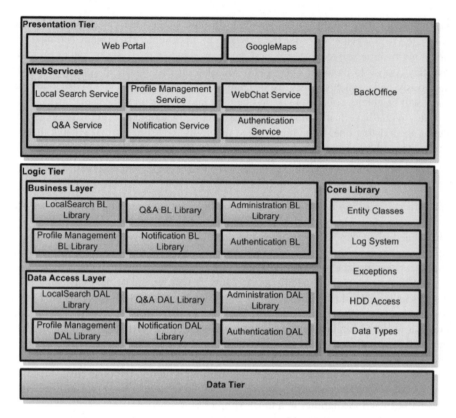

Fig. 1. Logical architecture of the system

Logic Tier

The logic tier also known as middle-tier forms the core of the system. It's composed of two layers, the business layer and the data access layer. These two are linked together in a loosely coupled way, which means that the business layer is aware of its data access layer only in the moment that it needs to call it (late binding). This is accomplished because the business layer has a well defined interface containing the methods that its correspondent data access layer must implement. This way we can make changes into a data access layer library or even change it with a new one while the system is running, only by changing a parameter in a configuration file. This gives us the flexibility of adding, removing or switching system components without the need to stop the system.

The logic tier also includes a core library that has common functions needed for the implementation of the system libraries, like specific exception handling, cryptography, logging, data validation, etc.

Figure 2 shows the physical architecture of the system, representing the optimum layout for the different services and modules. Taking into account cost or organizational issues some modules of the system could be joined together.

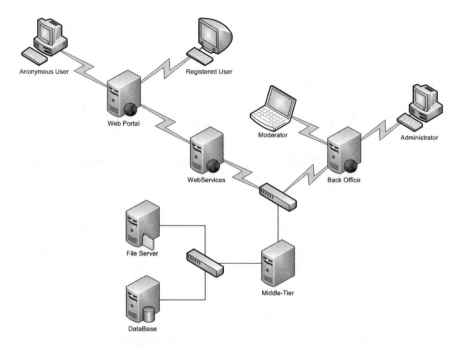

Fig. 2. Physical architecture of the system

Data Access Layer

In the data access layer, we will use a new .NET Framework component released in version 3.5, Language Integrated Query (LINQ), specifically LINQ to SQL [8]. This component allows us to query and manipulate a relational database like SQL Server in a simpler way because it maps the relational database structure into objects. This way we can work with objects making it very easy to query a database and manipulate its tables. Of course there's a compromise in performance compared to stored procedures, but we gain a lot more in terms of code structure and organization. Not to mention the easier it is to manipulate a data object than a long and complex stored procedure that mixes business logic with data access logic in the data tier.

Presentation Tier

The presentation tier is composed of the web portal used by the community, web services to access the modules of the system and the back office that let administrators manage the system. It will be developed using the Microsoft ASP.NET [9] web application framework. The web portal will consume the web services using asynchronous calls made in the client side using the JavaScript language and the lightweight JavaScript Object Notation (JSON) [10] data format to carry the information provided by the system. We use JSON instead of XML duo to its simplicity and ease of processing on the client side. By using asynchronous calls, we can change and update small parts of the web pages without refreshing the whole page, giving a more enjoyable navigation experience to the users. Another advantage of using this approach is that it leaves the server side of the system responsible for delivering the requested data, leaving the client side responsible to process the received data. This way we divide the computational power needed for the system to work to all its users.

3 Final Considerations

This article describes a web system with the goal to help online users in sharing their knowledge of the local services that they use. With the appearance of the social web and related technologies, people are more receptive to this kind of collaborative online communities. People also want to know more from a local service than its location and contact. They want to know the opinions of their friends, neighbors and members of the community about the local services that they use. With the use of an n-tier architecture allied with the use of web services for publishing the functionalities of the system, it is possible to create a modular web system that is easy to update, where we can add new services and functionalities as the system grows. To validate the prototype of the proposed system, there will be performed various tests like usability, performance, stress and security.

References

1. Wade Roush Answers by the People, For the People in Technology Review (2006), http://www.technologyreview.com/Infotech/17039/
2. Knights, M.: Web 2.0. Communications Engineer 5(1), 30–35 (2007)
3. comScore Social Networking Goes Global, comScore, Inc. (2007), http://www.comscore.com/press/release.asp?press=1555
4. Ludford, P., et al.: Capturing, sharing, and using local place information. In: CHI 2007: Proceedings of the SIGCHI conference on Human factors in computing systems. ACM, New York (2007)
5. National Retail Federation - BIGresearch's Simultaneous Media Survey (SIMM 9), http://www.nrf.com/ modules.php?name=News&op=viewlive&sp_id=229
6. Kim, S., Oh, J.S., Oh, S.: Best-Answer Selection Criteria in a Social Q&A site from the User-Oriented Relevance Perspective
7. Sadoski, D., Comella-Dorda, S.: Three Tier Software Architectures. Software Technology Review. Software Engineering Institute, Carnegie Mellon University (January 1997), http://www.sei.cmu.edu/str/descriptions/threetier.html
8. LINQ Project, http://msdn.microsoft.com/en-us/netframework/aa904594.aspx
9. Microsoft ASP.NET, http://www.asp.net/
10. JavaScript Object Notation (JSON), http://www.ietf.org/rfc/rfc4627.txt

Shared Resource Availability within Ubiquitous Collaboration Environments

Kimberly García[1], Sonia Mendoza[1], Gustavo Olague[2],
Dominique Decouchant[3,4], and José Rodríguez[1]

[1] Departamento de Computación, CINVESTAV-IPN, D. F., México
[2] Departamento de Ciencias de la Computación, CICESE, Ensenada, México
[3] LAFMIA UMI 3175 C.N.R.S.-CINVESTAV, CINVESTAV-IPN, D. F., México
[4] Laboratoire LIG de Grenoble, UMR 5217, Grenoble, France
kimberly@computacion.cs.cinvestav.mx,
smendoza@cs.cinvestav.mx, olague@cicese.mx,
Dominique.Decouchant@imag.fr, rodriguez@cs.cinvestav.mx

Abstract. Most research works in ubiquitous computing remain in the domain of mono-user systems, which make assumptions such as: "nobody interferes, observes and hurries up". In addition, these systems ignore third-part contributions and do not encourage consensus achievement. This paper proposes a system for managing availability of distributed resources in ubiquitous cooperative environments. Particularly, the proposed system allows collaborators to publish resources that are intended to be shared with others collaborators and to subscribe to allowed resources depending on their interest in accessing or using them. Resource availability is determined according to several parameters: technical characteristics, roles, usage restrictions, and dependencies with other resources in terms of ownership, presence, location, and even availability. To permit or deny access to context-aware information, we develop a face recognition system, which is able to dynamically identify collaborators and to automatically locate them within the cooperative environment.

Keywords: Availability of distributed resources, perception, ubiquitous cooperative environments, human face recognition, automatic user localization.

1 Introduction

Previous works about ubiquity in the CSCW field highlight the working group members' need of accessing and sharing relevant information anytime and anywhere. Markarian et al. illustrate this requirement by means of the following scenario [14]: "the members of a group physically meet together to discuss about some particular subject. During the meeting, one of the collaborators remembers that he owns relevant information, which can be shared with his colleagues, but it is stored in his PC. Consequently, the non-mobile character of his PC forces him to go to his office either to print this information or to make a copy on an

R.O. Briggs et al. (Eds.): CRIWG 2008, LNCS 5411, pp. 25–40, 2008.
© Springer-Verlag Berlin Heidelberg 2008

USB flash drive". This way of sharing information during a face-to-face meeting unavoidably breaks the interaction flow among collaborators.

This kind of problems arising from statically sticking information on computers motivates us to deploy cooperative applications in mobile computing devices, so that collaborators can access relevant resources when required. In order to support resource sharing and human collaboration in ubiquitous environments, it is required to dynamically manage information about availability of human, physical and electronic resources. Our field of study is an organization whose human resources are inherently potential collaborators who can share electronic resources (e.g., multimedia and software) and whose physical resources are heterogeneous computing devices (e.g., PCs, servers, laptops, PDAs, printers, plotters, and interactive whiteboards) that can be distributed in private places (e.g., offices) and public places (e.g., waiting rooms and corridors).

From the resource availability point of view, it is impossible for each collaborator to have, within his office, all the physical resources (e.g., printers, whiteboards, and clusters) existing in his organization. On the other hand, placing a collaborator's physical resources in public places or giving free access to them can be inconvenient for any collaborator as physical resources are susceptible of abuse or wrong management and use. Likewise, it is infeasible to replicate all his electronic resources on every device with storage capabilities because some resources are costly (e.g., multimedia) or restricted (e.g., software).

The main contribution of this paper concerns the development of a groupware system to make the collaborators aware of the availability of persons themselves, and of their shared physical and electronic resources. To determine the availability of such kinds of resources, we rely on several parameters such as technical characteristics, roles, usage restrictions and relationships with other resources in terms of ownership, presence, location and even availability. These parameters have not been comprehensively considered in previous works.

In this paper, we first analyze the concept of "perception" and its negative effects concerning violation of privacy and intrusion in cooperative environments (Section 2). Afterwards, we describe our proposal in two steps: firstly, we explain the main mechanisms to determine resource availability based on the previously mentioned parameters (Section 3). Secondly, we describe the main modules of our face recognition system, which is able to identify persons who enter to specific places (Section 4). Thus, the proposed system can permit or deny access to context-aware information. After presenting related work (Section 5), we conclude this paper and give some ideas for future extensions (Section 6).

2 Perception Characterization

"Perception" has been generally associated with cooperative systems that support synchronous distributed interaction (same time-different place). Thus, as the physical distance drastically reduces communication among the group members, these systems require dedicated mechanisms to provide collaborators with information about contributions, intentions and work focus of their colleagues.

Two characterization axes of perception have been identified [15]. The first axe refers to the capture of awareness information about two types of objects: 1) group members (e.g., presence, location and activities) and 2) shared data (e.g., modifications). The second axe considers two forms of capturing such information: 1) implicitly by I/O devices (e.g., keyboards, mice, cameras, microphones, or infrared sensors) or 2) explicitly by users throughout applications or artefacts (e.g., calendars or published information such as "do not disturb").

This perception characterization does not take into consideration a type of information object, which can be relevant for Ubiquitous Computing (UC) systems: the physical resources. However, this characterization highlights an essential requirement for the design of UC systems: the implicit capture of information. Although the UC field tries to avoid the explicit introduction of information by users, this capture form can be required in cooperative systems to help solving the difficult problems of privacy violation and intrusion.

Privacy Violation and Intrusion
One of the main requirements commonly highlighted by UC systems concerns to track the location of mobile devices held by nomadic users. For instance, CHIS (Context-Aware Hospital Information Systems) [14] shown that the fulfillment of this requirement is important for the management of information within a hospital. However, this tracking capability could reveal identification problems (PDA identification instead of person identification). Moreover, location information must be controlled by the owner in order to avoid privacy and intrusion problems, well known in the field of mediaspace systems.

Thus, these later systems have allowed to understand some technological and social effects of proving audio and video connections to perceive people located in other places and to eventually strike up a computer-supported conversation. Some of these systems (e.g., VideoWalls [1] and Portholes [5]) capture video from cameras located not only in common areas (e.g., cafeteria) but also in private areas (e.g., offices). At the beginning of 1990s, some of the main concerns for system designers were the excessive technological requirements, e.g., video and audio connections, bandwidth and inter-media synchronization. Although, nowadays, some of these technological problems are relatively solved, privacy violation and intrusion remain open problems.

These social problems come from the lack of methodologies to determine the kind of information required by collaborators to facilitate the contact among them [10]. Thus, cooperative systems could provide either: 1) excessive information, which can cause privacy violation, intrusion, scalability and bandwidth problems as well as difficulties to select relevant information; or 2) reduced information, which can cause inappropriate contacts or lost of opportunities. In fact, during a cooperative process, resource sharing control and associated information privacy could change because the group's needs evolve over time [2].

Some important questions arise from the need to cope with this dynamism: Who should be in charge of these tasks? The system? The resource owner? The group? A particular collaborator acting as the group leader? We take these questions as the starting point to define a set of parameters that have influence over the management of a collaborator's perception about his colleagues, physical devices, multimedia and software.

3   Resource Availability Management System

The Resource Availability Management System (RAMS), proposed in this paper, aims at providing a collaborator with: a) functions to publish relevant physical and electronic resources that he can share with some of his colleagues, b) allowed awareness information (e.g., presence, location and availability) about his colleagues themselves and their resources on which he has access rights to use them or to access their contents depending on temporal and spatial restrictions; and c) functions to subscribe to (maybe unpublished) required resources.

3.1   RAMS Use Scenario

Let us suppose that within his physical working environment Mr. Brown owns, among other devices, a high-quality color plotter that is able to print large format technical graphics. At a moment, he decides to share it with his colleagues. Using the RAMS system functions, Mr. Brown can: a) publish his plotter by specifying its main technical characteristics (e.g., resolution, speed, and paper size) and 2) control the usage of this resource by specifying how it can be shared. Thus, Mr. Brown defines roles (e.g., authorized or denied printing) and usage restrictions (e.g., in terms of estimated access schedules) that are associated to his resource. As Mr. Brown's printer is located in his office, he can offer the printing service only during specific time slots in order to prevent potential clients for his plotter from disturbing him (see Fig. 1).

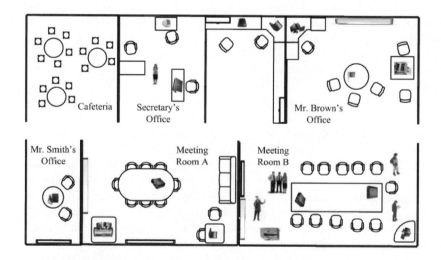

Fig. 1. Physical Cooperative Environment

Mr. Smith, who collaborates with Mr. Brown, needs to print a VLSI circuit design. To satisfy this requirement, the RAMS system provides Mr. Smith with three functionalities: 1) a view of all the types of resources on which he has access rights; 2) a view of the allowed resources belonging to a specific category (e.g., printers) selected by Mr. Smith, and 3) a set of templates to describe the required resource in terms of technical characteristics (e.g., resolution, speed, and paper size).

By means of these functionalities, the RAMS system informs Mr. Smith that Mr. Brown's high-quality color plotter is able to satisfy his requirements from a functional point of view. This information is determined from: a) the access rights attributed to Mr. Smith by Mr. Brown on his plotter, b) the resource category or the technical characteristics provided by Mr. Brown when he published it, and eventually c) the resource requirements specified by Mr. Smith. The RAMS system also informs Mr. Smith about the printer location, availability, and access schedule. However, despite the access schedule may authorize Mr. Smith to use the plotter during a given time slot, the RAMS system may indicate that it is temporarily unavailable because Mr. Brown is not yet arrived to his office!

Few times later, one of Mr. Brown's students, Miss White, goes to his advisor's office in order to test her still being developed application on whiteboard. She has no access restrictions to his advisor's office whenever he is absent. When Miss White comes into Mr. Brown's office, the RAMS system is able to recognize her and to automatically infer her new location. She intends to stay for two hours in his advisor's office and then declares herself available, which means that other persons can eventually disturb her anytime. Thus, although Mr. Brown remains absent, the RAMS system can infer and then notify Mr. Smith that Mr. Brown's plotter is now available. Of course, if Miss White would have declared herself unavailable, Mr. Brown's plotter also would remain unavailable.

Afterwards, Mr. Brown arrives to his office. The RAMS system notifies interested collaborators of his presence and location. Moreover Mr. Brown declares himself available. Automatically notified of these important changes, Mr. Smith sends his VLSI circuit design to the plotter. After printing completion, he goes to Mr. Brown's office in order to recover his printed sheets.

In order to highlight some important functionalities of the RAMS system, we study the three following cases (see Fig. 1):

Case 1: Mr. Smith is detected close to a technician's office
Mr. Smith is taking a tea in a colleague's office of another building. The RAMS system has to transmit an important document to Mr. Smith. Thus, it sends it to a technician's plotter that is located close to Mr. Smith's current location.

Case 2: Mr. Smith is detected within Mr. Brown's office
Some technical information becomes urgent to be diffused to Mr. Smith. Thus, the RAMS system asks Mr. Brown if he agrees to receive information intended for Mr. Smith in order to display it on his laptop screen and to show it to him. In the same time, Mr. Smith is notified of this event on his PDA. Thus, Mr. Smith's information is automatically displayed on Mr. Brown's high resolution

screen, whereas Mr. Smith can use his PDA to control the display (e.g., scrolling). Misters Smith and Brown can even establish a technical discussion based on the complementary use of the laptop (controlled by Brown) and PDA (controlled by Smith).

Case 3: Misters Smith and Brown establish a cooperative working session with their colleagues

If Misters Smith and Brown needs to establish a cooperative working session with other colleagues (e.g., by teleconference) to analyze such a technical drawing, the RAMS system (which is in charge of managing the set of collaborators' work contexts) locks the meeting room A (as the meeting room B is already reserved). Then, Misters Smith and Brown join up with the working session supporting by the interactive whiteboard, Smith's PDA and Brown's laptop.

This scenario requires the management of different kinds of resources: the building map where Misters Brown and Smith are located, the computer resources, and finally Misters Brown and Smith themselves.

3.2 RAMS Software Architecture

The RAMS system relies on the publish/subscribe asynchronous model [9] to allow collaborators: a) to publish resources that may be shared with others in a controlled way, and b) to subscribe to (not yet published) resources in order to eventually access or use them. Thus, using specific applications, collaborators can play the role of publishers and/or subscribers of awareness information (e.g., presence, location and availability) about people and computer resources. State information is transmitted via inter-application events (see Fig. 2).

An *event* is an information unit automatically produced each time an agent (i.e., a user's application instance) performs actions on resources. The RAMS system manages two different sets of agent clients: 1) *producer agents*, which generate events and transmit them to the RAMS system for diffusion, and 2) *consumer agents*, which subscribe to the RAMS system in order to receive events.

In contrast to conventional publish/subscribe systems where producers and consumers are decoupled from each other, cooperative systems need to identify agents in order to support resource sharing and relative tasks (e.g., activity coordination, contribution identification and change notification). Thus, the RAMS system provides each agent with a unique identifier, which allows to designate not only the active entity that executes actions on resources (i.e., user), but also the source of these actions (i.e., site and application).

This agent identification allows for: 1) controlling actions applied on shared resources, and 2) defining sets of roles by means of which agents can act. The role notion is essential to express the social organization of the group work, and constitutes the basis for protecting resources from unauthorized actions.

In this way, we can adapt information filtering and notification functions of the publish/subscribe model to the requirements of cooperative systems. More precisely, these functions have to take into account: 1) agent's identification and 2) agent's roles on shared resources.

The RAMS system is composed of three main functional parts: a *broker*, an *inference engine*, and a set of *sensors* (see Fig. 2).

The RAMS Broker

The *broker* consists of the *publication* and *subscription modules*, a *filter* and a *notification system*.

The *publication module* allows producer agents to describe resources in terms of their technical characteristics (see Fig. 2 ref. #1). Also, producer agents can define usage restrictions and attribute roles to consumer agents on shared resources. Likewise, the *subscription module* allows consumer agents to describe relevant technical characteristics of the resources (see Fig. 2 ref. #2).

Typically, consumer agents do not receive all the published events but a subset of them. Before being forwarded to them, events pass through by a *filter*, which organizes producer and consumer information respectively by topic and content (see Fig. 2 ref. #3). These modules are detailed in section 3.3.

The *notification system* is in charge of delivering to consumer agents not only dynamic awareness information, but also static information about the most suitable resources, e.g., presence and location of physical and computer resources (see Fig. 2 ref. #4) . This information is provided by producer agents whenever they publish their resources.

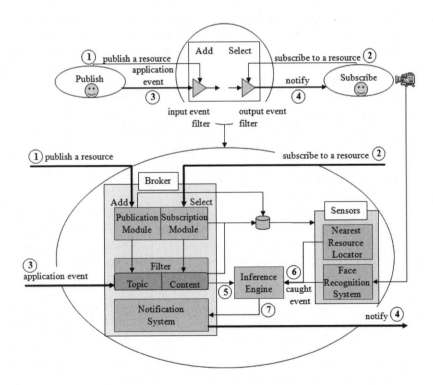

Fig. 2. Software Architecture of the RAMS System

The RAMS Inference Engine

For each consumer agent, the *filter* provides the *inference engine* with a potential set of shared resources. This set is selected from predefined and relatively static information (e.g., consumer agent's roles, resource usage restrictions and compatibility of the technical characteristics provided by the producer agent and eventually by the consumer agent).

Based on static and dynamic information (see Fig. 2 ref. #5 and #6), the *inference engine* establishes dependencies among resources from which the set of the most suitable resources is inferred (see Fig. 2 ref. #7). For instance, a collaborator's printer located in his office is available for others if: a) the owner is in his office; b) he is available; and c) the access schedule authorizes the printer usage. The *inference engine* is detailed in section 3.4.

The RAMS Sensors

In order to select the most suited resources from the potential set and to automatically infer new events, the *inference engine* relies on a set of sensors.

The first sensor is a *face recognition system*, which is in charge of identifying collaborators, whose faces are captured by cameras located in specific places. This system allows not only to inform about a collaborator's presence and location in a place, but also it is the basis to manage context-aware information. To cope with privacy and intrusion problems, the RAMS system also allows collaborators to handle their appearance within the cooperative environment by declaring themselves invisible for some colleagues.

The second sensor is a *resource locator*, which is responsible for determining the closest physical resource (relatively to the requester's current location) from the set of technically suitable, available and accessible resources.

3.3   Resource Description

Whenever producer agents publish resources in order to share them with their colleagues, the RAMS system allows to describe these resources using dedicated functions. Such a description depends on the resource type:

- A **human resource** description defines: a) his social information, e.g., name, position, and affiliation, b) his default location, e.g., office, and c) his office schedule. In contrast to physical resources, a collaborator's location can regularly change as he can move from one place to another.

Mr. Brown's User Definition

Name	Mr. Brown	Mr. Brown's Office Schedule		
Affiliation	University of Calgary	Everybody	Monday - Friday	9:00 - 19:00
Research Area	HCI, Groupware	Prof. Meeting	Monday	12:00 - 14:00
Location	office B 201			

- A **physical resource** description defines: a) its technical capabilities, e.g., resolution, double vs. single side, color vs. monochromatic, b) its default location, and c) its access schedule.

Characteristics of HP Printer

Description	Plotter
Brand and model	HP DesignJet 4500
Color	yes
Resolution	Up to 2400 x 1200 bpi
...	...
Owner	Mr. Brown
Location	office B 201

Usage Constraints for HP Printer

Brown's students	Mon to Sun	-
Mr. Smith	Mon to Fri	9:00-13:00

– An **electronic resource** description defines: 1) its technical characteristics, e.g., format, size and duration, b) execution requirements, e.g., viewer, c) its location, e.g., site address, port and path, d) the required communication protocol, e.g., HTTP, and e) eventually some data for access control, e.g., login and password.

Characteristics of Video: "VLSI"

Description	Video on VLSI Circuits
Title	"VLSI Circuits and Systems"
Author	Mr. John P. Uyemura
Format	Mpeg
Duration	2 hours
...	...
Owner	Mr. Smith

Usage Constraints for "VLSI" video

Mr. Brown	Everyday	10:00-18:00
Researchers	Wed to Fri	9:00-13:00
Smith's students	Wednesday	14:00-16:00

Subscriber's Roles on Shared Resources

The RAMS system also allows producer agents to attribute roles to consumer agents on their resources. Like resource usage restrictions, roles can be attributed to a specific collaborator or group. The set of roles varies from a resource type to another. For instance, a collaborator may consult, review, or modify a Web document, whereas he may remotely use, configure or download a software.

Resource Usage Restrictions

Two types of physical resources can be distinguished: the public and private ones. Public resources are generally owned by a non-human resource, e.g., an organization department. Some public resources might have usage restrictions defined in terms of time, e.g. a group of collaborators must search a free slot in order to use a meeting room. In the case of other resources (e.g., printers) such a restriction could be irrelevant as the use time per person is relatively short, but it could be required to limit the use for a specific role (e.g., students).

By contrast, private physical resources belong to a collaborator or a group. Thus, the RAMS system allows a producer agent to define usage restrictions based on different criteria. For instance, they can be expressed in terms of time (e.g., cluster usage schedule) and/or in terms of results (e.g., maximal number of printed sheets per month). Like roles, usage restrictions may also vary according to the collaborator or group to which they are associated.

Awareness Information Filtering

Resource awareness information (e.g., presence, location and availability) is notified by means of events. Consumer agents generally receive a subset of the published events. Thus, filtering is the process of selecting events for processing and transmission. Two forms of filtering have been identified:

- **By topic:** events are published on logically designated channels. Thus, consumer agents will obtain all the events published on the channels to which they are subscribed. Producer agents are responsible for the definition of event classes, which correspond to channels. For instance, let us consider the following topics: Printers, Displays, Scanners and Videos. If a consumer agent is subscribed to the Printers topic, then he will receive information about all published printers (e.g., high and low resolution as well as monochromatic and color printers) even if he looks for a high resolution and color printer.
- **By content:** events are notified to a specific consumer agent only if their attributes or contents fit in with his defined requirements. Consumer agents are responsible for event classification. For instance, if a subscriber specifies some attributes of the required printer e.g., high resolution, color and output device, he will receive information concerning all published high resolution color printers, but also high resolution color displays, PC screens, etc.

The proposed RAMS system combines the previously presented forms: producer agents publish events by topic, whereas consumer agents subscribe by content to one or more topics. We select this filtering approach as combining the advantages of two forms, i.e. it relieves consumer agents of information classification (by topic) and filtering is more fine (by content). For instance, if a consumer agent is subscribed to the Printers topic and also defines some attributes (e.g., high resolution and color) then he will only receive events about the published high resolution color printers.

3.4   Resource Dependencies

The definition of resource dependencies is based on three types of relationships:

- The **ownership relationship** establishes a m to n association between producer agents and their resources. Producer agents can also attribute roles to consumer agents and define usage restrictions on their resources.
- The **location relationship** establishes: a) a m to n association between collaborators and places; b) a m to n association between electronic resources and sites; and c) a m to 1 association between physical resources and places. A computer resource location is fixed by the owner and can be punctually modified. By contrast, as a collaborator's location can change over time, it is regularly computed by the *face recognition system* (cf. Section 4).
- The **collaboration relationship** establishes a m to n association between human resources and groups. In addition, an office schedule is associated to each human resource.

Presence and Availability of Resources

In the case of physical resources, the RAMS system relies on networking functionalities (e.g., in the case of a network printer) or application events (e.g., in the case of a meeting room) to determine whether a resource is present or not. Thus, a physical resource is considered either: a) *present* whenever it is reachable, e.g., an on-line printer or an usable meeting room, or b) *absent* otherwise, e.g., the printer is out of order or the meeting room is closed to be repaired.

When a resource is present, the RAMS system is able to inform the consumer agent whether the resource is available or not. As some physical resources (e.g., a cluster) can be remotely exploited, resource availability mainly depends on the usage restrictions defined by the producer agents. However, when physical resources are located in restricted access places (e.g., offices), the RAMS system relies on the *inference engine* to determine resource availability.

The *inference engine* uses application and caught events (see Fig. 2) to infer new events. Thus, by means of the defined ownership and location relationships, the *inference engine* is able to determine whether a producer agent and a given resource are currently co-located. Moreover, several persons (e.g., administrators) may be authorized to provide access to a room. It is important to notice that location of a room owner is required only if this room needs his presence to be open. This characteristic must be specified when declaring a room resource.

Nevertheless, when a resource is located within a restricted area, resource availability is inferred not only from the usage restrictions and the correspondence between the resource and the owner current locations, but also from the owner's availability. Thus, the RAMS notifies the consumer agent that the resource is available if the following conditions are true: a) the current context, e.g., time, is in accordance with the resource usage restrictions, e.g., current time is within the time slots during which the printer may be used or the meeting room is free (C_1); b) the owner is currently located within the resource room (C_2); c) the owner is available (C_3); d) the resource is functional, e.g., the printer is online and ready to print (C_4); and e) the resource is free (C_5).

Information about C_1, C_2 and C_5 are implicitly captured by the RAMS system sensors, while that of C_3 and C_4 are explicitly captured by users. Thus, the RAMS system allows the producer agent: a) to declare himself either available or unavailable and b) to declare his resource unavailable when it is not functional. The *inference engine* implements a sophisticated approach to infer resource availability, which relies on collaboration relationships among users. For instance, according to our scenario (cf. Section 3.1), Mr. Brown owns a plotter, which is located in his office. If Mr. Brown is absent, but Miss White is in his office and declares herself available, then Mr. Brown's authorized colleagues may access his devices located there.

If one of these conditions is false, the RAMS system notifies the consumer agent that the resource is unavailable. For instance, if the *face recognition system* (cf. Section 4) detects Mr. Brown within the secretary's office, the *inference engine* will notify his authorized colleagues that Mr. Brown's plotter is temporarily unavailable.

4   Face Recognition System

As previously noticed in our scenario (cf. Section 3.1), Mr. Brown is working within a distributed environment in which he is able: 1) to move from a PC to another and b) to use mobile devices such as laptops or PDAs. However, Mr. Brown has also to be considered as a (human) resource that other colleagues may want: a) to physically meet (informal or formal meeting with co-presence of the members), b) to establish a multimedia teleconferencing for discussing a special point, or c) to localize in order to send and show him some information he asked or might be interested in (e.g., relatively to his current location, printing of a document on the closest printer).

Localizing the User Instead of His Device

The RAMS system is particularly focused on better localizing users to make information closer to them and, by this means, allowing users to easily localize each others within the ubiquitous cooperative environment. The goal is then to follow each collaborator rather than to localize his mobile device, from which he may be regularly separated for a while! By this proposal, we do not want to negate the real interest in locating mobile devices by other approaches, e.g., triangulation of WiFi signals [14], which constitutes a complementary way. Rather, we aim at highlighting the possibility that an efficient face recognition system may provide to support cooperative mobile work.

A computer vision-based system for the recognition of human faces requires a learning phase before the testing phase, i.e., the effective real-time face recognition. The learning phase is carried out only once, while the testing phase takes place every time a face is captured by a camera. In order to develop a robust face recognition system, we have combined several techniques as described below.

The learning phase implements a specialized algorithm that is able to differentiate a human face from other thing. For every person that we want to identify, several pictures of his face in different positions are required, e.g., full-face portrait, profile, right or left three-quarter portrait. These pictures may be customized with accessories, e.g., glasses, hat, moustache, earrings, different hair colors (see Fig. 3 ref. #1). A picture database is generated using the OPENCV (Open Source Computer Vision) library and then analyzed with an Eigenface approach [16], which helps creating a classification model (see Fig. 3 ref. #2) that includes information to differentiate one person's face from another.

The OPENCV library is intended to facilitate the implementation of computer vision applications. This library offers functions to support human face detection. Particularly, we use a face recognition tool that supports object detection

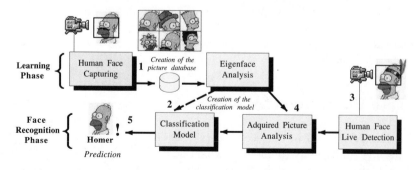

Fig. 3. Learning and Testing Phases of the Face Recognition System

based on the principle of Haar-like features [17]. This principle consists in codifying existing contrasts in some areas of the image (e.g., eyes, nose, mouth, forehead, chin and cheeks) to obtain particular relationships among similar objects (i.e., human faces). The used OPENCV implementation has been trained with hundreds of human face samples in order to facilitate detection.

The Eigenface analysis is based on appearances, i.e., the visual aspect of persons. As a result of applying the Principal Component Analysis (PCA) [12] on a representative set of face images corresponding to the persons to be identified, the Eigenface analysis creates low-dimensional representations, called *eigenfaces*, which contain the most significant variations in the face images of the testing set. To calculate such eigenfaces, the *face recognition system* first determines the mean of the testing set to calculate the covariance matrix. Then, the system gets the eigenvalues and eigenvectors as well as the deviation from the mean in order to obtain a set of eigenfaces. In fact, only the eigenfaces that present the most significant changes of the testing set are selected. Finally, based on the calculated eigenfaces and on the testing set of images, the *face recognition system* obtains the most representative weights, which serve as input data to the LIBSVM (Support Vector Machine) library [3]. This library creates a classification model that determines whether the captured picture belongs to one of the registered users. The *face recognition system* carries out these steps only once when the learning phase takes place.

Once the learning phase is achieved, the real-time face recognition (testing) phase takes place. This phase first captures a picture (see Fig. 3 ref. #3) using the OPENCV library, when a human face is detected. This picture is then analyzed according to the information produced by the Eigenface analysis (see Fig. 3 ref. #4) during the learning phase. Finally, the resulting information of this analysis combined with the classification model (see Fig. 3 ref. #2) allow the LIBSVM library to establish a correspondence (see Fig. 3 ref. #5) between a person and his recent captured picture. Installed as a main part of the ubiquitous cooperative environment, the *face recognition system* has to run in continuous mode, thus periodical updates are required.

5   Related Work

ProxyLady [4] is an application for PDA devices that supports casual interaction, i.e., a collaborator first informs ProxyLady about his intention to communicate with a specific colleague. When the requester and the requested collaborator become physically close: a) they are notified of their respective nearness and b) the requested collaborator is informed of the requester's intention. ProxyLady also considers privacy and intrusion problems as it allows collaborators to activate an "invisible mode" in order to be unnoticed by their colleagues. To scan the proximity area, ProxyLady relies on the bluetooth technology, whose signal range remains very limited and costly. Thus, each collaborator's presence is notified to the other, only if the two collaborators' PDA devices are "close". However, information about collaborator's availability and location is not provided.

Although ProxyLady allows collaborators to share private electronic resources (e.g., documents), it does not offers any awareness information about them. Support for physical resource sharing is out of the scope of this application.

CHIS (Context-Aware Hospital Information System) [14] allows managing information within a hospital using PDA devices. This application relies on WiFi signal triangulation to determine the PDA physical location. Instead of users, PDA devices are associated to roles, e.g., doctor, nurse, and anesthetist. Thus, based on the PDA location and role, CHIS offers context-aware medical information, e.g., when arriving to a patient's room, a nurse is informed of this patient's medication. CHIS provides users with role-based presence and PDA location information, but it does not take people availability into consideration. This application also supports physical resource sharing, but it is limited to public resources, i.e., non-subjected to user-defined usage and access restrictions. Moreover, presence and location information about physical resources is explicitly captured (by users) throughout a map.

In spite of the fact these projects propose interesting solutions that focus on providing some relevant features (e.g., PDA proximity detection using the bluetooth technology and determination of PDA location based on WiFi signal triangulation), they offer few, limited and non systematically updated information about the availability and accessibility of the shared resources distributed within the ubiquitous cooperative environment. On the other hand, it is also important to situate our proposition relatively to some relevant service discovery systems such as Jini [6], Ninja SDS (Service Discovery Service) [13], or SLP (Service Location Protocol) [11]. These systems are designed to reduce resource management, while increasing service use [7][8]. However, an important challenge of these systems is the suited integration of computer devices and human users. In fact, the majority of these systems only support interaction between computers and applications acting as clients.

The SLP protocol mainly supports client applications although it provides some additional facilities to support humans. These facilities allow users to briefly describe (by means of templates) physical device services (e.g., printing) in order to register them in the system or to use them for searches. In addition, SLP implements the concept of "scope" that allows to create service management groups, e.g., network managers attribute a set of services to a scope in order to create groups of services, which can be easily identified.

The Ninja SDS system provides a more flexible way to support humans because they can describe services and searches by means of XML documents. Finally, Jini offers a template-based mechanism to capture information about services (e.g., physical devices and applications), which are represented by means of Java objects. Although Jini also provides a better support for humans, the attributes required to describe services are reduce to the minimum. Consequently, client-side filters must be implemented in order to carry out finer searches.

Although these service discovery systems facilitate service sharing, none of them is focused on supporting human-human interaction nor management of awareness information about physical, electronic and human resources.

6 Conclusion and Future Work

The main objective of this work is to define principles and mechanisms to efficiently manage and provide up to date awareness information about the different resources in order to support distributed, mobile and ubiquitous cooperative work. Thus, all resources are parts of the cooperative environment in which their state may evolve depending on other resources. Relations between resources are managed to infer the availability property of each of them. For example, if a professor owns a scanner that is located in his office and gives to a student the right to use it; however, as the professor is currently within a meeting, this resource is presented within the student's cooperative environment as temporarily unavailable. A collaborator is also considered as a regularly moving (human) resource with which his colleagues can initialize a cooperative working session that uses other physical and/or electronic resources.

Thus, in addition to all common pre-declared resources (e.g., rooms, whiteboards, servers), collaborators can publish the different resources (technical characteristics, roles, usage restrictions) that they propose to share with some others. In a symmetric way, potential consumers can subscribe to some resources. Using a suited inference engine, the RAMS system processes all these resource declarations and inter-resource relations to provide collaborators with actual and updated awareness information.

In order to allow the RAMS system to offer such functionalities to collaborators, we have proposed a sensor-based architecture that includes two sensors: the first one is a *resource locator* which is in charge to determine the set of the closest resources that are suitable, available and accessible for a given requester and it depends on his current location. The second sensor is a *face recognition system* based on the Eigenface analysis that allows to identify and locate collaborators within the cooperative environment.

The developed work for designing and implementing the second sensor takes its source from a main observation: as we can see examining existing projects, the proposed solutions remain limited. In fact, the central mobile and dynamic collaborator remains a confusing and implicit notion or a bit better he is centered on the detection of his mobile devices. This is not enough to provide required efficient resource management functions. Mobile devices cannot be considered as sticked to collaborators and by this way, we propose a complementary *face recognition system* that is able to efficiently locate collaborators even when they are separated from their mobile devices. Improvements of this tool are still in development, especially to avoid or reduce confusing predictions that, for example, come from some appearance differences (e.g., hat, moustache, rings). To provide a more complete localization function, we also plan to develop a third sensor able to detect mobile devices. Of course, locating collaborators implies privacy and intrusion problems that are treated considering the users as resources.

A prototype of this ubiquitous cooperative environment has been successfully developed, but we plan to install and validate it in a widely distributed real environment including several buildings.

References

1. Abel, M.: Experiences in an Exploratory Distributed Organization. In: Intellectual Teamwork: Social Foundations of Cooperative Work, New Jersey (USA), pp. 489–510. Lawrence Erlbaum Associates Inc, Mahwah (1990)
2. Berard, F., Coutaz, J.: Awareness As an Extended Notion of Observability. In: CHI 1997 Workshop on Awareness in Collaborative Systems, Conference on Human Factors in Computing Systems SIGCHI 1997, Atlanta GA (USA), vol. 29(4). ACM Press, New York (1997)
3. Chang, C.-C., Lin, C.-J.: LIBSVM: A Library for Support Vector Machines (2001)
4. Dahlberg, P., Sanneblad, J.: The Use of Bluetooth Enabled PDAs: Some Preliminary Use Experiences. In: Proceedings of IRIS 23 Laboratorium for Interaction Technology (2000)
5. Dourish, P., Bly, S.: Portholes: Supporting Awareness in a Distributed Work Group. In: Proceedings of Conference on Human Factors in Computing System, CHI 1992, Monterey, CA, USA, pp. 541–547. ACM Press, New York (1992)
6. Keith, E.W.: Core Jini, 2nd edn. Prentice Hall, USA (2000)
7. Keith, E.W.: Discovery Systems in Ubiquitous Computing. IEEE Pervasive Computing 5(2), 70–77 (2006)
8. Zhu, F., Mutka, M.W., Ni, L.M.: Service Discovery in Pervasive Computing Environments. IEEE Pervasive Computing 4(4), 81–90 (2005)
9. Gamma, E., Helm, R., Johnson, R., Vlissides, J.: Design Patterns: Elements of Reusable Object-Oriented Software. Addison-Wesley, Reading (1995)
10. Greenberg, S., Johnson, B.: Studying Awareness in Contact Facilitation. In: CHI 1997 Workshop on Awareness in Collaborative Systems, Conference on Human Factors in Computing Systems (SIGCHI 1997), Atlanta, GA, USA, vol. 29(4). ACM Press, New York (1997)
11. Guttman, E., Perkins, C., Veizades, J., Day, M.: Service Location Protocol Version 2. IETF RFC 2608 (June 1999)
12. Hernández, B., Olague, G., Hammoud, R., Trujillo, L., Romero, E.: Visual Learning of Texture Descriptors for Facial Expression Recognition in Thermal Imagery. Computer Vision and Image Understanding 106(2–3), 258–269 (2007)
13. Hodes, T.D., Czerwinski, S.E., Zhao, B.Y., Joseph, A.D., Katz, R.H.: An Architecture for Secure Wide-Area Service Discovery. Wireless Networks 3(2/3), 213–230 (2002)
14. Markarian, A., Favela, J., Tentori, M., Castro, L.A.: Seamless interaction among heterogeneous devices in support for co-located collaboration. In: Dimitriadis, Y.A., Zigurs, I., Gómez-Sánchez, E. (eds.) CRIWG 2006. LNCS, vol. 4154, pp. 389–404. Springer, Heidelberg (2006)
15. McDaniel, S.E.: Providing Awareness Information to Support Transitions in Remote Computer-Mediated Collaboration. In: Proceedings of the Conference Companion on Human Factors in Computing Systems: Common Ground (CHI 1996), Vancouver, BC, Canada, pp. 57–58. ACM Press, New York (1996)
16. Turk, M., Pentland, A.: Eigenfaces for Recognition. Journal of Cognitive Neuroscience 3 (1), 71–86 (1991)
17. Viola, P., Jones, M.: Rapid Object Detection Using Boosted Cascade of Simple Features. In: Proceedings of the Conference on Computer Vision and Pattern Recognition (CVPR 2001), Hawaii. IEEE, Los Alamitos (2001)

Context Awareness and Uncertainty in Collocated Collaborative Systems

Roc Messeguer[1], Pedro Damián-Reyes[2], Jesus Favela[2],
and Leandro Navarro[1]

[1] Department of Computer Architecture, UPC, Spain
{messeguer,leandro}@ac.upc.edu
[2] CICESE, Mexico
{preyes,favela}@cicese.mx

Abstract. Context awareness is a necessary feature for mobile collocated collaborative learning. In this paper we describe how requirements for context-aware cooperative learning activities are derived from the jigsaw technique augmented with the use of mobile devices, applications to support the activities of groups, and tools to provide context-awareness to detect group formation. The emergence of groups is detected based on the location of the students within the classroom, but this information has to be careful filtered to evaluate the degree of uncertainty and protect from erroneous estimations. A three-phase strategy to manage uncertainty by identifying possible sources of uncertainty, representing uncertain information, and determining how to proceed under the presence of uncertainty is used for this propose. These requirements are validated and confirmed in experiments with students working together in the classroom, measuring neutral or positive effects on learning and the usefulness of introducing mobile devices, group support applications, and context aware-ness. The ratio of unwanted interruptions to users made by the system is used to evaluate the utility of the system. Results show that by managing uncertainty, location estimation becomes more reliable, thus increasing the usefulness of the learning application.

1 Introduction

In traditional learning environments, students are generally regarded as passive learn-ers. Assessments of student learning are generally based on their individual work such as quizzes, examinations and tests. Each student competes with his/her peers to obtain the highest score. Thus, in this method of teaching and learning, educational content is teacher-directed and learning is individualistic. In this context, the content is deliv-ered to the learners by the teacher and the students rely mainly on the teacher, the knowledge expert, for their knowledge and information.

In contrast, cooperative learning is an instruction method based on students working together in small groups to accomplish shared learning goals [10]. Students work and cooperate among themselves, helping each other to achieve the learning goals. This learning mode is student-centered and encourages students to cooperate and collaborate with each other in achieving their learning outcomes. There are many collaborative and

R.O. Briggs et al. (Eds.): CRIWG 2008, LNCS 5411, pp. 41–56, 2008.

cooperative learning methods, which also can be considered as group learning methods and used in both classroom-based and web-based environments. One of the methods adopted for achieving Cooperative Learning is The Jigsaw.

As computer-supported collaborative learning (CSCL) applications that support group tasks are introduced in collocated cooperative learning settings, the overhead imposed by these tools on students and the instructor also increases. In particular, introducing tools to support groups imply that the instructor has to manually assign the groups, and students have to wait until the instructor completes this task. This is not convenient as it stresses the already overloaded instructor and introduces delays and additional burden to any participant in synchronous collaboration that occur in real time, progressing in parallel. Although there are many tools for providing some degree of automation and support for activities within a group, we haven't found specific tools to automatically provide the applications with awareness of environmental changes, particularly on the organization of the classroom: mainly knowing in real time the structure of the classroom in groups and the roles assigned to each participant. The changing absolute or relative location of every significant element in the workplace (people or artifacts) is a rich source of information to understand the structure and performance of the collaborative task.

We intend to use this information to automatically infer contextual information [2] (groupware context) that facilitates CSCL support. In this sense, by context we refer to any information that can be used to characterize an entity, where the entity can be a person, place or object that is considered relevant for the interaction between a user and an application, including the user and the application itself.

A context-aware system has the capacity to perceive and capture the world surrounding the user with the goal to adapt its behavior to provide information and services that are useful and relevant to that place and time [1].

The context is characterized in different levels of abstraction: low level and high level. Low-level context is obtained directly from physical sensors, such as temperature, luminosity, noise, movement, air pressure, environmental humidity, among another. Whereas high-level context is abstract and inferred from low-level context. For example, the activity, or the state of mind of the user is context that can be inferred using low level context [15].

However, these low level mechanisms can provide uncertain information, because the contextual information can be imperfect, due to a flaw in the devices of perception or due to an error in the estimation or treatment of the contextual information. If this uncertainty is not considered and appropriately managed, a context-aware application might become unusable.

The uncertainty in the context can originate from different sources, among those that highlight the incomplete, wrong or ambiguous information, for this reason a reasonable doubt must always be maintained on all the available information. Some methods have been suggested to deal with uncertainty in the context. Mainly Bayesian networks and ontologies have been used to tackle this problem [7, 16, 18], some others base on the creation and chaining of rules [19], using mediation, a dialogue between the user and the system [3], and the use of fuzzy logic [8, 16].

In our case we use a strategy for the administration of the uncertainty following three main steps: identifying and measuring the uncertainty and establishing actions to carry out in the presence of uncertainty.

The substantial contributions of this paper are: First, a list of requirements for a CSCL application in a mobile and collocated collaborative learning environment, and second the analysis of a strategy to provide the requirement "The system should automatically form real and virtual groups of students using the current context. (dynamic teams)." Finally, we propose to use the unwanted interruptions to users made by system and their cost [9] as a metric to evaluate the utility of the system.

2   A Brief Review of Collaborative Learning and the Jigsaw Methodology

Cooperative learning is an instruction method based on students working together in small groups to accomplish shared learning goals [10]. To make collaborative learning a success, there must be some kind of "glue" that holds the group together. Group members must feel they need one another, must want to help each other learn, and must have a personal stake in the success of the group. They also must have the skills necessary to make the group work effectively and be able to regularly analyse the group's strengths and weaknesses to make adjustments as needed. [12] Those experienced in successful small group work have found five essential components (i.e., the "glue") that are necessary: Positive interdependence, Face-to-face promotive interactions, Individual accountability and personal responsibility, Teamwork and social skills and Group processing.

Successful collaborative learning requires effective implementation of student groups. Subdivision of the class into formal groups require more planning as to the size and have the same group members throughout its existence. In general, groups should be heterogeneous, so that in each group the different levels are represented, as well as both sexes and different socio-cultural backgrounds. For instance, students can be chosen randomly from an attendance roster or they can count off. For a class of 50 students working in teams of 5 students, count off from 1 to 10 and then have each number meet in a specified place in the room [12].

2.1   The Jigsaw Methodology

The collaborative learning methodology followed in the experiments was a jigsaw, applied with success in technical courses in the literature [4] and in our classes.

The basic premise of a jigsaw is to divide a problem into sections. Home groups are formed (fig. 2), with each team member taking responsibility for one section of the problem in question. Each student receives resources to complete only his part and becomes expert in this subject. Expert groups are then formed. Students who are responsible for the same section join together and form a new, temporary focus group whose purpose is for the students to master the concepts in their section, and to develop a strategy for teaching what they have learned to the other students in their original collaborative learning group. After the expert groups have completed their work, the home groups, the original collaborative learning groups re-assemble. The students then teach one another the sections they have worked on. To ensure individual accountability, the students can be evaluated on all sections of the task.

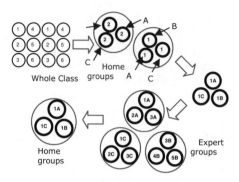

Fig. 1. The Jigsaw methodology

The stages of the jigsaw activity we followed are (fig. 1): 1.- Introduction of the topic (whole class), 2.- The teams go over the problem in question and assign a section to each member. (by group), 3.- Individual work of each section of the problem (by student), 4.- Expert groups work to master the concepts of their section (by group), 5.- Home groups work to connects the various section to answer the problem in question (by group), and 6.- Evaluation (by student and/or by group).

The jigsaw structure promotes positive interdependence and also provides a simple method to ensure individual accountability [11]. Positive interdependence means that team members need each other to succeed. In the Jigsaw, it relies on the fact that each individual possess specific resources needed for the group as a whole to succeed. This can be arranged by giving specific resources to different individuals in the group [12]. Individual accountability and personal accountability: to accomplish this, students must contribute their fair share. The instructor must structure the groups so that individuals do not have an opportunity to "hide". To ensure individual accountability, the students can be evaluated on all sections of the task [11]. Most cooperative learning experts agree that the approach works best if team grades are adjusted for individual performance. If this adjustment is not made, students who do little or no work may receive the same credit as those who do a great deal of work, which is unfair and works against the principle of individual accountability. We use peer ratings as a basis for team assignment grade adjustment [13].

The jigsaw activity implies several reorganizations of groups during the activity that is a critical and demanding task for the teacher which may require detailed planning and taking notes during the activity to track deviations from plan to facilitate further evaluation. Therefore dispensing the teacher from most part of this overhead is a goal for our work.

3 Our Case Study

The purpose of this study was to analyse and validate the requirements for the mobile collocated collaborative learning scenario, in particular the requirements for context awareness in terms of the detection of the structure of groups, and the consideration of the quality and uncertainty of that information used to immediately detect the organization of students in groups.

This experiment has been done at the EPSC campus of UPC, an engineering school built and designed to be based in the collaborative learning and project based learning models. Classrooms are equipped with tables and chairs having collaborative learning in mind. Experiments have been done in a subject where each student has a laptop connected via a WiFi network and they interact with one or more collaborating peers to solve a given problem.

This was applied to a course with 28 Telecommunications Engineering students. The students and instructors have experience in using the jigsaw methodology in different courses, only with paper and pencil, and also with laptops just to take notes, but not with additional support for communication and collaboration.

4   Requirements for Mobile Collocated Collaborative Learning

In this section, we determine the requirements for an application in a mobile/dynamic collocated collaborative learning environment by describing two typical use cases and pointing out how this application would improve the learning process. The first use case describes a student activity (see section 4.1) while the second describes the same activity but viewed by the instructor (section 4.2).

4.1   The Viewpoint of the Student

Pepe arrives to the classroom and, as all other students goes to collect a laptop to perform the activity. When all students are seated the instructor describes today's activity and the expected deliverables at the end.

The instructor creates the groups assigning a number to each student. Pepe has number 2 and looks for other students with the same number. All of them are going to form a working group for that activity. After students move around the classroom, Pepe finds his group companions. As a group, they grab a pair of tables and 3 chairs to "create" their own work space. The instructor delivers 3 exercises (A, B and C) in the campus web site. Pepe and his companions take a rapid look at the exercises and distribute one for each of them. Pepe has got the exercise C. In this moment, the instructor informs them they have 20 minutes to, individually, understand the assigned problem and attempt to solve it. Pepe starts to solve it, although he is not able to reach a conclusion. He asks the instructor for help. The instructor tells him to write down the steps followed in the development of the question and the results, together with the doubts he has.

After 20 minutes of work, then comes the creation of the groups of experts. The 4 students that have worked on exercise C form an expert group. The instructor informs they have 20 minutes more to solve the doubts, agree upon a possible solution for the exercise and prepare a few notes on how to explain it to his initial group colleagues. Pepe and his new companions share notes and discuss how each has solved the exercise, trying to clarify and resolve their doubts. They discuss on the potential conclusions but no agreement is reached. The instructor helps them to understand the exercise and reach to a correct conclusion that they write down as a summary.

Afterwards, the instructor informs that the time is over, and now they have to return to the original group to explain each other the exercises. Pepe has to come back to his home group. Finally the instructor tells them to put together the three exercises. The home group starts to write up the final report for the activity with their results and their conclusions on the final complex problem. The three of them work together on the same laptop to write up this report taking fragments from the previous documents.

To finish the activity, each of them individually has to perform a small test on the campus web site about concepts exercised during the activity.

A CSCL application that allows a mobile and dynamic collocated collaborative learning could have supported Pepe by enabling him to: 1.- Know which group he belongs to 2.- Write diaries or notes of his activity or experiments (individual document space) 3.- Share their diaries or notes (group document space) 4.- Use a collaborative group editor to assemble the final report, and 5.- Manage report and test deliverables.

4.2 The Viewpoint of the Instructor

Juan, the instructor, helps students to distribute the laptops inside the carriage. He explains the activity for today's class. He insists on the individual accountability of cooperative learning. Each student has to be accountable for the group results and that the individual mark depends on the marks of the group companions.

Juan is going to organize the groups. Students are numbered. He tells them to form groups by numbers. Juan waits a couple of minutes until all groups get together. As the groups organize, he goes around each table and delivers the forms for each part (A, B and C) of the jigsaw. Juan tells them to distribute the exercises and that reminds they have 20 minutes to work individually on them. A couple of students more call the instructor. The first has difficulties to understand the exercise, Juan does the necessary clarifications. The second has problems with his computer, there is not enough battery. During the 20 minutes of individual work, Juan continues to resolve the doubts of the students with the exercise, doubts with the use of the tools and interpretation of the exercise values. There are some students who stand up and clarify their doubts with students from other groups. 5 minutes before the end of the class he asks everyone to annotate their doubts, the steps followed on the solution and the results obtained.

Now Juan informs that the time is over and creates the groups of experts. He indicates to create 6 groups (two A, two B and two C). He assigns 20 minutes to jointly solve the doubts, the solution and conclusions. During the work in groups of experts, Juan solves a few doubts. The students inside the same group solve most of them. There is almost no movement of students asking or talking with other groups. However, Juan continues to handle technical problems.

Juan announces the time is over. He instructs the students to go back to the home groups. As the groups reorganize Juan delivers a list of questions and conclusions to resolve regarding the combined problem, sum of the three parts. In the classroom there is quite amount of movement, essentially people needing data or any conclusion from the work of experts and they don't have them properly written down. 10 minutes before ending the activity, Juan asks them to leave the reports and prepare to perform a short individual test.

A CSCL application working on a mobile and dynamic collocated collaborative learning could have supported Juan by enabling him to: 1.- Create and manage automatically and dynamically groups 2.- View and record for later inspection a real time classroom/activity snapshot, and 3.- Manage report and test deliverables

4.3 Summary of Requirements

The list of requirements obtained from the previous cases can be summarized as follows:

- R1: A student should be able to write and read his own documents, diaries or notes of learning activity (personal workspace)
- R2: A student or a group should be able to submit deliverables, and the instructor should be able to access and evaluate them (repository of deliverables)
- R3: A student should be able to move around the classroom, even further, with a computer: a mobile device connected to a wireless network (Mobility of students)
- R4: Students need an adequate space to work in groups (tables, chairs, etc)
- R5: Team-mates should be able to share their (digital) documents, diaries or notes (team repository)
- R6: Team-mates should be able to work together with the same text or document (collaborative workspace)
- R7: The system should automatically form real and virtual groups of students using the current context. (dynamic teams)
- R8: An instructor should have a real-time view or snapshot of the classroom to provide awareness of the activity going on and to support evaluation.

Most of these requirements have already been worked and proved in other studies. The 4th requirement is not technological. It is a prerequisite for working in groups at the classroom. It is covered with a classroom adapted to it. Electronic learning or E-learning is a general term used to refer to computer-enhanced learning. These systems cover the requirements R1 and R2 perfectly. Mobile learning or M-learning happens across locations, or that takes advantage of learning opportunities offered by portable technologies. In the technological evolution chain (Fig. 1), it corresponds with the second steep, Mobile Computing; so R3 is covered by mobile learning. Moreover, electronic and mobile learning with the right computer applications can become a Computer Supported Collaborative Learning (CSCL) environment that focus on the socio-cognitive process of social knowledge building and sharing. The CSCL systems provide tools to satisfy the R5 and R6 requirements.

A goal for this work was to satisfy requirements R7 and R8 in collocated collaborative learning.

5 Effects of Supporting Learning in the Performance of the Class

To assess the importance of some requirements, we performed experiments with groups of students from a course to distinguish the contribution of three aspects that we want to evaluate with respect to the requirements that were identified: Computer support (R1, R2) support for mobility (R3, R4), computer-support to the activity of the group (R5, R6), context-awareness (R7, R8). For that, we have defined four different situations:

a) Students use desktop (not mobile) PC, no computer-support for collaboration. This is the control group.

b) Students and groups use laptops (thus adding support for mobility), but no computer-support for collaborative tasks.
c) (b) adding a collaborative workspace software providing screen sharing for groups (thus adding support for collaboration), no automatic detection of groups (done manually by the teacher).
d) (c) adding location awareness to provide automatic context-awareness: groups with laptops, shared group folder application and automatic group detection.

5.1 The Process of Observation / Evaluation

Our observation and evaluation are focused on the impact of the technology in the learning process. In [14], they propose different methods of process-oriented evaluation. In our case study, we evaluate a real activity. The students must perform their work and then be evaluated. Therefore, we use this information assessment to complement with other methods.

The observation and evaluation of the experiments is based on four sources: 1.- Individual assessment, an individual quiz on the topics covered in the class, 2.- Group's outcome assessment, evaluating the activity group report, 3.-The opinions of students obtained from the Critical Incident Technique or CIT questionnaire [6] used for collecting direct observations of human behaviour that have critical significance[1], and 4.- The observations from the instructors (an ethnographic technique: Direct, first-hand observation of daily behaviour).

5.2 Results and Findings

Table 1 shows the results obtained from experiments in all scenarios measured against the individual assessment (average and standard deviation) obtained from a quiz on the topics covered during the activity. The knowledge acquired by each individual does not change, or at least not clearly (variations up to 10% in averages), with the scenario (different levels of support) where the activity was performed.

Table 2 shows the results from experiments in several scenarios measured against the group assessment (average and standard deviation) obtained from evaluating the activity group report. The assessment obtained by the group (table 2) does have a clear dependency (variations up to 40% in averages) with the scenario where the activity was performed.

Table 1. Individual accounting (scores from a quiz on the topics covered in the class)

	Control group	Mobility	Mobility + collaboration	Mobility + collaboration + context
Average (0-10)	8	7.4	7.9	8.2
Std Deviation	2.5	2.3	2.6	2.5

---

[1] A critical incident can be described as one that makes a significant contribution - either positively or negatively - to an activity or phenomenon.

Table 2. Group's outcome accounting (scores from evaluating the activity group report)

	Control group	Mobility	Mobility + collaboration	Mobility + collaboration + context
Average (0-10)	6.25	7.0	8.4	9.1
Std Deviation	1.2	2.3	2.6	1.9

We have found that deriving group membership information from location information based on WiFi networks is technically viable, can be incorporated in CSCL applications and the use is beneficial for group participants using CSCL applications. The effect can be perceived in terms of improvement on the learning outcomes and thus in student's qualifications. As we add further support to the scenarios the outcomes of the work in groups improve, and therefore collaborative groups become more efficient.

The results from the CIT questionnaire support these findings. The great majority of students affirm that either the use of laptops (21/28 75%), or the use of a collaborative system (17/28 60%) have been useful in the activity. However, when more details are requested, some responses supporting that statement seem to be less reliable as the higher motivation seems to come from the technological novelty. Other responses highlight that these scenarios are more adapted to the activity. A few of them observe an improvement of group work when a collaborative system was used, or an improvement in the work in groups and in mobility in the scenario using laptops. Among the negative opinions, the duration of the activity appears at the top. Some students claim that both laptops and the collaborative software only contributed to spend time by using unknown programs or computers. This was probably due to the lack of experience of the instructor and students who had the additional work of learning to use the tools as this didn't appear in further sessions.

Finally, among the observations from the instructors, several technical problems are reported, unrelated to the planning of the activity: load and change of laptop batteries, lack of experience in using the collaborative application, technical problems to access to the wireless network, etc. These were also solved in further sessions.

5.3 Lessons Learned and the Need for Context Awareness

The initial lessons and obstacles we faced were related to the problems with the location mechanism (errors in the estimation due to the movement of students). The estimation errors are disturbing when they imply a change in group membership and then an interruption in the activity of the students involved.

These estimation errors and uncertainty that cause interruptions in the activity should be appropriately managed otherwise a context-aware application might become unusable. For example when the system gives a wrong estimation, the user is changed to another group or context, interrupting his work, breaking the user's attention on the current task to focus on the interruption temporarily [17] and requiring an action to return to the correct group. There are many types of unwanted interruptions. In this paper we focus on the interruptions generated by the own system and their cost [9] as a metric to evaluate the utility of the system.

Comparing the collected location data from what it really occurred and annotated during the experiments, we learned that: a) students were always part of a group, even when the location data incorrectly indicated they were far from any group or outside the classroom, and b) there are not many frequent group membership changes around the classroom during the activities, even when the location data indicated a student was rapidly moving to several groups: some students just stand up and ask their doubts to colleagues that belong to other groups, the majority of them carrying their laptop. This is useful for the dynamic management of groups, but does not signal a formal group change.

Finally, we find the following requirements useful but they are not yet fully addressed or resolved:

R7: The system should form automatically real and virtual groups of students using the current context. (dynamic teams)

R8: An instructor should view a real time classroom snapshot for activity log and evaluation support.

6   Context-Awareness and Management of Uncertainty for Collocated Collaborative Learning

Our group membership service calculates group membership based on location in-formation with some rules to filter out misleading or unreliable information. This in-formation on group membership enables applications to automatically allocate tools, resources, etc. to students depending on the group to which they belong.

6.1   Formation and Estimation of Working Groups

Our mechanism for determining group membership uses the location of the user to determine the group to which it belongs. For that it collects several data about each student: location (calculated from the received power from the WiFi network using fingerprinting from multiple WiFi access points), identity (calculated from the MAC network address of the laptop assigned to each student) and the time (in fact an event counter calculated from the timestamps of the centralized logging system). Figure 2 presents the general structure of the mechanism of group assignment.

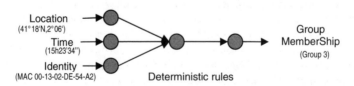

Fig. 2. Group estimation

The first step to test or use our service is to map the classroom: this is done by tak-ing multiple measures of power levels from multiple access points (fingerprinting) at many places within the classroom. With that collection of measures we need to statis-tically test the reliability of the location estimations at each point (a very similar or the same pattern of power can appear at a distant place in the room).

We need to determine the criteria to decide on membership: it can purely be the proximity to some coordinates, the proximity to a certain object (e.g. a table), or to a person (to group leaders). This is therefore very specific to the physical room (but ideally reusable for any activity performed at that room) and the activity performed (regarding the criteria for membership).

After that, the actual activity can take place and our service will be able to guess the group to which a student belongs on the basis of his location within the classroom.

6.2 Management of Uncertainty

In our scenario, we have found that uncertainty appears due to low resolution and lack of accuracy of the contextual information, precisely on the location of every student.

To be able to handle uncertainty firstly we must be able to identify the existence of it. For that, we must create a representation of the uncertainty, with the aim of creating rules that can signal its presence or absence.

Spatial uncertainty occurs when the location information of a laptop points to a incorrect place, the "forbidden zone" (e.g. far away from any groups, outside the room) as in our activity students were always assigned to a group. A rule detects these cases that are signaled as "uncertain".

Temporary uncertainty in group assignment is signaled by a rule every time there is a change in location that leaves the participant in a new group as in our activity students do not change group too often. This rule allows us to evaluate the presence or absence of uncertainty in an assignment to a group (if the change of location and thus group membership is confirmed with further location samples, then that uncertainty disappears).

These two rules help us to know about the presence of uncertainty in the assignment to a group of one student.

This way we classify all the estimations as true or uncertain, following the schema shown in Figure 3.

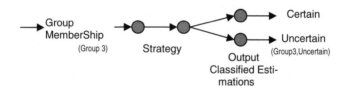

Fig. 3. Management of uncertainty

The utility of this classification is double, first it allows identifying unsafe and probably erroneous estimations, and it also allows to process and to mask these uncertain estimates to improve the precision of the system.

For the case of original uncertainty (uncertainty in the location data: incorrect or multiple possible locations) (Fig. 4), a rule combines that information with other contextual elements. This is, if original uncertainty is detected, the service looks for the nearest student to the current location and assigns the same group to it. The establishment of this mechanism of group assignment is based on the fact that a student is relatively close to his group partners, and there is strong evidence that the closest student to the student with uncertainty is usually a member of that same group.

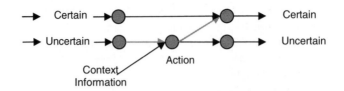

Fig. 4. Action of combination with other contextual elements in original uncertainty

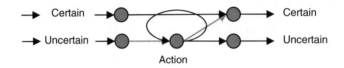

Fig. 5. Action of re-estimation in derivated uncertainty

In the case of temporary uncertainty (Fig. 5), the action to carry out is to wait or perform a re-estimation of location. This is, when the temporary uncertainty is detected the application repeats the process based on a new location sample with the aim of confirming the change of group. It has been found that two consecutive assignments to the same group usually implies that the change of group is correct.

6.3 Simulation, Results and Findings

Using the log data collected in the experiments with students in the mobility+collaboration+context scenario we have evaluated our mechanism (the rules) comparing to what it really happened (the ideal or "true logs").

Table 3. Estimations classified and improved with the strategy of management of uncertainty

Estimation	Correct (%)	Uncertain	Wrong (%)
Home	1546 (95%)	44	30 (1%)
Expert	1579 (97%)	27	14 (<1%)

Table 4. Relationship of wrong estimations among those marked as uncertain

Estimation	Wrong/Uncert
Home	41 / 41
Expert	25 / 27

Table 3 shows the accuracy of this mechanism to detect uncertainty (measured group membership estimations considered correct, considered uncertain with respect to those we knew were wrong). A pattern that we have observed in this application specific scenario is that the great majority of the estimations marked as uncertain, correspond to wrong estimations in reality, as it can be observed in Table 4.

6.4 Effect on the User: Interruptions and Notifications

It has to be noted that not every erroneous or uncertain estimations implies an interruption in the attention of the student. For example, two consecutive erroneous or uncertain estimations are not two interruptions but just one. The first one interrupts the student and changes its focus away from the main activity towards the change of context, tools, group, etc. decided by the system, but the second one does not interrupt. The student already has left his main activity.

For that reason we defined a burst as a sequence of erroneous and/or uncertain estimations. The end of the burst, the return to normality, is identified with two correct consecutive estimations. From the log data we identified the bursts with erroneous and/or uncertain estimations. In the case of an uncertainty estimation burst, without erroneous estimations, the system would not carry out any action: it would just inform (notification) the user of this state of uncertainty, without interrupting the activity of the student.

In table 5 we present the total number of bursts with erroneous and uncertain estimations that can be observed during a concrete activity. We also show the total number of bursts composed by uncertain estimations. Table 6 shows the impact on the activity of the student, measured in interruptions and notifications.

Table 5. Total of bursts of erroneous estimations and in the case with uncertainty management also the bursts with uncertain estimations

Uncertainty management	Without	UIT	
Group/Burst	Erroneous & Uncertain	Erroneous & Uncertain	Uncertain only
Home	211	34	20
Expert	200	18	11

Table 6. Average of interruptions for student and activity with the strategy of management of uncertainty and the average of uncertain information for student and activity

Group	Interruptions	Information
Home	1.1	1.6
Expert	0.5	1.0

6.5 Comments and Discussion

Looking at the requirements presented before:

"R7: The system should form automatically real and virtual groups of students using the current context. (dynamic teams)."

In table 6, we can clearly see that with the original algorithm the student had an average of 7 to 8 interruptions while performing an activity of 30 minutes of duration. We think that this value is too high as it distracts too much the operation of groups and student activities. With uncertainty management, these interruptions go down to one, on average, for each student during the same activity. We believe this value is acceptable and has little impact in the operation of the group or the student.

Therefore we find that a mechanism for the administration of uncertainty, precisely the labelling of the uncertain estimations, is very useful for the design of context-aware applications that assisting the user with automatic group membership detection.

This group membership information can be further exploited as pointed out by R8 *"An instructor should view a real time classroom snapshot for activity log and evaluation support"*. This application has not been built but it would be possible to build and probably very useful for the teacher as a record of the activity (it contains the history of location, identity, time and all estimations of location certainty and groups) for evaluation.

7 Related Work

In [21], working on supporting collaborative activities amon learners, they focus on supporting social interaction, and they design three levels of social interaction functions: Encounter, Communication and Collaboration. In this paper, we focus on the encounter functions of social interaction, while applications use this information to provide the later functions. At the encounter process, we propose to dynamically form groups by adding location information to context learner information. Moreover, we propose to dynamically assign roles to learners.

Previous work that integrates Ubiquitous Computing in learning surroundings with Computer-supported collaborative learning includes [20], which focuses on a context aware environment to support the needs of peer-to-peer collaborative learning in virtual communities. A difference with this paper is that we focus on face-to-face cooperative learning with computer support environments.

[6] discusses on contextual information about groups (team learning context). They focus on workspace and social awareness and they even comment on team formation support: closed and opened teams, joined and left manually and dynamic teams formed automatically by the system based on context and metainformation, but there is no evaluation of it. [22] proposes the use of self-organization in CSCL with the use of macro-scripts, a pedagogical method to organize activities. In some way the design of learning activites can be seen as macro-scripts. In fact they present as an example the "Jigsaw-script family" [23] describes how to integrate tools, individuals and learning material in a flexible manner as computational grids, called "learning grids". In our work contextual information is used to build these grids dynamically. [17] discusses about interruptions on team awareness. It claims that the problem with interruptions are not on the data on in the processing but on the collection and representation and they propose a novel interface.

8 Conclusions

In this paper, we have explored the use of context-awareness in collaborative learning scenarios. Therefore we set up a ubiquitous learning scenario in a mobile and collocated collaborative learning environment. Based on several experiments performed in real classrooms and classes we have reported on lessons learned from such an experience and a list of requirements for a CSCL application.

Automatic derivation of contextual information is particularly needed to satisfy the important requirement of "An automatic and dynamic group manager that supports a student interface with collaborative group tools and document spaces".

In the light of this, we have implemented a system or service that based on location information of the laptops used by the students in the classroom connected to multiple WiFi access points is able to automatically estimate group membership. This information has to be careful filtered to evaluate the degree of uncertainty and protect from erroneous estimations that cause undesirable interruptions to the students. For this propose we used the strategy for uncertainty management based on a structure which consists of three main stages: identification, measurement and dealing with uncertainty. The utility of this strategy is double, first it allows identifying unsafe and probably erroneous estimations, and it also allows to process and to correct these uncertain estimates to improve the precision of the system. Finally we have evaluated the utility in terms of the rate of unwanted interruptions to users' activity made by system. The quality of the filtered location estimates has been found good enough to reliably detect the formation of groups. These results enable the construction of group support applications that effectively assist group members to automatically share, communicate and coordinate as they move and reorganize in synchronous and collocated collaborative learning activities.

Acknowledgments. This work was supported in part by the Spanish Ministry of Science and Innovation under grant TIN2007-68050-C03-01 and by CONACYT through a scholarship provided to Pedro Damian-Reyes.

References

1. Abowd, G.D., Ebling, M., Hunt, G., Lei, H., Gellersen, H.-W.: Context-aware computing. IEEE Pervasive Computing (2002)
2. Borges, M.R.S., Brezillon, P., Pino, J.A., Pomerol, J.C.: Bringing context to CSCW. Computer Supported Cooperative Work in Design (2004)
3. Dey, A.K., Mankoff, J.: Designing mediation for context-aware applications. ACM Transactions on Computer-Human Interaction (2005)
4. Felder, R., Rebecca, B.: Cooperative learning in technical courses: Procedures, pitfalls, and payoffs: ERIC Document Reproduction Service ED 377038 (1994)
5. Flanagan, J.C.: The critical incident technique: Psychol Bull. (1954)
6. Ferscha, A., Holzmann, C., Oppl, S.: Team Awareness in Personalized Learning Environments. In: International Conference on Mobile Learning,(MLEARN) (2004)
7. Gu, T., Pung, H.K., Zhang, D.Q.: A bayesian approach for dealing with uncertain contexts. In: International Conference on Pervasive Computing (Pervasive) (2004)

8. Guan, D., Yuan, W., Gavrilov, A., Lee, S., Lee, Y.-K., Han, S.: Using fuzzy decision tree to handle uncertainty in context deduction. In: Huang, D.-S., Li, K., Irwin, G.W. (eds.) ICIC 2006. LNCS, vol. 4114, pp. 63–72. Springer, Heidelberg (2006)
9. Horvitz, E., Apacible, J.: Learning and reasoning about interruption. In: International Conference on Multimodal Interfaces (ICMI) (2003)
10. Johnson, D.W., Johnson, R.T., Smith, K.A.: Active learning: Cooperation in the college classroom, 2nd edn. Interaction Book Co., Edina (1998)
11. Millis, B.J., Cottrell Jr., P.G.: Cooperative learning for higher education faculty. The Oryx Press, Phoenix (1998)
12. NISE. Collaborative learning (retrieved 04/06/2008),
    http://www.wcer.wisc.edu /archive/cll/CL/doingcl/DCL1.asp
13. Oakley, B., Felder, R.M., Brent, R., Elhajj, I.: Turning student groups into effective teams. Journal of Student Centered Learning (2004)
14. Oliver, M., Harvey, J.: What does 'impact' mean in the evaluation of learning technology? Educational Technology & Society (2002)
15. Prekop, P., Burnett, M.: Activities, context and ubiquitous computing. Computer Communications (2003)
16. Ranganathan, A., Al-Muhtadi, J., Campbell, R.: Reasoning about uncertain contexts in pervasive computing environments. IEEE Pervasive Computing Journal (2004)
17. Röcker, C., Magerkurth, C.: Privacy and interruptions in team awareness systems. In: International Conference on Universal Access in Human-Computer Interaction (UAHCI) (2007)
18. Truong, B.A., Lee, Y.-K., Lee, S.-Y.: Modeling and reasoning about uncertainty in context-aware systems. In: IEEE International Conference on e-Business Engineering (ICEBE) (2005)
19. Xu, C., Cheung, S.C.: Inconsistency detection and resolution for context-aware middleware support. European software engineering (2005)
20. Yang, S.J.H.: Context aware ubiquitous learning environments for peer-to-peer collaborative learning. Educational Technology & Society (2006)
21. Zhang, G., Jin, Q., Lin, M.: A framework of social interaction support for ubiquitous learning. In: International Conference on Advanced Information Networking and Applications (AINA) (2005)
22. Tchounikine, P.: Directions to knowledge learners' self-organization in CSCL macro-scripts. In: Haake, J.M., Ochoa, S.F., Cechich, A. (eds.) CRIWG 2007. LNCS, vol. 4715, pp. 247–254. Springer, Heidelberg (2007)
23. Harrer, A., Lucarz, A., Malzahn, N.: Dynamic and flexible learning in distributed and collaborative scenarios using grid technologies. In: Haake, J.M., Ochoa, S.F., Cechich, A. (eds.) CRIWG 2007. LNCS, vol. 4715, pp. 239–246. Springer, Heidelberg (2007)

Information Needs for Meeting Facilitation

Adriana S. Vivacqua[1], Leandro Carreira Marques[2],
Marcos S. Ferreira[2], and Jano M. de Souza[2,3]

[1] ADDLabs/UFF – Active Design Dcomentation Labs - Fluminense Federal University
[2] PESC-COPPE/UFRJ – Graduate School of Engineering - Federal University of Rio de Janeiro
[3] DCC-IM/UFRJ – Institute of Mathematics - Federal University of Rio de Janeiro
avivacqua@addlabs.uff.br, leandrom@cos.ufrj.br,
marcos@cos.ufrj.br, jano@cos.ufrj.br

Abstract. In many group work settings, meetings take up a reasonable amount of time and often do not achieve satisfactory outcomes. One of the techniques that has been introduced to ensure meetings run smoothly and reach their goals places an individual in the role of meeting facilitator. Facilitation involves putting together the meeting agenda, designing meeting dynamics and overseeing the meeting at run time, to ensure goals are met. This may involve intervening or adjusting meeting structure to produce desired results. Thus, a facilitator should be able to act according to perceived group dynamics or problems. In this paper, we investigate information needs during facilitation activities. Our goal is to be able to construct systems that provide information to the facilitator so he or she can decide when to act and what to do.

Keywords: meeting facilitation, information provision.

1 Introduction

Meetings take up a significant amount of time, especially for upper level management and information workers [12]. Complex problems frequently demand exploratory discussion and decision making by groups of people who must come together to discuss and explore possible solutions, focusing on the best ones. Large projects frequently require a team working together and discussing possible alternatives, their evaluations and tradeoffs and selecting the most appropriate ones. A number of factors may reduce production in meetings [6], and certain techniques have been introduced to improve meeting productivity [2]. One strategy involves employing facilitators, professional meeting coordinators to assist the group in reaching their goal. Facilitators act before, during, and after the meeting happens.

Facilitators design the meeting process and its activities in order to ensure the desired goals are reached, sometimes using pre-defined group dynamics patterns such as thinkLets [3]. Facilitators also run the meeting, making sure the group is proceeding as expected, the necessary issues are being covered and the schedule is being followed. Thus, one of the facilitator's roles is monitoring the meeting and correcting any deviation in group dynamics. If the facilitator notices the group is steering away from its objectives, he or she should try to get the group back on track.

R.O. Briggs et al. (Eds.): CRIWG 2008, LNCS 5411, pp. 57–64, 2008.
© Springer-Verlag Berlin Heidelberg 2008

This involves a decision on the facilitator's part of whether, when and how to act. To make this decision, a facilitator requires information. Experienced facilitators should have no difficulty reading situations and finding appropriate actions to correct any problems. However, a less skilled or inexperienced facilitator may find it difficult to read a situation and understand different dynamics. Regardless of the case, a computer-based meeting support system should provide information to support the facilitator in his or her task of analyzing group dynamics and deciding when to act either through intervention or through restructuring of the meeting. In this paper, we take the first step towards such support, investigating information needs of facilitators. The paper is organized as follows: in the next section, we introduce background theories upon which our work is based. In section 3 we present considerations on information needs for facilitation, followed by a discussion in section 4.

2   Related Work

The main role of a facilitator is to oversee the meeting process and make sure that the group reaches their goals [4]. He or she must be someone who understands group processes and can assist the group, identifying problems and looking for solutions, and acting within the process when necessary. Facilitation is a process through which a person intervenes to help improve the way the group solves problems and makes decisions [8]. The benefits of facilitation have been recognized in face-to-face as well as distributed meetings, and poor facilitation can lead to ineffective meetings and weak outcomes [6].

The facilitator is the person responsible for setting the contexts, the norms, the agenda, providing recognition, prompting, weaving and meta-commenting [5]. It is his or her job to assure that, at the end of the meeting, everything that was in its agenda has been accomplished. He or she assists the group members with their tasks [4]. This involves understanding group processes and the complexity of meetings. Guiding the meeting requires an understanding of individual, social and political issues [7].

Some of the tasks of the facilitator are: encouraging the participants to take responsibility for their decisions; keeping the group focused on the objectives; managing conflicts; ensuring equal member participation; leading and managing meetings; using the agenda to guide the group; preparing the meeting and its activities beforehand and reinforcing meeting rules [10].

ThinkLets have been devised as a tool to assist the facilitator in planning and conducting meetings. A thinkLet is the smallest unit of intellectual capital required to create one repeatable, predictable pattern of thinking among people working toward a goal [1]. It is composed of three elements: a tool (the specific technology used), a configuration (a way to prepare the tool) and a script that determines what the facilitator should do or say during the execution of the thinkLet [3].

ThinkLets are organized into seven classes corresponding to the following group though patterns: generate, reduce, organize, clarify, evaluate and build consensus [11]. Each thinkLet is classified under one or more of these classes, which are associated to group dynamics that happen during a meeting. Creating a meeting process involves putting thinkLets together to achieve the desired outcomes, and conducting the meeting is a matter of taking the group through the thinkLets, using the scripts provided on their definitions. However, meetings may still require actions to adjust course. The script provided, for instance, does not provide instructions on how to deal with problematic situations.

The need to act during the meeting means that the facilitator must make a conscious decision to do so. Classic decision making theories [13][14] picture the individual as a rational actor, which, given a set of choices, will try to choose in order to maximize gains. This applies to the facilitator, who, given a perceived situation, will try to take the action most likely to steer the meeting to the desired direction. As with any decision, this one demands information. The more information a facilitator has about the meeting dynamics, the better the decisions he or she will make.

2.1 Typical Meeting Problems

An action taken by a facilitator to adjust the course of the meeting is called an intervention. One of the tasks of the facilitator while keeping track of the meeting is intervening to adjust direction, change strategy or prompt group members [4]. Interventions are prompted by the perception of certain cues of group behavior. Five generic problem syndromes have been identified in previous research [4] [9], each with telling cues and possible interventions:

The *Multi-Headed Beast* syndrome happens when there is no agreement on the agenda, the process design or when the problem-solving strategies are mixed. In these cases, the cues are usually digressions, interruptions, multiple topics appearing, individuals not listening to one another are saying, and little integration of the generated ideas. Possible interventions are suggesting a round robin to clarify the task, listing perceptions of the task, seeking synthesis or (re)formulating the agenda.

The *Feuding Factions* syndrome happens when there are hidden agendas, power struggles or when the participants fear change. In these cases, individuals start to repeat arguments or attack each other. Recommended interventions range from allowing individuals to list criteria privately or independently of alternatives, to measuring alternatives against criteria.

The *Dominant Spaces* syndrome can be perceived by passive-aggressive body language, unequal time or withdrawal. There are two possible causes for his syndrome: people become frustrated because they are not being heard or become insulated and afraid. Possible interventions are making a pool of under-participants or devise an activity to share perceptions (e.g., self-rating, commenting on other views).

The *Recycling* syndrome happens when there is no record of ideas or when there is confusion about the problem-solving process. The cues usually are "broken record" behavior, irritation with the lack of progress and/or failure to gain consensus. The facilitator might then (re)introduce the problem-solving process, identify which issues belong to which steps or identify where the participants are, where they were and where they are going.

Finally, the *Sleeping Meeting* syndrome can be found in meetings with long silences, absence of energy or ideas, or withdrawal. It is caused, most of the times, due to fear of volatile issues, hostility, depression or fatigue. The facilitator should point out the observation, suggest a mood-check and then take a break, address any underlying problems, decide an action plan to rectify and/or return to task, or leave the problem to the end of the meeting.

This short list shows that a facilitator must be alert to the dynamics of the meeting, and that there are many cues that can be observed to diagnose problems and come up with solutions.

3 Meeting Dynamics and Facilitation

Given the aforementioned aspects, we can see that facilitation involves active monitoring of the meeting and constant decision making, deciding when to act and what to do to ensure the meeting achieves its objectives. With the DynaMeeting project, we aim to better understand this decision-making process, verifying what information is needed for appropriate decisions to be made and how to provide it to facilitators, helping monitor the meeting. We consider two types of possible actions during the meeting:

- Intervening within an activity (intra-thinkLet)
- Restructuring the process (inter-thinkLet)

These are discussed in the following section.

3.1 Taking Action to Adjust Meeting Course

ThinkLets are the building blocks with which a facilitator can design meeting structures. The first step in this design process is putting together these blocks in a coherent way, in order to achieve desired goals. ThinkLets are put together in such a way that one thinkLet's output serves as another's input, and control flows from one thinkLet to another. Meetings may be represented as data flow or activity diagrams. Initial thinkLet selection is based not only on input and output compatibility, but also on goals, resources and expected outcomes [18]. Additionally, there are particular characteristics of the thinkLets (e.g., number of participants or level of agreement) that should be taken into account.

Meeting processes usually present a fixed set of thinkLets, since the meeting is designed beforehand. However, some situations may occur during the meeting that may render these choices less than optimal. One possible case is the number of ideas generated being smaller than the expected at the next step. In this case, it might be beneficial to run another generation thinkLet before moving on, or change strategy and select a different thinkLet. Thus, at the end of each step, a reevaluation is necessary to check the validity of the next step.

The same choice guidelines used to create the initial meeting structure can be used to decide about possible restructuration. While tools have been proposed to assist meeting process design (e.g., [17] [16]), these stop at the initial design. Criteria used for the initial selection can be applied to reevaluate each choice at the end of each particular thinkLet, and tools of this kind could be adapted for use during the meeting.

Problems may also occur during the execution of each thinkLet. When this is the case, the facilitator must intervene to steer the group in the right direction. This means that continuous evaluation within each thinkLet is also necessary, as certain behaviors may lead to undesirable outcomes and should be avoided. Therefore, information needed in this step has to do with the identification of these problems and possible courses of action. Different problems may occur, depending on thinkLet class.

These problems can be detected by checking thinkLet goals and verifying if the activity is moving in the right direction. For instance, a generate activity has as a goal to create more concepts or ideas. If the meeting stalls or the group is repeating itself, it may be necessary to intervene or change strategy. Each thinkLet class has particular problems, and the information needs should be similar for all thinkLets pertaining to the same class.

3.2 Meeting Indicators

With these analyses in mind, we designed a few indicators to help the facilitator in his or her job. Forms of automatically measuring each indicator are currently under research. A list of possible indicators follows. The correlation between thinkLet classes and indicators that could be useful for each class is shown in Table 1:

A. Group Participation Rate: members who are effectively contributing ideas and comments. Provides an idea of how representative the results being presented are of the group's opinions. It also allows the determination of whether the meeting is being dominated by a select few.
B. Distribution of contributions: number of contributions each member has provided. This shows the facilitator whether there are free-riders in the group. The facilitator can then take action, for instance, explicitly soliciting their contribution.
C. Individual Participation Rate: number of contributions per individual, given a timeslot. This measure denotes the level of interest a participant has in the task. If necessary, the facilitator can solicit interaction or change the dynamics to bring more energy to the meeting.
D. Idea Flow: time elapsed between two ideas. This indicator provides the facilitator with an idea of how well the meeting is proceeding, and at which stage participants are: in initial phases, lots of new ideas are generated. This rate usually goes down with time, so at later points fewer new ideas will be generated. If the group is becoming stagnant or runs out of ideas too early, the facilitator should take action.
E. Attention Allocation: number of responses to contributions versus new contributions, per member. This show how much attention has been devoted to analyzing others' contributions versus providing new ones (listening to others).
F. Idea Discussion: responses generated by each idea. This indicates the level of discussion generated by an idea. Controversial ideas should generate more discussion. When the group loses itself in discussions, the facilitator may decide to intervene. If members focus too strongly on one idea, new ideas may not be generated. The meeting resumes when group members have refocused on the subject.
G. Idea Distinction: indicates how different ideas are, and the level of divergence of the group. At the beginning of a generation activity, many ideas will be generated, later on this number should stabilize as the domain becomes better mapped. More repetition should be seen.
H. Interpersonal Agreement or Disagreement: pros and cons submitted by one user in relation to another allow a facilitator to detect interpersonal problems between two or more participants. This can be done by looking at the type of responses given to another's contributions. Should there be patterns of agreement or

disagreement, it might be necessary for the facilitator to intervene, to break the cycle participants have gotten into.

I. Individual Positioning: percentage of pros and cons submitted. This reveals "conforming" behavior (when individuals try to conform to the norm or avoid providing their opinions for fear of disturbing the meeting). In reverse, it also distinguishes individuals who have adopted a "devil's advocate" role, criticizing ideas. Facilitators may want to discourage conformity.

J. Divergence level: number of different ideas being generated. This shows if the group is converging on a topic or solution, or if it continues diverging, threatening to spin out of control and not reach a solution. If necessary, the facilitator may cut the divergent behavior short.

Table 1. thinkLet classes and indicators

thinkLet Class	Goals [11]	Useful Indicators
Generate	Increase number of concepts	D, G, J
Reduce	Decrease number of concepts (focus)	D, G, J
Organize	Increase understanding of relations between concepts	G
Clarify	Increase understanding of concepts	F, H
Evaluate	Increase understanding of relative values of concepts	E, F, H, I
Build Consensus	Increase commitment to proposals	E, F, H, I

Indicators A, B and C are generic and can be used to verify participation and decide when to stimulate participants into contributing.

4 Discussion

Understanding what is taking place is an important part of any facilitator's job, and these indicators should help the facilitator better decide when to intervene and what action to take. We expect that the syndromes described in section 2 may be identified using some of the aforementioned indicators, and that these will provide subsidies for the facilitator to act and direct the meeting according to the situation. The correlation between indicators and syndromes that can be identified through them is shown in Table 2.

Table 2. Indicator x problem syndromes

Syndrome	Indicators
Multi-Headed Beast	E, J
Feuding Faction	H
Dominant Spaces	B, C
Recycling	F, J
Sleeping Meeting	A

It is likely that these indicators could easily be detected by experienced facilitators. However, systems don't usually provide measurements to render this task easier, so that it might be performed as well by a non-professional facilitator. When selecting indicators, we tried to focus on those that could possibly be built into a system. We tried to design indicators that could be automatically calculated by a meeting support system, making inferences and providing suggestions to the facilitator.

We hope this research will provide a first step in the direction of automatic provision of meeting indicators. The indicators listed should provide the meeting facilitator with a wealth of information through which he or she can detect how the meeting is progressing and take action if necessary. These indicators are meant to assist the facilitator in his or her decision to intervene in the meeting, and are based on literature and on informal observations of meetings. Much work remains to be done, especially regarding information needed to decide about restructuring the meetings. Closer analyses and interviews with facilitators need to be conducted to elicit different types of information needs.

Meeting support systems should not only support meeting participants, but also the facilitator in executing his or her task. This means performing a real time analysis of the ongoing meeting. While reading cues in a meeting may be easy for an experienced facilitator, it may be hard for a novice.

Reading cues becomes even more complicated in completely electronic meetings, where individuals interact solely through the computer, and the facilitator cannot see their expressions or analyze body language. In these cases, it is even more important that a computer system help the facilitator with the determination of possible problems. During the meeting, the system should enforce selected dynamics (as specified by the thinkLets sequence established) and monitor group activity, providing information that may help the facilitator determine when to act. A proposed system architecture and a description of how these indicators could be transformed into computational calculations can be found in [19].

Even though different meeting support systems exist, most do not support the facilitator in his or her activities, especially with regards of evaluating the meeting progress and dynamics. A meeting support system should assist the facilitator by analyzing the group dynamics and providing information about cues that might have been missed. A system using these principles is in the final phases of implementation, and some experiments have already been scheduled.

Acknowledgements

This research is partially supported by CNPq.

References

1. Briggs, R.O., de Vreede, G.J., Nunamaker Jr., J.F., David, T.H.: Thinklets: achieving predictable patterns of group interaction with group support systems. In: Hawaii International Conference on System Sciences. IEEE Computer Society Press, Los Alamitos (2001)
2. Kolfschoten, G.L., Briggs, R.O., Appelman, J.H., de Vreede, G.J.: ThinkLets as Building Blocks for Collaboration Processes: A Further Conceptualization (2004)

3. Vreede, G.J., de Briggs, R.O.: Thinklets: Five Examples Of Creating Patterns Of Group Interaction. In: Ackermann, F., de Vreede, G.J. (eds.) Proceedings Of Group Decision & Negotiation 2001, June 4-7, La Rochelle France, pp. 199–208 (2001)
4. Viller, S.: The Group Facilitator: A CSCW Perspective. In: Proceedings of the Second European Conference on Computer-Supported Cooperative Work (1991)
5. Feenberg, A.: Network design: an operation manual for computer conferencing. In: Proc. Conference on Computer-Supported Cooperative Work, Austin, Texas. ACM, New York (1986)
6. de Vreede, G.J., Davidson, R.M., Briggs, R.O.: How a Silver Bullet May Lose Its Shine. Communications of the ACM 46, 96–101 (2003)
7. Macaulay, L.A., Alabdulkarim, A.: Facilitation of e-Meetings: State-of-the-Art Review. In: Proc. IEEE Int. Conf. e-Technology, e-Commerce and e-Service (EEE 2005), pp. 728–735 (2005)
8. Schwarz, R.: The Skilled Facilitator. Jossey-Bass Publisher (1994)
9. Westley, F., Waters, J.A.: Group Facilitation skills for managers. Management Education and Development 19, 134–143 (1988)
10. Antunes, P., Ho, T.: Facilitation Tool - a Tool to Assist Facilitators Managing Group Decision Support Systems. In: Ninth Workshop on Information Technologies and Systems, WITS 1999, Charlotte, North Carolina (December 1999)
11. Kolfschoten, G., Vreede, G.J.: The collaboration engineering approach for designing collaboration processes. In: Haake, J.M., Ochoa, S.F., Cechich, A. (eds.) CRIWG 2007. LNCS, vol. 4715, pp. 95–110. Springer, Heidelberg (2007)
12. Andriessen, J.H.E.: Working with Groupware. Springer, London (2003)
13. Simon, H.A.: A behavioral model of rational choice. The Quartely Journal of Economics 69(1), 99–118 (1955)
14. Simon, H.A.: Theories of Decision Making in Economics and Behavioral Science. The American Economic Review 49(3), 253–283 (1959)
15. Turban, E., Aronson, J.E.: Decision Support Systems and Intelligent Systems. Prentice Hall, Englewood Cliffs (1998)
16. Kolfschoten, G.L., Veen, W.: Tool Support for GSS session design. In: Hawaii International Conference on System Sciences. IEEE Press, Los Alamitos (2005)
17. Antunes, P., Ho, T.: The Design of a GDSS Meeting Preparation Tool. Group Decision and Negotiation 10(1), 5–25 (2001)
18. Kolfschoten, G.L., Rouette, E.A.J.A.: Choice Criteria for Facilitation Techniques. In: Hawaii International Conference on System Sciences (HICSS-39). IEEE Press, Los Alamitos (2006)
19. Vivacqua, A.S., Marques, L.C., Souza, J.M.: Assisting Meeting Facilitation through Automated Analysis of Group Dynamics. In: International Conference on Computer Supported Cooperative Work in Design (CSCWD 2008), Xian (2008)

Risk Assessment in Healthcare Collaborative Settings: A Case Study Using SHELL

Pedro Antunes[1], Rogerio Bandeira[1], Luís Carriço[1], Gustavo Zurita[2],
Nelson Baloian[3], and Rodrigo Vogt[2]

[1] University of Lisboa,
Department of Informatics of the Faculty of Sciences, Campo Grande, Lisboa, Portugal
{paa,lmc}@di.fc.ul.pt
[2] Universidad de Chile,
Department of Information System and Management of the Economy and Businesses
School,, Diagonal Paraguay 257, Santiago de Chile, Chile
gzurita@ing.puc.cl
[3] Universidad de Chile,
Department of Computer Science of the Engineering School, Blanco Encalada 2120,
Santiago de Chile, Chile
nbaloian@gmail.com

Abstract. This paper describes a case study addressing risk assessment in a hospital unit. The objective was to analyse the impact on collaborative work after the unit changed their installations. The study adopted the SHELL model. A tool aiming to support the inquiring activities was also developed. The outcomes of this research show the model is adequate to analyze the complex issues raised by healthcare collaborative settings. The paper also provides preliminary results from the tool use.

Keywords: SHELL, Risk Assessment, Collaborative Settings, Hospitals.

1 Introduction

Risk assessment in healthcare has been maturing over the last years in USA, Britain, Europe and Australia [1]. At its origins, the focus was on developing a framework for controlling litigation, which has been a major worry for clinicians and hospitals. Studies of medical error in healthcare have brought a growing awareness of the scale of the problems directly and indirectly causing harm to patients. Risk assessment is also at the heart of the concept of clinical governance [2], a management approach making those in charge of healthcare organizations accountable for the quality of care delivered.

Until the 1980s, a major goal of risk assessment was to evaluate the technical and human contributions to catastrophic breakdowns in high-hazard enterprises such as aviation, nuclear power generation and chemical production [3]. Accidents such as the ones that occurred at the Three Mile Island and Chernobyl raised much political and social concern. By contrast, medical mishaps tend to affect single individuals and thus received less attention.

R.O. Briggs et al. (Eds.): CRIWG 2008, LNCS 5411, pp. 65–73, 2008.

But since the mid-1980s research has begun on the technological and human factors affecting the safety of healthcare systems [4]. Much is already known today about human error, work environments, information overload, attention problems, and human-machine interfaces [5]. One outcome of this research is the widespread acceptance of the models of causation of accidents [3, 6]. We should expect an increased preoccupation with risk assessment as more technological advancements are brought into healthcare.

This paper is related with collaboration technology in two ways. The first one concerns the highly technological and collaborative nature of hospitals, since various types of professionals orchestrate their activities in coordinated and concerted ways. The collaborative setting is part of the problem when mishaps occur and thus should be involved in risk assessment. Research on Computer Supported Collaborative Work (CSCW) may contribute to risk assessment with insights on collaboration and technology use. Secondly, groupware technology may also support risk assessment. I.e., technology is not only part of the problem but also may become part of the solutions. In this paper we address these two facets of the problem. We illustrate our approach to risk assessment in a hospital, analyzing technology changes in a very rich collaborative setting: the intermediate care unit for newborns. Our approach is based on the Software – Hardware – Environment – Liveware – Liveware (SHELL) model [7], well known in the human factors field.

We also discuss a tool we have been developing to support risk assessment. The tool implements gesture-based data management functionality over Tablet-PC, adopting the SHELL model to support the interviewers' activities. The paper is organized as follows. In the remaining of this section we give a brief description of the adopted model. The section 2 describes the case study. The section 3 is dedicated to the SHELL tool. The section 4 discusses the obtained results and provides some conclusions.

1.1 SHELL

The SHELL model characterizes the socio-technical context of working environments, disentangling the relationships between humans, called liveware (L), and four other elements of the working environment [7]: Hardware (H), the physical sources; Software (S), including rules, regulations, procedures and practices; Environment (E), the physical, economic and social aspects influencing human performance; and liveware (L), the other humans operating in the working environment. This additional liveware dimension is fundamental to account for the communication, coordination and collaboration aspects of the working environment.

The interfaces between the SHELL elements define major areas of analysis: liveware-liveware (L-L), liveware-hardware (L-H), liveware-environment (L-E) and liveware-software (L-S). These interfaces define the underlying structure for risk assessment. SHELL has been extensively applied in manufacturing, nuclear power production, aviation, ship and railway operations, and maintenance. Many aspects related with human factors have been analyzed using SHELL, including requirements analysis, safety assessment, psychological issues, accident investigation and human operations.

Concerning the healthcare sector, the number of studies adopting SHELL is scarce. [8] studied the anesthetists' workload in operating rooms, with some emphasis on collaboration issues such as delegation and supervision. [9] used SHELL to define an instrument to evaluate work performance in an intensive care unit. [10] discussed the use of SHELL in developing a healthcare report system. None of these projects adopted the CSCW perspective to analyze collaboration issues. Therefore, one of the major goals of our research was to evaluate the applicability and usefulness of SHELL in the CSCW domain.

2 Case Study

The study was conducted on a hospital specialized in neonatal and pediatric services – Maternidade Alfredo da Costa. The specific target was the intermediate care unit for newborns (designated ECU – Especial Care Unit). This unit receives infants unable to live independently, usually in consequence of premature birth complications, who surpassed the most critical phase and do not require intensive care. Many infants residing in the ECU are in incubators, subject to extended electronic monitoring and receiving enriched oxygen mixtures, while others stay in open cribs and are essentially gaining weight. The unit contains 18 incubators and 13 cribs and most of the time operates very close to that limit (29 to 31 newborns).

The ECU is a rather complex organization. Besides the diversity of clinical cases and care services, the ECU involves multiple players with different goals, cultural background and attitudes. Besides the clinical staff, the parents are also present during long periods (usually between two to six hours a day), apart from the nursing assistants and secretariat. The overall maneuver entails the collaboration of all players. The nurses take one of the principal roles there. They have a constant presence in the infants' rooms. They support the detection on abnormal situations, the containment of their consequences and the restoring of the normality. They are also responsible for controlling the nursing assistants and interacting with the parents and doctors.

The doctors are always available in emergency situations to diagnose problems, prescribe treatments and coordinate the nurses' actions. The parents' presence is encouraged, particularly for the infants that already stay in open cribs. They collaborate in feeding their babies, holding them, etc. The nursing assistants are essentially responsible for hygiene and fetching and delivering materials.

The reasons for studying this unit were threefold. First, the unit handles collaboration at a reasonable pace that offers good opportunities for external observation. Secondly, it has a diversity of players and collaboration requirements. Third, the unit recently suffered a complete change of their installations and the hospital administration showed interest in assessing their impact. Furthermore, as the new ECU started operating in January 2008, the players still remember objectively the details of the antecedent situation.

The study was based on interviews and long visits to the premises with several stakeholders: hospital administration, unit's executive board, head of the hospital informatics department, unit's principal nurse and one of the chief doctors.

2.1 Assessment of the Environmental Changes

The interviews and visits to the ECU were framed by SHELL. The model provided the structure necessary to disentangle many aspects of work in the ECU, elucidating the fundamental drives behind the structural changes caused by the new installations. In this section we will also rely on the SHELL model to summarize our findings.

The liveware elements (L) collaborating in the ECU are doctors, nurses, nursing assistants and parents. Most of the work depends on the clinical staff. We found the nursing assistants are regarded quite distinctively from the other staff; presenting a lower level of education, being subject to different management rules and rotating a lot between units. The parents participate in the process but are mostly regarded as external entities.

The collaboration depends on many regulations and procedures, as well as practices and traditions (S). The most relevant hardware (H) found in the ECU includes incubators, medical equipment, and computers. We observed nurses most frequently handle this hardware, especially computers, which are seldom used by doctors. The nursing assistants emerge again very distinctively as they operate their own and specific equipment.

All liveware share the same physical environment (E), consisting of several rooms, offices and corridors. The complete renovation of the ECU introduced significant changes in this environment. Therefore, (E) should be considered the control variable in this study. The following main changes were identified:

- New automatic electric doors isolating the ECU from the other units.
- The previous infants' rooms had small windows so their interior could be seen from the corridor. Now they have glass walls and are completely visible from within the unit.
- New automatic electric doors isolating the infants' rooms from the corridor. In the previous condition these doors were permanently kept open.
- The unit has an office for the chief nurse and a doctors' room. In the previous conditions these rooms were located far away from the unit.
- New working and cleaning rooms. The incubators previously serviced in the corridor are now serviced in the cleaning room.
- As before, the computers are placed in the main corridor. But they are now in a different position, facing the corridor and with glass walls behind.

We now discuss these changes according to the areas of analysis proposed by SHELL:

L-E. The automatic doors contributed to reduce the ambient noise to more comfortable levels, with positive impact on the liveware and their activities. It was considered that the doors increased the parents' awareness and care for the work performed in the unit, which lead to a quieter attitude. Furthermore, the nurses now spend more time working in the infants' rooms rather than moving immediately to more private premises. The interviewees found two major reasons for this new attitude: the increased quietness stimulates the nurses to accomplish their tasks inside the infants' rooms; and the increased noise isolation refrains nurses from leaving the infants' rooms unattended.

The glass walls had a significant impact on the nurses, as they now have a clear view of the incubators and organize more swiftly their interventions. The nurse office and the doctors' room contributed to the longer presence of the principal nurse and the doctors in the unit. The working room was also welcomed, as the previous situation was characterized by the unpleasant coexistence of very different functions, such as cleaning, eating and writing.

L-H. The new position of the computers in the main corridor affords working on the computer and at the same time controlling the incubators through the glass walls. The location of these computers in the previous environment disallowed such level of awareness.

As in the previous setting, the ECU has an emergency incubator located in the end of the main corridor. However, since the corridor is much longer now, the doctors are considering the necessity to obtain a new emergency incubator to be located in the other end of the corridor, since more time is taken to respond to emergency situations.

L-L. According to the interviewees, the new ECU supports more structures work, more quietness, better awareness, and more fluidity and collaboration. The nurses reported lesser coordination problems and the same level of communication necessary to handle emergency situations. The doctors and the principal nurse spend more time in the unit, which was very positively regarded. There is less conflict between the nursing assistants and the other staff, because maintenance tasks have been relocated from the main corridor to a specialized room. One negative outcome of this new arrangement is that by the end of the day, when the staff is reduced, the nursing assistants leave the incubators' rooms unattended.

L-S. The new environment changed the relationships between staff and rules and procedures, although more time is necessary to detect more profound changes (the new ECU was operating for three months when the interviews were done). One change is related with the nursing assistants. Since the rotation of these resources is very high and there are strict rules about hygiene, disinfection, etc., there is a constant need to instruct the new personnel on those matters. While in the past the instruction was done at the corridors, it has now moved to the service room, with positive impact on the remaining activities.

The outcomes from this study showed the new working environment had a very positive impact on the unit's responsiveness and safety. The SHELL model facilitated establishing the causal relationships explaining the positive outcomes (Figure 1). From our point of view, the model served very well the set research goals.

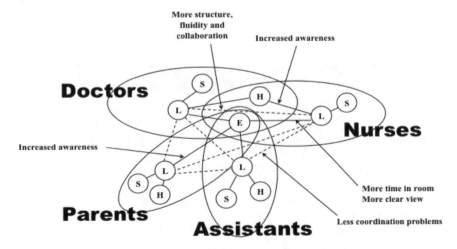

Fig. 1. Schematic view of the ECU unit according to the SHELL model

3 SHELL Tool

The SHELL tool aimed to facilitate the elicitation of the SHELL model elements, using the touch-screen features and mobility of Tablet-PC. The analysts used the tool to record findings during interviews and observations in the hospital. The tool offers additional features to those of the pen and paper: the annotations may be organized immediately and there are pen-based editing options making the manipulation of the information more comfortable. For this, the SHELL tool captures the strokes hand-written over the Tablet-PC screen, along with the recognition of predefined gestures for triggering edition functions. The tool enables analysts maintaining visual contact with the collaborative setting.

The user-interface consists of a main working screen, where the analyst may take notes about the subjects (liveware). These notes are recorded inside nodes labeled with the name of the subjects. The nodes are represented by rectangles created by gesturing an "⌐". This gesture is automatically recognized as a node creation (see Figure 2). The other SHELL model elements are specified inside each node. This is done by clicking with the pen over the node (Figure 2, right side), which makes an arrow pointing down to appear at the top-left corner. By clicking on that arrow, the node is "opened-up" and the whole screen shows four predefined sub-nodes corre-sponding to the SHELL elements (figure 3, left side). Each sub-node should be filled with the information elicited by the analyst. In this way the "father" node along with the four "son" nodes conform the relationships L-L, L-S, L-E, L-H over which the SHELL model is applied. This enables an easy analysis of the collaborative situation. When working inside a node, an arrow pointing up is always displayed and serves to leave the node (Figure 3, left side).

Fig. 2. creation and selection of a node

The recursive creation of model elements is not allowed. When entering text in a node, if the bottom of the screen is almost reached, then the working area is automati-cally scrolled up to give the user more space to enter information. To scroll up and down, a panning mode may be activated using an option in the menu. In this mode, gestures up and down will scroll the working area instead of writing a stroke. The editing functions are activated by strokes matching pre-defined gestures having cer-tain meanings, for instance: a) a double lace selects all the strokes inside it; b) a cross deletes all strokes touched by it or previously selected; c) selected objects may be moved by dragging them; f) a spiral gesture copies the previously selected strokes.

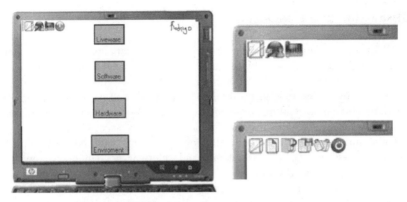

Fig. 3. Left side: The working area related with one interviewed person and the SHELL sub-nodes. Right side: Close-up of the menu. Above menu closed; below menu open.

3.1 Observations from the Tool Usage in the Case Study

The preliminary interviews and visits to the hospital were conducted with the traditional data elicitation tools, paper and pen, to avoid surprise and discomfort. The SHELL tool was only introduced in the process when it was considered that a good relationship was established between the analysts and the interviewees, the goals of the study were well established, and the purpose of the tool was understood.

The SHELL tool was then used as a substitute for the paper and pen, usually departing with empty pages, filled with hand-scribed text as the interviews progressed. The tool served to organize risk issues (problems, causes and effects) and focus the interview on the SHELL elements. The data elicitation always departed from a specific liveware element – the interviewee –, from where multiple L-L, L-S, L-E and L-H relationships were established. After these experiments, two informal interviews were conducted with the tool users. These users were not involved in any way in the tool development.

The following observations were drawn from the interviews. The interviews and visits usually took a long time and challenged the autonomy of the Tablet-PC. The connection of the power plug was not possible most of the times. The tool usage was thus accompanied with a discouraging level of stress. The hardware was considered more problematic than the software.

The software was considered simple to learn but not simple to master. Most of the problems concern the use of special gestures necessary to organize and edit information. Some of the gestures used to manipulate information were similar to the way one user was accustomed to write, thus resulting in unintended recognized gestures. Most difficulties were related to deleting information since, when it fails, users find an unwelcome drawing, which has to be deleted and so the problem is recursive. The failures had to be recovered in front of the interviewees, which increased the stress. These problems were somewhat mitigated by having two persons doing the interviews, being one more focused on annotating and the other on interacting with the interviewees. But even in these situations using the tool was regarded as problematic, as the user must keep up with the interview.

4 Discussion and Conclusions

The SHELL model allowed us to obtain very insightful data about the consequences of the installation changes done by the target organization. Of course, many of those changes were intended by design. For instance, the glass walls and the new rooms were intended by design to improve the ECU's structure and performance. However, as mentioned by the hospital management, there had not been any previous attempt to assess if those changes had the expected impact on performance. The results showed a remarkable improvement in the overall work structure, with positive impact on performance (less coordination problems) and safety issues (more awareness and presence from nurses and doctors).

The SHELL model allowed us to focus on the fundamental drives of change when inquiring about the changes, and highlighted the causal relationships between the installation changes and the L-E, L-H, L-L and L-S model elements. Therefore one outcome from this case study is the very positive role of SHELL elucidating the complexity of the collaborative work done in the hospital unit, and the causal relationships explaining what occurred after the installation changes. The SHELL model also demonstrated flexibility and plasticity to the varied situations that were encountered during this study.

One curious outcome of this study is that a small number of negative impacts and increased risks were found. Indeed, only two major issues were raised, one related with the increased distance between the emergency incubator and the infants' rooms, and another related with the lesser availability of the nursing assistants by the end of the day.

We observed that, beyond the changes intended by design, some unexpected consequences occurred. For instance, the more presence of nurses in the infants' rooms was not deliberately designed. It just occurred as an indirect consequence of having automatic doors separating the ECU and parents assuming a different attitude. From our point of view the SHELL model was invaluable pointing out these important consequences and the causal relationships explaining them.

The SHELL model was also invaluable disentangling the collaborative nature of the work done in the ECU. The model has a strong focus on the liveware element, which emphasizes the human aspects of the system under evaluation. But the model also emphasizes the L-L relationships, which were instrumental to analyze what was happening with the collaboration in the work setting. Most of the positive outcomes coming from this study were related with L-L relationships (better work structure, more awareness, more presence, less conflicts between staff), indicating a positive role of SHELL assessing the collaborative setting.

This research project thus had very positive results. From the hospital management point of view, the project was their first opportunity to address risk assessment with a focus on collaborative settings. From our point of view, this research was a preliminary step towards applying the CSCW view to risk assessment in the healthcare domain.

Currently, our experiments with the SHELL tool served to highlight the possibilities and problems of such a kind of tool. The obtained preliminary outcomes showed some resistance from the analysts towards using the tool. More work has to be done improving the functionality and, most importantly, increasing the capability to manage the model in a more comprehensive (e.g., integrating data from multiple liveware) and collaborative way (e.g., supporting multiple persons working on the same model, a functionality currently supported by the tool but that has not yet been experimented).

Acknowledgements. Special thanks to all people at Maternidade Alfredo da Costa who made this research possible. This paper was supported by the Portuguese Foundation for Science and Technology (PTDC/EIA/67589/2006) and Fondecyt 1085010.

References

1. Vincent, C.: Clinical Risk Management, Enhancing Patient Safety. BMJ (1995)
2. Walshe, K.: The Development of Clinical Risk Management. In: Vincent, C. (ed.) Clinical Risk Management, Enhancing Patient Safety, BMJ, London, UK (2001)
3. Perrow, C.: Normal Accidents, Living with High-Risk Technologies. Princeton University Press, Princeton (1999)
4. Leonard, M., Graham, S., Bonacum, D.: The Human Factor: The Critical Importance of Effective Teamwork and Communication in Providing Safe Care. Quality & Safety in Health Care 13, 85–90 (2004)
5. Redmill, F., Rajan, J.: Human Factors in Safety-Critical Systems, Butterworth Heinemann, Oxford, UK (1997)
6. Reason, J.: Understanding Adverse Events: Human Factors. Quality in Health Care 4, 80–89 (1995)
7. Edwards, E.: Man and Machine: Systems for Safety. In: Proceedings of British airline pilots Association Technical Symposium. British Airline Pilots Association, London, pp.21–36 (1972)
8. Leedal, J., Smith, A.: Methodological Approaches to Anaesthetists' Workload in the Operating Theatre. British Journal of Anaesthesia 94, 702–709 (2005)
9. Helmreich, R., Davies, J.: Human Factors in the Operating Room: Interpersonal Determinants of Safety, Efficiency and Morale. Baillière's Clinical Anaesthesiology 10, 277–295 (1996)
10. Rizzo, M., McEvoy, S.: Medical Safety and Neuroergonomics. In: Parasuraman, R., Rizzo, M. (eds.) Neuroergonomics: The Brain at Work. Oxford University Press, Oxford

Persuasive Virtual Communities to Promote a Healthy Lifestyle among Patients with Chronic Diseases

Eduardo Gasca, Jesus Favela, and Monica Tentori

Department of Computer Science, CICESE Research Center, Ensenada, México
{egasca,favela,mtentori}@cicese.mx

Abstract. The World Health Organization has declared obesity a world-wide epidemic. People with obesity have a higher risk to attain chronic diseases, high risk of premature death and a reduced quality of life. Recent studies have shown that persuasive technologies and virtual communities can promote healthy life-styles. In this article, we describe the development of a Persuasive Ecosystem aimed at promoting a healthy lifestyle in patients with a chronic disease that participate in a support group. The study was inspired in the results of a case study conducted in a hospital responsible for running this group. The results of a preliminary evaluation show an increased engagement of the patients with the program due to the use of the system.

Keywords: Persuasive Ecosystem, Virtual Community, pHealthNet.

1 Introduction

The increased consumption of energy-dense foods high in saturated fats and sugars, and reduced physical activity, are some of the main causes of increased obesity around the world [1]. Mexico ranks 2nd as the country with most people with obesity [1], where around 50% of the adult population is obese [2]. To cope with this, the Instituto Mexicano del Seguro Social (IMSS) recently implemented PREVENIMSS, a national four-week program for education and prevention of diseases caused by mal-nutrition and obesity. The success of the program, however, has been limited by the lack of continuous support once the patients complete the four-week workshop.

Virtual communities allow many individuals to collaborate and share experiences with people who are geographically distributed [3]. In the medical area, virtual com-munities have been successfully used in the care of patients [4]. Their benefits include reduced stress, social satisfaction, the availability of information relevant to their disease, and increased communication between patients and physicians [5]. Similarly, interactive systems have been designed to change users' attitudes and/or behaviors to achieve specific goals. This type of applications are called *Persuasive technologies* [6]. For instance, the UbiFit Garden system was designed to encourage regular physi-cal activity. The system uses wearable sensors to detect and track people's physical activities and displays them through an aesthetic image. This image is presented to the user in the form of a flower garden [7]. When the sensors detect a new physical activ-ity, it improves the appearance of the plants in the garden and adds a new element,

R.O. Briggs et al. (Eds.): CRIWG 2008, LNCS 5411, pp. 74–82, 2008.

such as a butterfly. If no physical activity from a user is detected, the flowers in the garden might perish. Indeed, several persuasive prototypes have been developed to improve people quality of life by successfully motivating them to make positive decisions in regards to their health [8].

In this paper by binding the ideas of persuasive computing and virtual communities we propose the development of a *persuasive ecosystem*. Our work aims to provide a design of a technological solution focused on supporting the interaction and communication between specialists and patients as proposed by the PREVENIMSS program while at the same time persuading them to maintain good nutrition and physical activity habits. We argue that patients that use this solution would feel more motivated to keep working on their programs, since they get a feeling of being personally attended while benefiting from the support of a group.

2 Studying PREVENIMSS to Promote a Healthy Lifestyle

For three months, we conducted a workplace study at IMSS to understand the PREVENIMSS support program. The study was conducted in three phases. In the first phase we studied how patients are canalized to the program and how it works. We conducted eight semi-structured interviews with the people involved in such process – two social workers, two nurses and four physicians. We also shadowed them for two complete shifts. The second phase of our study included a set of passive observations of a couple of support groups –including a total of eight sessions. Our observation helped us become involved with the groups and to identify our target informants. In addition, we conducted eight semi-structured interviews: three with the PREVENISS staff (including the nutritionist, psychologist and exercise trainer) and five with the patients. For the final phase we evaluated patients' perception of the program. We conducted thirty-two phone interviews with patients that had recently participated in the program. The patients interviewed were eight men and twenty four women with an average age of 44 year old. All patients interviewed have overweight problems and expressed interest in improving their lifestyle habits. The interviews were analyzed using a comparative verification of evidence resulted on the identification of major themes for each topic of inquiry.

2.1 SODHI Group: Issues and Opportunities for the Deployment of Persuasive Technology and Virtual Communities

PREVENIMSS works by organizing informational sessions about particular topics such as the risks of maintaining a sedentary life and bad feeding habits. This particular self-support group is called SODHi. In SODHi a group of health specialists (nutritionist, exercise trainer, physician and psychologist) assesses the patient's health to design a personalized diet and exercise plan. Specialists and patients attend four meetings, one per week, were their diet plans are handed over and a social support group is created. After the fourth and final session, patients are encouraged to keep up with the plan and assist to a quarterly appointment with specialists.

We found three major issues that impact patients' motivation and their healthy habits after or during SODHi.

- *SODHi does not help create a long-lasting support group among staff and patients.* Attending four sessions is not enough to connect with a community. For instance one patient made the following comment during an interview: *"Four sessions is a short time ... and now that I'm encouraged the group is over".* This lack of communication among patients and specialists after the group is finished result in a loss of motivation for patients to stick to the healthy habits acquired.
- *SODHi sessions demand physical presence.* Most of patients found it difficult to attend the weekly sessions. A patient said: *"Sometimes I can't keep my appointments because I live far from the hospital and when I miss them I tend to gain weight".* Missing one session impacts in the social relations within the group and the access to information relevant to keep track of their progress.
- *Limited access to relevant information.* During the sessions specialists provide them with didactic material. When the program finishes patients no longer have access to information on new ways to maintain a healthy lifestyle.

2.2 SODHi Impact in Patients' Healthy Habits

Information from the interviews was used to compare how patients' dietary and exercise habits change during and after SODHi. Results show that the amount of patients that conducted exercise and kept diet was higher during the program than after it was completed. For instance, while 75% of the patients did exercise during SODHi only 41% continue exercising once the program was completed –a 34% decrease. Similarly, 46% of the patients stop keeping their diet after SODHI. This could be partially explained because patients loose communication with specialists and other members of the group when the program finishes. For instance, during an interview a patient explained: *"I would like to attend again the group or persuade them to continue gathering because it is [impossible] to continue just by yourself".* Indeed, we found that the most relevant aspect to motivate users is to maintain communication with specialists and others members of the group. We found that 87.5% of patients lost all contact with the specialists and 91% with other members of their group once they completed the four-week program.

Despite of this, we also found that patients agree that participating in Control SODHi persuaded them to improve their health and their quality of life. 87.5% of the patients emphasized that they had achieved their goals while assisting the program, 91% stated that their quality of life improved after attending SODHi and 94% of patients qualified the program as a good motivator.

3   pHealthNet: A Persuasive Ecosystem for PREVENIMSS

We envision a persuasive ecosystem to help SODHi staff and patients maintain their diet and exercise programs beyond the four-week sessions. The persuasive Health Network system (pHealthNet) uses two devices (pedometer and mobile phone) and a virtual persuasive site (Figure 1). The pHealthNet *site* allow users to maintain community attachment, challenge friends about nutritional habits and physical activities, provides activity awareness and gives proper credit for user activities encouraged by the system. The mobileHealthNet client allows users to maintain a connection with

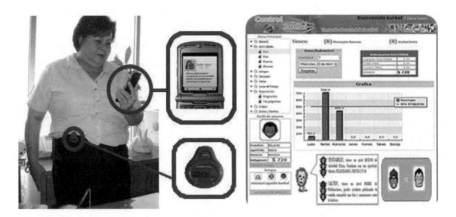

Fig. 1. pHealthNet (a) Maria receives a notification of a challenge on her mobile phone (b) Rita reviews Maria's goals through a timeline

relevant events in the site, as well as, easily and quickly upload the amount of steps walked during the day. Altogether the services of the site and both devices form a persuasive ecosystem for users to monitor their lifestyle habits while keeping a record of them. To better explain our systems' functionality we present a scenario of a patient currently attending the program and using the system.

Maria[1], a 50 years old woman with diabetes and overweight problems is attends a SODHi group. She meets the support group weekly and soon gets attached to Rita –a 48 years old woman with hypertension. During the sessions Maria and Rita support each other's dietary and exercise habits. Mr. Diaz, the social worker, recommends her the pHealthNet site. Maria registers herself to the site and adds Rita as a friend. Rita challenges Maria with the goal of completing 10000 steps a day for a week. While Maria is at work she receives in her mobile phone a message notifying her of the challenge. Maria logs in and accepts the challenge. As soon Maria arrives home she grabs her pedometer and walks off to the nearest park to exercise. At the park, she walks for an hour approximately completing 15000 steps. When she finishes her exercise she uses her mobile phone to upload the amount of steps just completed. After a couple of hours, Maria receives a message from the physical trainer congratulating her for the progress. When Rita is introducing her steps she consults Maria's timeline to assess her progress. She realizes that Maria has just crashed her record and creates herself a new goal of completing 20000 steps for the next day.

To provide this functionality the system uses a client-server architecture as a basis for its implementation. Connectivity between the server and the mobile phone is achieved through SOAP 1.1, WSDL 1.1 and HTTP 1.0/1.1. The mobile phone has a component that works in pair with its counterpart in the server to send and receive messages. The site is implemented on top of a web server with MySql, Apache and PHP. The system includes three main services as described next.

---

[1] All names have been changed to protect the privacy of our informants.

3.1 Supporting Community Attachment

The purpose of this service is to provide a set of tools for users to maintain a social network and a connection with other members with similar problems and that have attended the PREVENIMSS program. For instance, a patient commented: *"Having the support of somebody to lose weight keeps me motivated and helps me not to feel [that I am the only one]…"* Our system allows users to manage friends and specialists, as well as, send/receive email and SMS messages. Also, users may participate in collaborative games to persuade others to keep a healthy lifestyle or received encouraging messages. As shown in the scenario, the physical trainer sends Rita a message to keep her motivated.

3.2 Providing Activity Awareness

We found three important types of activity awareness to provide: a history of past behavior, current status, and activity level performance. During an interview a patient commented: *"I would like to keep contact with the people in my group, see them again to talk about our experiences or even assess their progress"*. Users can use PHealthNet's timeline to consult the physical activities executed by them or others, whom they have registered in their roster. In addition, a meter activity in the form of a traffic light shows the level of physical activity and the level of participation on the site –red being worst and green being best. Moreover, depending on the color of the traffic light a persuasive message is shown to the user. This type of awareness might help users to improve their lifestyle habits. As shown in the scenario, Rita challenges Maria resulting in an increase in the amount of steps walked by both.

3.3 Persuading a Healthy Commitment

This service allows patients to challenge friends about their dietary and exercise habits. A patient explained: *"By committing myself and with my partners, being close to them and being consistent … our habits could improve"*. The system allows users to challenge others by introducing goals such as an amount of steps for a period of time, a level of glucose and the amount of weight lost. For instance, as shown in the scenario Rita challenged Maria to complete 10000 steps a day for a week. In addition, the system gives patients' proper credit for their activities to maintain users motivated the system recognizes patients' efforts by giving prizes to them in the form of electronic money called "SaluPesos". Patients can win money every time they introduce a healthy activity to the system. For instance if they increase their amount of steps or complete a goal.

4 Preliminary Evaluation

We conducted a preliminary evaluation to assess the system's core characteristics and users' intention to use the system. We deployed the system for one month during one complete SODHI group –four sessions in total. Six patients attending the group voluntarily participated in the evaluation. The patients were women with obesity, diabetes or hypertension between 40 and 60 years old (Figure 2).

The dynamics of each SODHi session was changed to incorporate pHealthNet. In each session an amount of time was set aside to allow patients to use the system and learn new features. During the first session pHealthNet was introduced to each patient –they were given a pedometer, internet access and their userID to log into the system. At this point, participants learned how to enter recipes, steps, diet plans, comments, testimonies, and how to earn salupesos. In the second session we focused on the community services offered by the tool. In this session, patients learned how to add friends and send them messages –including SMS. Moving to a persuasive approach, in the third session patients learned how to use salupesos, and games to challenge each other. In the final session, we conducted a focus group and a brainstorming session to gather feedback and capture their experiences with pHealthNet. We logged some of the user events specifying which features were used and at what time.

Fig. 2. Using pHealthNet in PREVEINMSS

4.1 Results and Discussion

Overall, patients' viewed the application as useful, efficient, and generally appealing. pHealthNet was qualified by patients as the main motivator of the program. Here we present the results obtained through interviews and system's usage.

Logs of computer behavior and self-reported measures. The use of the system has been rather heterogeneous due patients' computing skills and computer access. All patients had a mobile phone; however, some of them didn't have internet access or even knew how to use a computer. For instance, in four weeks Maria published nine diets, twenty three recipes, two comments and one testimony earning a total of $2160 salupesos. She has registered nineteen times her footsteps and updated her weigh seven times. In contrast, Carmen has only published three recipes, one goal and she has registered fourteen times her footsteps earning a total of $180 salupesos. In contrast with Maria, Carmen didn't have access to a computer at home. However, she kept a paper-based journal of their activities and brought it along with recipes to the session to be assisted in recording the data. In total, patients recorded their footsteps seventy four times, added four comments, nine diets, twenty-seven recipes, ten goals, one question to the specialist, one testimony and updated their weight twelve times.

Promoting an individual commitment. Indeed, the lack of internet access was the main obstacle in using pHealthNet. Despite of this, patients discovered new ways to interact with the system and participate in the community. For instance, while some participants (such as Carmen) participated "offline" others asked third parties to upload their information for them. These "assistants" ended up registering themselves to the system and participating with it as well. We also found that patients where persuade by stages in accordance with the system's features that were revealed. For instance, at the beginning the use of the pedometer engaged the patients while at the end it was the shared challenges that motivated them. Revisiting Maria's case we observed that for the first week she focused on uploading her personal information such as diets or recipes while at the end she only registered her footsteps.

Using collaboration to persuade. Finally, we observed that collaboration and social support was the main feature to engage users to achieve goals. Going back to Maria's case, while for the first week she scored at most 6167 steps, once Rita challenged her she increased her footsteps by 300% -up to 18150 steps (Figure 3). By doing so she completed a 10000 challenge resulting in a loss of 5 kilos. Patients repeatedly expressed that the system would keep them motivated by connecting their goals and problems with those of others. For instance, a patient made the following comment: *"you are used to skip exercise at home, but with the [site] you know that you aren't alone and that motivates you"*. Another patient stated that: *"A [rivalry] in a good way exists because you could see the amount of exercise others have been doing but that cheers you up because you don't want to be left behind"*. As a result of the use of our system, three of the participants decided to continue with the program to become part of a larger community with the group that started the following month.

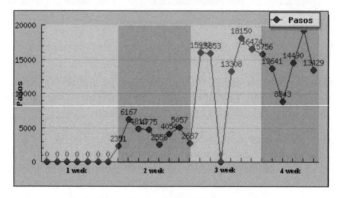

Fig. 3. Maria's timeline –footsteps activity

5   Related Work

Recent research has focused on the use of virtual communities and persuasive technologies to promote healthy lifestyles in social groups. An example of this is *Houston*; a software application that promotes healthy lifestyles in social groups, allowing users to register physical activities and send instant messages[9]. As in our system, *Houston* uses a pedometer to measure the physical activity and mobile phones to support the social group. Houston's main function is to record physical activity, goals, message exchanges with friends and activity time lines. In contrast, pHealthNet works with virtual communities which include not only friends, but also healthcare specialists, and allows sharing additional information such as receipts and exercises.

UP Health is a computer system based on desktop computers and mobile phones to share information to promote physical activity and smoking cessation in social groups [10]. The application uses persuasive tools to accomplish these objectives. However *Up Health* offers limited social support because it only shares records and goals among their members, but there are no collaborative activities such as the games included in pHealthNet. An additional difference between these systems and pHealthNet is that the latter was created to support a social program established in a hospital to assist patients with a chronic disease.

Several virtual communities for healthcare have been created. A good example is *HutchWorld* [5], to assist cancer patients through social interaction tools, information, and entertainment activities. Although these systems have proved to be very successful in improving patients' quality of life, their focus is only on community support, and not on persuading patients to change their living habits.

6 Conclusions

We conducted a case study of a program aimed at promoting a healthy lifestyle among patients with chronic diseases. We identified the following issues: poor communication between participants (patient-specialist), no buildup of personal goals, loss of interest after completing the course, and high desertion rate. With information gathered from interviews with the patient, we designed and implemented a persuasive ecosystem to motivate patients to do exercise and diets. A preliminary assessment of the system provides evidence that the patients increase their trust in themselves and their physical activity and have begun to abide by their plans.

Acknowledgements

We thank Jorge Fisher for his support in deploying the system at IMSS and his helpful comments throughout the development of the system.

References

1. OECD, OECD in Figures, OECD publication. 10, Paris (2005)
2. Secretaria de Salud and INSP, Encuesta Nacional de Salud y Nutrición, 2nd edn., Méxco. 131 (2006)
3. Joon, K., et al.: Encouraging participation in virtual communities. Commun. 50(2), 68–73 (2007)
4. Jadad, A.R., et al.: Are virtual communities good for our health? BMJ 332(7547), 925–926 (2006)
5. Shelly, F., et al.: HutchWorld: clinical study of computer-mediated social support for cancer patients and their caregivers. In: Proc. CHI 2002. ACM, Minneapolis (2002)

6. Fogg, B.J.: Persuasive Technology: Using Computers to Change What We Think and Do. In: Jonathan, G., Jakob, N., Stuart, C. (eds.) Science & Technology Books, vol. 224 (2002)

7. Consolvo, S., Paulos, E., Smith, I.: Mobile Persuasion for Everyday Behavior Change. In: Eckles, E.F.a.E. (ed.) Stanford Captology Media, p. 166 (2007)

8. Intille, S.S.: A New Research Challenge: Persuasive Technology to Motivate Healthy Aging. IEEE Transactions on Information Technology in Biomedicine 8(3), 235–237 (2004)

9. Sunny, C., et al.: Design requirements for technologies that encourage physical activity. In: Proc. CHI 2006. ACM, Montreal (2006)

10. Misook, S., Jeunwoo, L.: UP health: ubiquitously persuasive health promotion with an instant messaging system. In: CHI 2007 extended abstracts. ACM, San Jose (2007)

Supporting the Social Practices of Distributed Pair Programming

Till Schümmer[1] and Stephan Lukosch[2]

[1] FernUniversität in Hagen, Department for Mathematics and Computer Science,
58084 Hagen, Germany
`till.schuemmer@fernuni-hagen.de`
[2] Delft University of Technology, Faculty of Technology Policy and Management,
Systems Engineering Department, Jaffalaan 5, 2628BX Delft, The Netherlands
`s.g.lukosch@tudelft.nl`

Abstract. When considering the principles for eXtreme Programming, distributed eXtreme Programming, especially distributed pair programming, is a paradoxe predetermined to failure. However, global software development as well as the outsourcing of software development are integral parts of software projects. Hence, the support for distributed pair programming is still a challenging field for tool developers so that failure for distributed pair programming becomes less mandatory. In this paper, we analyze the social interaction in distributed pair programming and investigate how current technology supports this interaction. We present XPairtise, a plug-in for Eclipse that allows instant pair programming in distributed development teams. In addition, we report on experiences and findings when using XPairtise in a distributed software development setting.

1 Introduction

Agile software development practices [1], especially the eXtreme Programming [2] methodology, most importantly differ from other software development practices in the way how they address collaboration among participants. In the agile manifesto [3], the authors state 12 general principles that all highlight the importance of flexibility and collaboration. With respect to group interaction, principles 4, 5, 6, and 11 are most relevant:

> "(4) Business people and developers must work together daily throughout the project. (5) Build projects around motivated individuals. Give them the environment and support they need, and trust them to get the job done. (6) The most efficient and effective method of conveying information to and within a development team is face-to-face conversation. (11) The best architectures, requirements, and designs emerge from self-organizing teams." [3]

Taking these principles seriously would imply that a distributed application of agile methods, especially the application of distributed eXtreme Programming

R.O. Briggs et al. (Eds.): CRIWG 2008, LNCS 5411, pp. 83–98, 2008.

(DXP), is a paradoxe predetermined to failure. In the same sense, global software development and outsourcing could not go together with agile approaches.

On the other hand, researchers have proposed several tools to better support distributed agile software development. The first notable publications that related distributed collaboration with agile methods were presented at the first international conference on eXtreme Programming. The *Team Streams* system [4] provided support for asynchronous interaction in XP while the TUKAN system [5] had a focus on partner finding and synchronous interaction. Both of these tools mapped the social practices to groupware applications in order to improve the interaction between the participants. In the following years, additional tools were presented that again mapped social processes of XP to groupware solutions. These include tools for distributed pair programming and tools for better supporting the planning process in XP.

Eight years later, we still see the need for additional research on tools and processes for DXP, especially for solutions that extend the most obvious solution of providing a shared code editor. For that reason, we have revisited well-known assumptions for tool support in DXP [5] and extended these assumptions with novel interaction settings. These interaction settings focus on knowledge transfer and testing which are integral parts of most agile processes.

Our findings are presented in this paper: We first summarize the social practices of pair programming before we present XPairtise, yet another but different tool for distributed pair programming. We describe the interaction metaphors used in XPairtise and present first observations from a long term evaluation where two software development teams used XPairtise during a 6 month project. Our experiences show that XPairtise can be a valuable component in a DXP practitioner's tool suite and thus contribute to making DXP reality at the end.

2   The Social Practice of Pair Programming Its Technology Implication

In this section, we briefly summarize the interaction that takes place in pair programming, i.e. coordination, coding, communication, teaching, and testing. Our assumptions are based on findings reported in [5,6,7]. As in [5], we take a look at the interaction between developers in a pair programming setting and discuss possible design alternatives for mapping this interaction to a computer-mediated setting by using a collaborative application. We make use of design patterns for computer-mediated interaction (P4CMI) [8] to describe the core design considerations. These patterns capture commonly used collaborative system design solutions and thus allow us to describe a hypothetic DXP system. We also use the patterns to compare existing solutions with the hypothetic solution by identifying the presence of patterns in the existing solutions. More details on the individual patterns can be found on the CMI patterns repository web site at `http://www.cmi-patterns.org/`. Note that we will use SMALL CAPS to identify a pattern name.

2.1   Coordination

The traditional setting: eXtreme Programming employs a lightweight planning metaphor using index cards as a main planning artifact. The use cases of the system under development are written down in the form of user stories on index cards (non-digital paper). In the planning game, these stories are discussed and sorted according to their importance. User stories are further decomposed to development tasks. Again, developers use index cards to store task details.

Before a pair programming session can start, a developer picks a task (from a set of shared task cards) and looks for a peer. In co-located settings, finding a partner is easy. The developer looks for other people who are currently finishing their tasks or work alone on other tasks. During daily planning meetings (the daily stand-up meetings), teams can be assigned for the day.

The computer-mediated setting: In distributed settings, both, the handling of story and task cards as well as the formation of pairs is much more challenging. Cards need to be stored in a light-weight planning environment. For group formation it is not as easy to detect the current status of remote users. Time shifts may make a stand-up meeting for coordination difficult or impossible. It is thus required that the developers become aware of one another, e.g., by having ACTIVITY INDICATORS, i.e., peripheral status views communicating the other users' current actions. Task cards, in addition, need to be available as shared objects, e.g., by organizing them in a SHARED FILE REPOSITORY or by means of a planning wiki.

2.2   Coding

The traditional setting: Once the team is formed, the team members sit together in front of the same screen and discuss a possible solution for the task. They maintain task awareness by placing the card next to the screen. One developer takes the role of a driver (the user having the keyboard) while the other user acts as a navigator. The navigator's task is to comment on the possible solutions for the task and check the quality and understandability of the created code.

The computer-mediated setting: In the distributed case, this should be supported with a SHARED EDITOR. The editor should be part of the integrated development environment and automatically open for the navigator when the driver opens it on his screen (whenever they join the same COLLABORATIVE SESSION). The content of the driver's and the navigator's editor should be coupled so that both developers can see the same file at the same time (as described in the SHARED BROWSING pattern). The complexity of the shard editor can be reduced by following the FLOOR CONTROL model proposed by XP. Since only one developer should have the keyboard at a time, this developer also controls the current scrolling and cursor position. Without FLOOR CONTROL driver and navigator would be able to type at the same time (note that the users may then produce edit conflicts so that the solution would require mechanisms for CONFLICT DETECTION and means for resolving conflicting changes, e.g., using an OPERATIONAL TRANSFORMATION approach). However, synchronous editing blurs the ROLES in the pair programming session, leading to two developers that lose a common focus.

One could think of using an APPLICATION SHARING approach for supporting this kind of interaction. The problem is that this takes reasonable bandwidth and that it is sometimes difficult to focus communication, which brings us to the next interaction in XP.

2.3   Communication

The traditional setting: Communication between developers and between developers and customers is the core of any agile method. By developing in pairs, communication naturally takes place between the developers. By changing partners frequently, knowledge is distributed epidemically. Communication is focussed by the shared display and gestures. Quick sketches on a sheet of paper can further support the communication. Having all developers in the same room allows other pairs to overhear conversations and thereby dynamically react to issues discussed in another pair.

The computer-mediated setting: For distributed interaction, communication poses the biggest problem in agile methods. On one hand, we would benefit from a media-rich communication channel, such as a video channel. On the other hand, the communication channel should not consume too much network bandwidth, be stable enough, and establishing connections needs to be quick and easy.

The simplest communication means would be an EMBEDDED CHAT. In addition, scribbles could be drawn in a synchronous graphical SHARED EDITOR and gestures could be conveyed in forms of REMOTE SELECTIONS in the source code that allow the navigator to point out relevant sections in the code. These kinds of communication have the advantage that they can be easily kept persistent and thereby enrich the comments of the discussed software artifacts.

The biggest disadvantage of textual communication for DXP is that the developers need their hands to produce code. Normally, coding and talking goes hand in hand. Thus, we expect that textual chats will not be the most important communication medium.

Embedding an audio channel is another valid option that allows parallel communication and coding. The biggest disadvantage of this solution is that audio communication is typically transient. This becomes important when communication logs should be used for teaching.

2.4   Teaching

The traditional setting: The intensive communication fosters peer-to-pear learning. By pairing a strong developer with a novice, the novice will learn best practices and gradually take more responsibilities. The expert will learn by making his knowledge explicit. We propose to extend this model for classroom education where an additional number of students can participate as spectators (following the SHOW PROGRAMMING pattern of [9]).

The computer-mediated setting: Supporting the MENTOR interaction in distributed settings may be much easier. Due to the fact of global distribution, the opportunity for finding an expert for a specific problem domain may increase. An almost unlimited number of SPECTATORS [10] can be added to the application by distributing the actions of driver to all of them. Since learning becomes more effective when learners actively interact with the subject, we envision that SPECTATORS become active learners who comment and analyze the activities of the driver and the navigator. The computer-mediated setting allows parallel communication channels for the audience and the pair of developers (multiple EMBEDDED CHATS). The developers can perform their pair interaction and communication and the audience can in parallel perform a meta discussion (a comparable interaction has been applied in several scientific conferences where the presentation was complemented with a textual chat channel, e.g. in [11]).

Capturing the interaction would further allow that students REPLAY interesting pair-programming episodes and thereby better understand the evolution of software artifacts (this is an argument why textual communication may be more suitable than audio in some cases).

2.5   Testing

The traditional setting: eXtreme Programming advocates a test-first approach. This means that no code is written as long as no test fails. In a co-located setting, developers first think about how to test a feature that is requested in a specific task. They then create test code that tests the functionality of the feature. This test is executed by a test automation tool such as jUnit [12]. Usually the test fails since the feature is not yet implemented. In a next step, the developers create code that fixes the broken test.

Driver and navigator use debugging tools that allow stepwise execution of the developed software. Additionally, they can inspect and modify variables and provide input values for the software.

The computer-mediated setting: This practice shows that it is not sufficient to have a SHARED EDITOR when considering tool support for distributed XP. Instead, the developers need support for collaborative execution of tests. In a first approach, the system would create DISTRIBUTED COMMANDS for triggering unit tests. This would allow the developers to execute the tests locally at their machines and inspect the results. In a next step, the system would allow coupled debugging including collaborative inspectors of application data and collaborative stepwise execution of a program. To enable collaborative use of debugging tools, one could capture and replicate the commands performed by the driver to control the debugger (DISTRIBUTED COMMAND). Additionally, break points can be modeled as shared objects and views of the variables could be shared. The challenge with such an approach is to keep both client machines (or even more in a setting with spectators) synchronized.

APPLICATION SHARING may ease the technical problems at this stage. However, the application will not be collaboration aware, which makes it again difficult to, e.g., point at specific artifacts.

3 Related Work

As briefly mentioned in the introduction, existing approaches for supporting distributed pair programming either use an APPLICATION SHARING approach to enhance an existing tool suite or provide customized tools that include various groupware features such as SHARED EDITORS or SHARED BROWSING support. As Hanks [13] pointed out, customized groupware tools do often "not provide all of the features used by a particular software developer" and thus "limit her ability to successfully accomplish her work." On the other hand, APPLICATION SHARING solutions lack process support and are thus not collaboration aware. Examples for systems with an APPLICATION SHARING APPROACH are JAZZ and MILOS.

The JAZZ system [14] is an extension of eclipse that supports the whole XP life cycle. Its main focus lies on supporting the workflows in asynchronous interaction. With respect to synchronous interaction, users can stay aware of co-workers and initiate chat sessions with co-workers who are logged in at the same time. Using an INTERACTIVE USER INFO, available users can also be invited to a synchronous pair programming session using an APPLICATION SHARING system.

MILOS [15] aims at supporting the coordination between software developers in an XP team. As in JAZZ, MILOS provides awareness of co-present users and allows users to initiate pair programming sessions using APPLICATION SHARING.

Both JAZZ and MILOS make use of existing IDEs and thereby provide the full functionality that developers know from single-user development environments. Examples for customized solutions are TUKAN and Moomba.

Providing awareness is one of the central aspects of the TUKAN system [5]. The main focus lays on partner finding for pair programming. Developers working on related artifacts are identified and the system proposes them to create a pair for distributed pair programming. SHARED EDITORS are provided for manipulating code together. Users can highlight important code using a REMOTE SELECTION. Unfortunately, TUKAN was built as an extension of a relatively unpopular development environment, namely ENVY for VisualWorks Smalltalk. This is one of the main reasons why it has not gained high popularity.

Moomba [16] extends the awareness model of TUKAN and translates it to a Java environment. Developers are made aware of other developers who work on related tasks. Once they decide to join closer collaboration, they can launch a collaborative Java IDE where the developers can use a SHARED EDITOR. Although Moomba supports Java development, it is still built as a proprietary tool and thereby can not provide the same domain-specific tool support as it is present in modern IDEs.

Solutions that combine the two approaches mentioned above extend professional IDEs with collaboration facilities so that they become collaboration aware. Examples for IDEs that are extended this way are TogetherJ and Eclipse.

Cook [17] created the CAISE architecture to allow users of the Together Architect for Java to share different tools of the IDE. Unfortunately, this IDE does not propagate key-strokes to the CAISE system which has the effect that code changes can only be shared on a per save basis. Eclipse is a more open environment that allows closer coupling of the developers' IDEs.

To our knowledge, there are three collaborative code editors for Eclipse that integrate distributed editing in the context of the IDE. The oldest plug-in that we are aware of is the Sangam system [18]. It allows developers to couple editors. Commands are replicated between the editors so that all connected developers can see and edit the same code. In our tests, we were able to create inconsistencies between the different editor instances. This means that the developers could end up with different data in their editors.

The Saros plugin [19] supports driver-navigator interaction in Eclipse. Once users decide to start a distributed pair programming session, the system synchronizes the code base of both developers and provides awareness on files that are opened at the driver's site. A SHARED EDITOR allows collaborative code creation and REMOTE SELECTIONS allow the navigator to point at relevant pieces of code. Saros is available under an open source licence at http://dpp.sourceforge.net/.

Finally, the XecliP plugin for Eclipse provides to a large extent a comparable functionality as the XPairtise system that we will present in the next chapter. The reason for this is that XecliP was developed in competition in another subteam of our research group. The developers had the same goals as those who developed XPairtise. However, there are slight differences with respect to project sharing, where XPairtise provided the simpler solution. This is the reason why we present XPairtise in this paper. More information on XecliP can be found on its project home page at sourceforge: http://xeclip.sourceforge.net/.

Debugging support is available in several distributed development tools. One of the oldest references is the FLECSE system, that supports users in stepwise execution of text-based software [20]. More recently, Moomba [16] allows developers to share the textual output of a program and to collaboratively execute jUnit tests using a DISTRIBUTED COMMAND approach. The Jazz system [14] uses an APPLICATION SHARING approach for supporting shared debugging. We are not aware of any system that allows loosely coupled interaction in the debugging context (e.g., independent exploration of variables). This is still an open research problem.

In summary, we can observe a trend to better integrate support for SHARED EDITING in professional IDEs. Not surprisingly, the Eclipse IDE becomes more popular for such developments. However, we still see the need for better support of the interaction, especially with respect to the roles in distributed pair programming. None of the tools explicitly addresses learner interaction. Saros seems to be one of the most promising plug-ins so far, but even this plug-in lacks a sophisticated role support.

4   Approach

In this section, we present XPairtise, our approach for supporting distributed pair programming. XPairtise is an Eclipse plugin that offers shared editing,

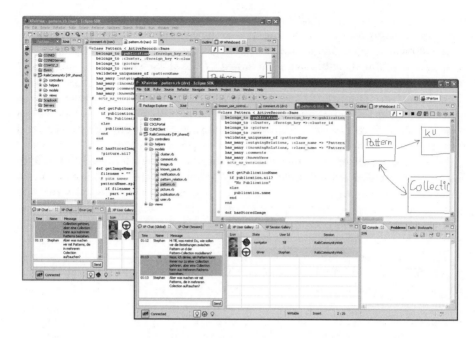

Fig. 1. XPairtise

project synchronization, shared program and test execution, user management, built-in chat communication, and a shared whiteboard.

Figure 1 shows two instances of the XPairtise plugin for Eclipse. In the following, we present the functional and user interface properties of XPairtise in more detail and relate them to the identified social practices.

4.1   Coordination

Story card management is not part of the XPairtise system. The reason for that is that we can employ traditional web-based solutions for capturing and editing the user stories and the task cards and use eclipse's embedded web browser to access the cards. In our setting, we used the CURE wiki [21] that provides templates for story cards and thus helps to ensure that all required information for a story card is present.

Concrete distributed pair programming sessions are modeled as COLLABORATIVE SESSIONS in XPairtise. When users feel the need for a pair programming session, they create a new collaborative session in INTERACTION DIRECTORY that is visible for all other XPairtise users (see Figure 2.1).

COLLABORATIVE SESSIONS have a name (typically the name of the task card) and an Eclipse project that is going to be shared via the proprietary XPairtise server. Once such a session has been created, it is listed in the INTERACTION DIRECTORY (see Figure 2-1). Each user who is currently connected to the XPairtise server can now join the session. When joining the session users can decide to join as navigator or driver. Of course, this is only possible when no other user has joined with the selected role so far.

Users can browse for other users who are connected with the XPairtise server via the USER GALLERY (see Figure 2-2). This view includes the remote users' current AVAILABILITY STATUS and thereby eases the selection of an appropriate partner. By sending an INVITATION they can invite another user to the pair programming session (see Figure 2-3). This opens a request at the invited user's site and the user can decide whether to join or not (see Figure 2-4).

The local workspace of a joining user is stored and the project for the distributed pair programming session is retrieved from the XPairtise server. This ensures that driver as well as navigator have synchronized workspaces when starting the session. Additionally, this approach makes XPairtise independent from code repositories like CVS or SubVersion and allows to establish ad-hoc distributed pair programming sessions.

1) XPairtise INTERACTION
DIRECTORY

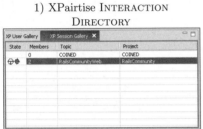

2) XPairtise USER GALLERY

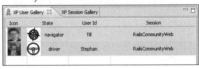

3) Driver invites another user

4) Dialogue to accept or reject an
invitation

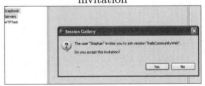

Fig. 2. Setting up a distributed pair programming session with XPairtise

4.2 Coding

Once a distributed pair programming session is established, driver and navigator can cooperate in a SHARED EDITOR (see Figure 3). All actions of the driver are also performed at the navigator's site. This, e.g., includes opening source files, scrolling the window, marking text, moving the text cursor, highlighting lines, editing text, as well as refactoring source code.

In co-located settings, driver and navigator switch roles by passing the keyboard among each other. To reflect this in a distributed setting, XPairtise makes use of a FLOOR CONTROL technique. Driver as well as navigator can request to switch roles (see Figure 4.1) by pressing a role change button. This request is highlighted in the user interface at the other user's site (see Figure 4.2).

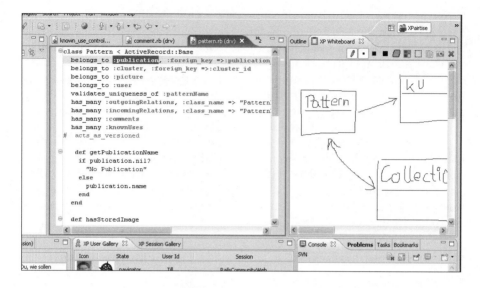

Fig. 3. Shared Editor in XPairtise

Fig. 4. Changing roles in XPairtise

A role change cannot be forced. It only takes place when the other agrees by also pressing the role change button.

4.3   Communication

XPairtise supports multiple communication channels: driver and navigator can use the integrated shared whiteboard and a graphical SHARED EDITOR (see right part of Figure 3) to exchange ideas. They can use the EMBEDDED CHAT for textual communication (see lower left corner of Figure 1). And they can use an integrated Skype control to establish audio connections.

To allow the navigator to raise the driver's attention to specific parts of the source code, the shared editor also supports REMOTE SELECTIONS, i.e. navigator can select text in their editor and the same text is highlighted in the driver's editor.

4.4   Teaching

With the above functionality, XPairtise already enables distributed pair programming. As it is also possible to create ad-hoc distributed sessions via the

XPairtise server without the need of project-specific code repositories, a novice can easily invite an expert to a pair programming session. The expert can then act as a MENTOR and teach the novice on how to solve current problems.

To widen the audience of a pair programming session, XPairtise furthermore supports the additional role of a SPECTATOR. Users who join an ongoing pair programming session as SPECTATOR can watch the interaction among the driver and the navigator. For that purpose, XPairtise also retrieves the project of the session from the XPairtise server. When a SPECTATOR joins as a latecomer and the driver already performed some changes, XPairtise still ensures that the workspaces are synchronized. Thereby, XPairtise can easily be used to teach a group of novices in a specific problem domain when the driver is an expert in that domain. Additionally, this also allows to teach distributed pair programming, when novices join an ongoing pair programming session among two experts in eXtreme Programming.

4.5   Testing

When the driver performs run actions or starts tests, these are started at the navigator's site as well. Thereby, XPairtise enables basic collaborative testing.

However, since testing is more than the execution of JUnit tests, we have recently added an XPairtise extension that supports collaborative debugging. Break points are modeled as shared objects as well allowing all participants to stop the program under test at the same place. In the same way as editor inputs were shared among the members of a collaborative session, the use of the Eclipse debugger can also be coupled: Eclipse commands at the driver's client such as stepwise execution of code or inspection of variables are monitored by the XPairtise plugin and distributed to all other clients. FLOOR CONTROL is an important issue here again since the it needs to be ensured that only one client at a time is able to continue the execution of the program under test.

The main reason why our debugging support for XPairtise is not yet part of the official open-source release is that we are still working on synchronizing the effects of external influences on the execution of the program under test. This includes that all input (e.g., files, streams, mouse movements or hardware signals like timer values) used by the tested program needs to be identical to ensure the same execution. To what extent this can be solved is still an open issue.

4.6   Implementation of XPairtise

XPairtise makes use of a client server architecture. On a technical level, XPairtise clients communicate with the XPairtise server using a JMS infrastructure (Java Messaging Service), namely the ActiveMQ messaging server (see Figure 5).

During the bootstrap phase of the XPairtise infrastructure, the XPairtise server connects to the message bus and creates a message channel that allows clients to request information on shared objects stored in the HSQL database that acts as XPairtise's object repository (CENTRALIZED OBJECTS).

Whenever clients register by sending a registration message to the server's message queue, the XPairtise server creates a message queue for the client. The

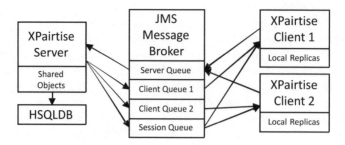

Fig. 5. XPairtise conceptual system architecture

client subscribes to this queue and from then on receives updates on changes to shared objects (REMOTE SUBSCRIPTION). COLLABORATIVE SESSIONS are also modeled as shared objects. In addition, each collaborative session has a message queue to which the server can add updates needed by all participants in the session. These updates are either sent as state updates (e.g., when users add drawings to the shared whiteboard) or as DISTRIBUTED COMMANDS. When a client, e.g., changes the selection in the SHARED EDITOR of Eclipse, the XPairtise plugin captures the selection command and sends this to the server. The server in turn adds the command to the session's message queue so that it is received by all members of the COLLABORATIVE SESSION. Each client executes the selection command locally with the effect that all clients can see the REMOTE SELECTION.

5 Experiences

XPairtise was developed in our 2006/07 lab course on cooperative system development. As our university is a distance teaching university, the team members only met at the beginning and the end of the lab course. In the meantime, the team members collaborate at a distance. Once the team had a first running prototype, the team used XPairtise for distributed pair programming. This allowed the team to identify problems early in the development cycle and address such problems directly. At the end of the lab course, all team members reported that XPairtise simplified their collaboration a lot. Since then, XPairtise is available at sourceforge.net.

A more formal evaluation was performed in our 2007/08 lab course. Again, students were asked to develop a collaborative application. We observed two teams of 5-6 students for a period of 18 weeks. During that time, we recorded the JMS messages exchanged in 52 XPairtise sessions. This allowed us to make a detailed analysis on actions performed in the sessions. In addition, we recorded the audio communication for seven (randomly selected) sessions where XPairtise was used. Complementary to these observations, we conducted semi-structured interviews in order to get feedback on the perceived usefulness of XPairtise.

5.1   Results

Based on this observation, we report first anecdotal results in relation to the social practices presented in the previous sections. We do not explicitly report on coordination issues since the existence of the sessions is already an indication that users were able to meet in COLLABORATIVE SESSIONS or join existing sessions as spectators (in 17 of 51 sessions).

Coding: In all 52 observed sessions, code was changed. The shared editor was used as expected. Surprisingly, we could observe a less agile interaction between driver and navigator. While in co-located pair programming sessions, role switches are expected to happen every 20 minutes [22], we could observe only 21 of 52 sessions where a role change took place at all. And even where a role change took place, there were only 4 sessions where the navigator was active for more than 30 % of the time. An ideal pair programming session would have frequent role switches and lead to an equal participation of both partners. This could not be confirmed in our observations. One reason for this was that in many cases experienced developers interacted with novice navigators (see section on teaching below).

Communication: When designing XPairtise, we expected that audio communication would be the most prominent communication channel but that the EMBEDDED CHAT would also be used frequently. However, there was almost no use of the text-based chat. Only 9 of 52 sessions had any chat entries in the session chat. Even fewer sessions had entries in the global chat that was intended as a meta communication channel. Using chat logs for augmenting code comments would thus not be possible in the observed groups.

Only 4 sessions utilized the whiteboard. These sessions were not used for pair programming but for creating sketches for the final project presentation. The interviews did not provide any further clues why the whiteboard was not used more frequently. Actually, students reported that they liked the feature of the whiteboard, which is in contrast to the log data that shows that the students did not use the whiteboard frequently.

The REMOTE SELECTION was used in all sessions. In all but 9 sessions, the navigator also selected code to focus the communication on a specific part of the code. However, there were much fewer occasions of remote selections than expected. In average, the ratio between driver selections and navigator selections was 92 to 8. The interviews on the other hand indicated that the users perceived the remote selections as a very important feature of XPairtise.

Teaching: To our surprise, there were fewer than expected sessions with SPECTATORS (17 of 52 sessions). In most of these cases, there was exactly one SPECTATOR (12 cases) who joined the XPairtise session for a short time. From the developers' feedback, we can say that in some cases, the guest was an expert who joined the session with the goal of explaining a specific part of the code. Instead of the driver educating the SPECTATORS, these were directing the driver in this case.

In addition to these sessions, we could observe teaching to a large extent: Many pair-programming sessions brought together expert developers as drivers with novices as navigators (approx. 40% of the sessions). In these settings, there were almost no role switches (only 36% of all observed sessions included a role switch). The driver kept his role throughout the session (between 1 and 2 hours). Looking at the audio logs of these sessions, we could observe that the driver was speaking much more than the navigator (in four of the seven observed sessions, the driver talked more than 85% of the time). All these observations indicate that the driver was presenting his code to a rather passive navigator. However, especially the unexperienced developers who participated as navigators in these sessions reported that observing the expert was very helpful for them.

Testing: Since the observed version of XPairtise only provided collaborative unit test execution, we cannot provide proofs on the usability of further testing support. Interestingly, the developers reported that they would prefer testing alone. This is in contrast to the XP methodology that puts collaborative test execution in the center of all development activities. The main reason for this judgement of the observed developers could be that none of them was experienced in testing.

5.2 Summary

The above evaluation shows that XPairtise supports the social practices for distributed pair programming. However, compared to pair programming in a traditional setting, our observations highlight some interesting differences. For coding, it is interesting to note that role switches did not occur as often as expected from a traditional setting. When considering communication, the EMBEDDED CHAT was used less than expected and almost all communication was handled via an audio channel. It is interesting to note that the whiteboard as well as the REMOTE SELECTION feature were rarely used but still considered as an important feature by the users.

Concerning teaching, we also expected a different behavior. Instead of laymen in the role of the SPECTATOR, experts were using this role. Still, this feature enabled the transfer of knowledge which is the basic task of teaching. Finally, when considering testing, the basic support was not used at all. But, this is mainly due to the fact that developers preferred testing on their own as well as developers not being experienced in testing at all.

6   Conclusions

In this paper, we have discussed the main social practices for distributed pair programming, i.e. coordination, coding, communication, teaching, and testing. We analyzed the technology implications when transferring these practices to distributed settings and provided guidance for developing technology support. We have discussed to what extent existing systems support these social practices and presented a tool that integrates support for the the practices in Eclipse.

First experiences, when using our tool during its development and during two long term development projects indicated that XPairtise supports the social practices for distributed pair programming. The evaluation revealed some interesting aspects on the tool usage. We plan to continue the evaluation of XPairtise in further lab courses as well as in commercial distributed development projects. In these evaluations, we aim to better understand the impact of the plugin on distributed team performance and to identify additional functionality to further support distributed pair programming.

Independent from these findings, we will in the next future improve the functionality for pair formation. For that purpose, we will include ACTIVITY INDICATORS or EXPERT FINDER mechanisms as they were, e.g., present in TUKAN [5] that allow to retrieve experts from the registered users for specific problem domains. This retrieval could, e.g., be based on source code analysis or activity analysis of pair programming session. For improving the teaching functionality, we plan to include a REPLAY mechanism which records pair programming session and allows to review them afterwards.

References

1. Boehm, B., Turner, R.: Balancing Agility and Discipline – A Guide for the Perplexed. Addison Wesley, Boston (2004)
2. Beck, K.: eXtreme Programming Explained. Addison Wesley, Reading (1999)
3. Beck, K., Beedle, M., van Bennekum, A., Cockburn, A., Cunningham, W., Fowler, M., Grenning, J., Highsmith, J., Hunt, A., Jeffries, R., Kern, J., Marick, B., Martin, R.C., Mellor, S., Schwaber, K., Sutherland, J., Thomas, D.: Manifesto for agile software develqopment (2001)
4. des Rivieres, J., Gamma, E., Mätzel, K.U., Moore, I., Weinand, A., Wiegand, J.: Team Streams: Extreme Team Support. In: Extreme Programming Examined, pp. 333–353. Addison Wesley, Reading (2001)
5. Schüumer, T., Schüumer, J.: Support for distributed teams in extreme programming. In: Succi, G., Marchesi, M. (eds.) eXtreme Programming Examined. Addison Wesley, Reading (2001)
6. Braithwaite, K., Joyce, T.: Xp expanded: Patterns for distributed extreme programming. In: Longshaw, A., Zdun, U. (eds.) Proceedings of the 10th European Conference on Pattern Languages of Programs, EuroPLoP 2005, Konstanz, Germany, UVK, pp. 337–345 (2006)
7. Kircher, M., Jain, P., Corsaro, A., Levine, D.: Distributed extreme programming. In: Proceedings of XP2001 - eXtreme Programming and Flexible Processes in Software Engineering, Villasimius, Sardinia, Italy (2001), http://www.kircher-schwanninger.de/michael/publications/xp2001.pdf
8. Schümmer, T., Lukosch, S.: Patterns for Computer-Mediated Interaction. John Wiley & Sons, Ltd., Chichester (2007)
9. Schmolitzky, A.: Patterns for teaching software in classroom. In: Hvatum, L., Schümmer, T. (eds.) Proceedings of EuroPLoP 2008, Konstanz, UVK, pp. 37–54 (2008)
10. Lukosch, S., Schümmer, T.: The role of roles in collaborative interaction. In: 13th European Conference on Pattern Languages and Programs, EuroPLoP 2008 (2008)

11. Rekimoto, J., Ayatsuka, Y., Uoi, H., Arai, T.: Adding another communication channel to reality: an experience with a chat -augmented conference. In: CHI 1998: conference summary on Human factors in computing systems, pp. 271–272. ACM, New York (1998)
12. Beck, K., Gamma, E.: Junit cookbook (2002)
13. Hanks, B.F.: Empirical Studies of Distributed Pair-Programming. Dissertation, University of California Santa Cruz (December 2005)
14. Hupfer, S., Cheng, L.T., Ross, S., Patterson, J.: Introducing collaboration into an application development environment. In: CSCW 2004: Proceedings of the 2004 ACM conference on Computer supported cooperative work, pp. 21–24. ACM Press, New York (2004)
15. Maurer, F.: Supporting distributed extreme programming. In: Wells, D., Williams, L. (eds.) XP 2002. LNCS, vol. 2418, pp. 13–22. Springer, Heidelberg (2002)
16. Reeves, M., Zhu, J.: Moomba – a collaborative environment for supporting distributed extreme programming in global software development. In: Eckstein, J., Baumeister, H. (eds.) XP 2004. LNCS, vol. 3092, pp. 38–50. Springer, Heidelberg (2004)
17. Cook, C.: Towards Computer-Supported Collaborative Software Engineering. PhD thesis, University of Canterbury, Christchurch, New Zealand (December 2006)
18. Ho, C.W., Raha, S., Gehringer, E., Williams, L.: Sangam: a distributed pair programming plug-in for Eclipse. In: Eclipse 2004: Proceedings of the 2004 OOPSLA workshop on eclipse technology eXchange, pp. 73–77. ACM Press, New York (2004)
19. Djemili, R.: Entwicklung einer Eclipse-Erweiterung zur Realisierung und Protokollierung verteilter Paarprogrammierung. Master's thesis, Freie Universität Berlin (2006)
20. Dewan, P., Riedl, J.: Toward computer-supported concurrent software engineering. IEEE Computer 26(1), 17–27 (1993)
21. Haake, A., Lukosch, S., Schümmer, T.: Wiki-templates: adding structure support to wikis on demand. In: WikiSym 2005: Proceedings of the 2005 international symposium on Wikis, pp. 41–51. ACM Press, New York (2005)
22. Hanks, B.: Tool support for distributed pair programming: An empirical study. In: Proceedings of XP/Agile Universe 2004: 4th Conference on Extreme Programming and Agile Methods, Calgary, Canada (2004)

Using Second Life in Programming's Communities of Practice

Micaela Esteves[1], Ricardo Antunes[1], Benjamim Fonseca[2], Leonel Morgado[3],
and Paulo Martins[3]

[1] Polytechnic Institute of Leiria, Ap. 4163, 2411-901, Leiria, Portugal
[2] UTAD/CITAB, Ap. 1013, 5001-801, Vila Real, Portugal
[3] UTAD/GECAD, Ap. 1013, 5001-801, Vila Real, Portugal
{micaela,antunes}@estg.ipleira.pt,
{benjaf,leonelm,pmartins}@utad.pt

Abstract. This paper presents a novel approach to teaching and learning computer programming, using the three-dimensional virtual world Second Life® to develop a programming community of practice. Our students have developed their programming projects as part of this community as an alternative way of learning. The learning of programming is a difficult process, with many students experiencing difficulties which result in high levels of failure in introductory programming courses. In this paper, we describe and analyse how this approach spurred students' motivation and interest in learning programming. We also present observations on the difficulties felt by both students and teachers in the development of projects and activities, and discuss the approaches taken to overcome those difficulties.

Keywords: Communities of practice, Collaboration, Programming learning, Virtual worlds, Second Life.

1 Introduction

Learning how to program a computer is a hard task, and a diversified set of skills must be learned for one to become a good programmer. Typically, when students initiate the study of computer programming, they usually come across several difficulties, which are then reflected in highs levels of failure in entry-level courses (commonly called "Computer Science 1/2" or "Computer Programming 1/2"). Several research efforts have sought to find the causes of this failure (*e.g.*, [1], [2]). Amongst the reasons pointed out by research are: lack of contextualization of the learning process [4]; the nature of traditional teaching method, based on lectures and specific programming language syntaxes [3], and difficulties in understanding the basic concepts of programming, such as variables, data types or memory addresses [2-3], described as abstract concepts without an equivalent representation in real life.

Compounded with these factors, we have a new generation of computer science students for whom computers have been a constant presence in their lives, an important tool, but don't feel themselves motivated to learn computer programming [5].

R.O. Briggs et al. (Eds.): CRIWG 2008, LNCS 5411, pp. 99–106, 2008.

They often don't understand why they should write code, since there is a world of complexity to be mastered just by combining applications and settings, by fiddling with configuration files and formats. Also, typical computer environments and applications students employ as users are of a visual complexity and appeal far beyond what students typically achieve on entry-level programming courses, a factor that does not support self-motivation. On the other hand, the stereotype of a programming student as someone that is alone, programming all night long, without social contact, contributes to hinder student's personal view of programming subjects and even shed aside possible future careers related with it [5].

All the aspects mentioned above make the students feel and experience some disorientation and lose interest in learning. Although students belong to a community – the academic community – they "learn lonely and alone are tested" [4]. The vision of having the students all connected as a network node, each contributing to another's learning while building personal knowledge[4], drove us to create a programming community of practice in Second Life. The practical applications of the acquired knowledge in the community, its reflection and exchange are some of the strategies suggested by Fleury [6] and Dillenbourg [7].

In this paper, we present the result of two years of observations using Second Life as a platform for teaching and learning computer programming, with the purpose of identifying practical issues that teachers and students face in such approach, and ways to overcome those issues. It was not our goal to compare this approach to others, since we believe that any such evaluation depends on the educational methods and practices; and that the establishment of methods of practices requires educational practitioners to be aware of the practical issues that may hinder or disrupt the educational process. In the following section, we give an overview of the concept of communities of practice, and in the second section, we present the research activities and analyse the results (identified issues and approaches to overcome then). Finally, we present some conclusions based on reflection upon the results.

2   Communities of Practice

"Communities of practice are everywhere and we are generally involved in a number of them – whether that is at work, school, or in civic and leisure interests. In some groups we are core members, in other we are more at the margins", [9]. Within these communities to which we belong, people share a common interest and join each other in its pursuit, developing and learning practices and world-views in the process. The practices may reflect activities, but also social relations.

These communities may have a formal or informal organization (formal communities of practice being those with regular meetings with predefined work, informal ones all others, including those that may not even see themselves as a community). Typically, communities are organized around some particular area of knowledge / activity that provide members a sense of joint enterprise and identity [9].

As stated by Wenger [8], there are three elements involved in defining a community of practices (CP). One is the domain: the community must have a subject to talk about. The second is a community of people that interact and thus facilitate the development of relationships regarding the domain. A Web page is not a CP, or if there are seventy managers that never talk with each other, they are not a CP, even if they have the same functions. There must be a community of people, a sharing and construction of knowledge. The third element is the set of practices (the "practice"): the community must have a practice and not just a common interest that people share. They learn together how to do the things they do (or want to do). And that learning involves participation in the community. A participation that refers not just to local events of engagement in certain activities with certain people, but to a more encompassing process of being an active participant in the practices of social communities and constructing identities in relation to these communities [8].

According to Wenger [8], a community of practice is a good way to promote learning and good practices, not only because it develops knowledge in a living and experimental way, but also because it helps participants reach solutions to possible problems, with significant connections leading individuals to higher creative levels than they could reach on their own [9]. A typical community is made up by different levels of participation: central, active and peripheral. Initially, people join communities and learn at the periphery. As they become more competent they move more to the "centre" of the community. According to Wenger [8], in order for a community of practice to be successful it needs to motivate the participants' involvement at the different levels, establishing the dialogue between the internal and external perspectives of the community. The participation of external elements in a community is extremely useful for the development of practices in that community as well as the integration of the community itself in other groups.

3   Developed Activity

The main objective of this study is to find out if and how could SL be used as a platform for teaching / learning the imperative programming language paradigm that is commonly taught in college level computer science courses. For this purpose, we have employed action research methodology. For this purpose, we create a community of practice for teaching and learning computer programming in the Second Life virtual world (SL), we provided to students elective alternative assignments on some compulsory college-level subjects. This took place at two Portuguese Higher Education institutions: the University of Trás-os-Montes e Alto Douro (UTAD) and the Higher School of Engineering and Management of the Polytechnic Institute of Leiria (ESTG). The subjects' main aim is to allowing students to develop semester-long projects, to improve programming skills.

3.1   Methodology

In this study we have employed action research (AR) methodology, a cyclical process approach that incorporates the four-step processes of planning, action, observing and reflecting on results generated from a particular project or body of work [12].

The first action research cycle started by planning a model for teaching introductory programming concepts in SL. To that end, it was necessary to make a pre-exploratory research and pre-observation with the goal of identifying problems and planning actions [13]. Initial plans were formulated, and actions for their prosecution were devised and implemented. While the action (teaching-learning) took place, results were monitored for reflecting later on.

The data collected for the reflection step of the action research methodology was based on daily session reports, classroom images and questionnaires. The reports, written down by the teacher-researcher at the end of each session, describe what happened during the class, indicating all the critical incidents and its implications. Classroom images (screenshots) have been taken in order to assist the teacher to review the lesson when necessary, such as when a critical incident had happened. Questionnaires with open questions concerning the learning / teaching method were presented to students at the beginning, middle and end of the process, to provide further information on the learning process. These elements are used as a tool to adjust and improve the learning / teaching approach.

The final step in the first research cycle was reflection upon the outcomes and based on them planning the next cycle. This goes on until the reflection of a cycle showed that the problem was then solved or level of knowledge achieved is fixed. At this point the study was concluded and a report was produced.

3.2 Programming Environment

The programming environment was SL itself, not any offline editor. SL is a persistent on-line 3D virtual world conceived by Philip Rosedale in 1991 and is publicly available since 2003 [10]. It allows large numbers of users to connect, interact and collaborate simultaneously at the same time and in the same (virtual) space. Figure 1 shows a typical programming session in this research: we can see 6 avatars on black rugs (students programming) and two other - teachers' avatars.

Fig. 1. Typical programming session

SL programming is currently done with a scripting language named *Linden Scripting Language* (LSL), which has C-style syntax and keywords. 3D objects created in SL can receive several scripts that are executed concurrently. Each script has its own state machine: program flow is sequential, using common methods from imperative programming, such as procedures and flow-control primitives, and structured in the traditional way, via function definition and function calls, but also by triggering events and responding to them (events can be raised either by environment interactions such as object collision or programmatic components such as requesting a server-based service). The programmer defines the states of each state-machine and explicitly specifies when to switch state. The language's programming libraries include functions and events both for SL-based results programming and for communication with external servers: sending and receiving e-mail, accepting XML remote procedure calls, and handling HTTP requests and responses.

SL enables synchronous collaboration among students because the system permits two or more avatars to edit the same object and include their own scripts, which act concurrently on the object (and may exchange messages). Also, it is possible to share scripts, so that students can access and edit the same piece of code while programming it. By default, only the creator of an object or script has full access to it. Thus, to share an object or script it's necessary for the creator of the object to explicitly set its permissions adequately. Figure 2 presents two avatars editing the same object (a car): the left window shows the car's contents, one being a script that is opened by double-clicking.

Fig. 2. Two avatars sharing an object

One particular aspect of script sharing is that although several avatars may read and change it, saving the script overwrites the current version. Initially, this is not really a problem, because scripts are shared by the teacher and there is only a student editing the script. But as the community evolves, students are able to contribute more often and in larger numbers, and so coordination among participants is required. Chat channels can be used to coordinate who is accessing and changing the script.

Asynchronous collaboration is also supported because the SL world is persistent. Students and teachers may access and leave in-world objects and messages to the other members (group messages and privates messages are supported). When a user logs in all his/her messages are shown, and he/she can see any objects left in the world by others (and edit them, if adequate permissions have been set).

3.3 Community Structure

Teachers were the community coordinators, so they defined the projects to be developed, encouraged and motivated the periphery students through the exchange of opinions between members of the community, as well as sharing experiences of active participants that were once in periphery.

The community had meetings about two hours per week in Second Life (SL), where they developed their programming work and kept track of community's progress, exchanged ideas and made suggestions. Face-to-face meeting took place only once a month, because the teachers were in Leiria and the students in Vila Real, 270 km apart: once a month there was a meeting at Vila Real to discuss the projects and the details of on-line cooperation.

The first students that participated in the community (2nd term of 2006/2007), 50% had little experience in both programming and SL, so they were in periphery level. The remaining 50% already had some experience in programming, although this was their first contact with SL, and thus while at the periphery of the "programming in SL" community they are already active members of our community if seen just from a programming perspective. The teacher's task was to motivate students at periphery to reach the active stage [10]. At the beginning of the subsequent term, 80% of students were at the periphery and the others within the active level.

3.4 Analysis of Results

It is possible to distinguish two phases in these activities: the first consisted in building objects with the modelling tools of SL, and is devoid of programming (robots, trains and dogs); and the second one consisted in the development of programs in LSL, to provide behaviours to the objects created previously.

During the first phase there weren't significant disparities between students at different levels of participation, the difficulties felt by both students were identical. For example: how to link objects with each other, how to make a copy or to line up objects. This is consistent with the previously-mentioned fact that while some students were at the periphery and others within the active level, regarding programming expertise, all began at the periphery regarding SL use.

In the second phase, some differences were observed among students. Students from the active level didn't have great difficulties in understanding how LSL works. Although they had already worked with event-based programming in other courses, these students weren't familiar with the concept of state machines or their programming. The major difficulty they faced consisted in selecting which library functions and events to use, and how to use them to implement specific functionalities. The teachers guided them, by showing alternative ways of creating identical object behaviours, so that they could ponder which would be more adequate.

Students at the periphery, programming-wise, weren't used to self study or autonomous computer programming development, so closer guidance from the teachers was needed. It began with simple examples that students would experiment with and modify. Whenever they had difficulties in understanding the examples, some explanations were provided for that specific part of the code. This way, students could understand what these small programs could do and the goal of each one. Based on the personal experience of the programming teachers and the scientific literature in this field, it is known that this level of understanding is difficult to reach when students are learning to program using traditional environments, such as C command-line compilers, where the students generally feel great difficulty in understanding the programming objective [2].

A particular important aspect in programming learning is the students' reaction to compilation errors [11], which are inevitable in the learning process. Students from active level corrected the compilation errors whenever they happened without the teachers' help, whereas the periphery students found themselves without knowing the reason why they occurred or how to correct them. When students had some difficulty about the code they had implemented, they shared it with the teachers, so they could observe and at the same time find out what was wrong and follow the teachers' indications/instructions. This way, the students corrected the code and went on.

Execution errors occurred more often with active-level students. These students tested more programs by their own initiative and noticed more frequently that these didn't execute as they expected. It was observed that the students were not less motivated because of this; on the contrary, they corrected the programs and tested them until they behaved has they wanted them to.

One of the projects set forth by teachers to the community consisted only in data manipulation and very little graphic interaction (one of SL's differentiating factors). This resulted in difficulty for teachers to motivate periphery students to strive to reach the active level. In order to go overcome this, community leaders had to involve active students from the previous semester and external elements, in order to motivate and increase the activity inside the community. In order to assess the work done by students, it was observed that it was difficult to manage the attribution by students of access privileges for teachers to their scripts, leading to situations that rendered assessment impossible without contacting the student and requesting correction of wrong privileges (for example, when a student would send the teacher an object without conferring the necessary permissions to access the scripts).

One of the difficulties felt in the community development was the lack of a common space for impromptu presentation of schematic ideas and reflections: a "blackboard" as it was. Another issue we came across was the absence of a mechanism that would inform the teachers, by email or some other non-SL system, what the students had achieved throughout the week, i.e., what was reached, what had caused more delays, which difficulties had been felt, and which attempts had been made to try and overcome them.

4 Conclusions

In this paper we presented a study that has been conducted using the action research methodology. In this study we created a programming community with the aim to explore the viability of using SL as platform for teaching and learning a computer programming language.

This study is not finished yet but we can conclude that:

- SL has characteristics that make it a platform suitable for teaching / learning a computer programming language but it is necessary to use it in association with another platform where the teacher can supply students with teaching materials.
- Students learning how to program by programming physical interactions in SL (e.g., making a dog follow you and obey your voice command) are typically motivated. Students who focused primarily on non-visible techniques such as data structures and string processing, benefiting from the environment just for enhanced context and not as a source of feedback for programming behavior, did not seem to exhibit any motivational advantage over students who employ a

traditional console-oriented (text-only) approach. Thus, teachers must pay special attention when conceiving students' assignments, particularly if the students are novice in programming because they need projects that stimulate their imagination.

References

1. Gray, W.D., Goldberg, N.C., Byrnes, S.A.: Novices and programming: Merely a difficult subject (why?) or a means to mastering metacognitive skills? Review of the book Studying the Novice Programmer. Journal of Educational Research on Computers, 131–140 (1993)
2. Miliszewska, I., Tan, G.: Befriending Computer Programming: A Proposed Approach to Teaching Introductory Programming. Journal of Issues in Informing Science & Information Technology 4, 277–289 (2007)
3. Lahtinen, E., Mutka, K.A., Jarvinen, H.M.: A Study of the difficulties of novice programmers. In: Proceedings of the 10th annual SIGSCE conference on Innovation and technology in computer science education (ITICSE 2005), Monte da Caparica, Portugal, June 27-29, 2005, pp. 14–18. ACM Press, New York (2005)
4. Figueiredo, A.D., Afonso, A.P.: Managing Learning in Virtual Settings: the Role of Context. Information Science Publishing (2006)
5. Lethbridge, C., Diaz-Herrera, J., LeBlanc, Jr., Thompson, B.: mproving software practice through education: Challenges and future trends. In: Future of Software Engineering (FOSE apos;2007), pp. 12–28 (May 2007)
6. Fleury, M., Oliveira Junior, M.: Gestão do Conhecimento Estratégico – Integrando Aprendizagem, Conhecimento e Competências. Editora Atlas, São Paulo (2001)
7. Dillenbourg, P.: Learning In The New Millennium: Building New Education Strategies For Schools. In: Workshop on Virtual Learning Environments (2000),
   http://tecfa.unige.ch/tecfa/publicat/dil-papers-2/
   Dil.7.5.18.pdf
8. Wenger, E.C., Snyder, W.M., McDermott, R.: Cultivating communities of practice: a practitioner's guide to building knowledge organizations. Harvard Business School Press Book (2002)
9. Lave, J., Wenger, E.: Situated Learning: Legitimate Peripheral Participation. Cambridge University Press, Cambridge (1991)
10. Esteves, M., Antunes, R., Morgado, L., Martins, P., Fonseca, B.: Contextualização da aprendizagem da Programação: Estudo Exploratório no Second Life. In: Proceedings of IADIS Ibero-Americana WWW/Internet 2007, Vila Real, Portugal, Outubro, 7–8 (2007)
11. Esteves, M., Mendes, A.: A Simulation Tool to Help Learning of Object Oriented Programming Basics. In: Proceedings of the 34th ASEE/IEEE Frontiers in Education Conference, Savannah, Georgia, USA, October, 20–23 (2004)
12. Lessard-hébert, M., Goyette, G., Boutin, G.: Investigação Qualitativa: Fundamentos e Práticas, Lisboa, Instituto Piaget (1994)
13. Dick, B.: A beginner's guide to action research (2008),
   http://www.scu.edu.au/schools/gcm/ar/arp/guide.html

Integrating Collaborative Program Development and Debugging within a Virtual Environment

Hani Bani-Salameh[1], Clinton Jeffery[1], Ziad Al-Sharif[1], and Iyad Abu Doush[2]

[1] University of Idaho
[2] New Mexico State University
{hsalameh,ziada}@vandals.uidaho.edu,
jeffery@cs.uidaho.edu,
idoush@cs.nmsu.edu

Abstract. A collaborative integrated development environment enables developers to share programming-related tasks. This paper presents the design and implementation of a collaborative IDE named ICI (Idaho Collaborative IDE). ICI enables developers in different locations to collaborate on a variety of software development activities in real-time. It supports software development in C, C++, Java, and Unicon. ICI combines a synchronous collaborative program editor and a real-time collaborative debugger within a 3D multi-user virtual environment. ICI reduces cognitive context switches between tools inside the IDE and between IDE tasks and virtual environment activities, allowing developers to share, in real-time, the process of editing, compiling, running, and debugging of their software projects.

Keywords: collaborative environment, debugger, run-time debugging, Integrated Development Environment (IDE).

1 Introduction

Integrated Development Environments (IDEs) are one of the most heavily used tools in programmers' everyday activities. For this reason, IDEs are a primary venue for adding collaboration support for software development and software engineering. A collaborative IDE is a place where developers work together to design, solve coding problems, and share development knowledge.

Much collaboration in software development is accomplished using asynchronous technologies such as e-mail or revision control systems that do not require real-time interactions. The IDE described in this paper augments those facilities with interactive collaboration tools for n users that are needed for applications such as remote pair programming, distance education, and distributed code reviews. Alternative technologies for performing these tasks are available but frequently require too much coordination and setup efforts to be comfortably ubiquitous, especially for developers who interact with many different teams and projects.

As Churchill and Bly observed, communication in collaborative tasks is always about the tools of collaboration [1]. A difficulty appears when, for example, developers discuss

R.O. Briggs et al. (Eds.): CRIWG 2008, LNCS 5411, pp. 107–120, 2008.
© Springer-Verlag Berlin Heidelberg 2008

changes to a program's code in a chat window (or in a phone conference), while at the same time trying to view the same code using a separate application. There are generic tools that provide shared views of an entire PC desktop, but they are bandwidth-intensive, have distracting session setup and teardown costs, and may provide more access and less control than is intended for a given collaboration.

2 Overview of ICI

Among current collaborative software development tools, most are limited to specific software development tasks, such as source code editing. Contexts such as computer science distance education need a more integrated collaborative IDE that 1) supports real-time collaborative compiling, linking, running, and debugging sessions, and 2) provides an environment where developers can communicate easily either by text or voice; all from within the same tool [2], [3].

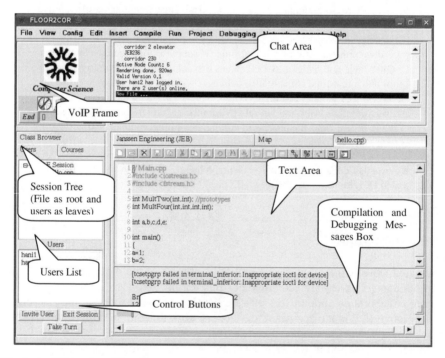

Fig. 1. An ICI session. Tabs allow easy switching between virtual world and IDE tasks. Users can text and voice chat with each other; see who is logged in and how busy they are; invite others to watch or help; and edit, compile, test, run, and debug programs.

This paper presents the design and implementation of a collaborative IDE named ICI (Idaho Collaborative IDE). ICI is integrated inside the CVE virtual environment (cve.sourceforge.net). Fig. 1 shows ICI inside CVE. CVE is a multi-platform collaborative virtual environment where users can interact with each other within a 3D virtual world. CVE's graphical environment is cartoon-like (similar to popular games), rather than aiming at being photorealistic. Fig. 2 shows an example scene developers might see in this environment. The collaborative virtual environment provides developers with a general view of other users and what they are doing. It allows developers to chat via text or VoIP with other team members and with developers from other teams in real time. Users may invite one another into collaborative IDE sessions, where they work together on tasks such as program design, coding or debugging. The integration of the IDE within a virtual environment makes searching for collaborators, seeing who is available, or queuing for the attention of an expert less difficult and less intrusive, especially for the busy developers who serve as architects, chief surgeons, or instructor/mentors. These users are often on the receiving end of a large proportion of collaboration/assistance requests, motivating special attention in the user interface.

While logged in to the collaborative virtual environment, developers can use ICI for their normal IDE tasks, in between collaborative sessions. The fact that it is online, videogame-like, and chat-enabled is a potential distraction, but enriching the sense of online presence while programming is what allows ICI to make it easier to support multiple collaborations or to switch between tasks. For the languages C, C++, Java, and Unicon, ICI provides interactive collaboration for compilation, linking, error messages, and debugging.

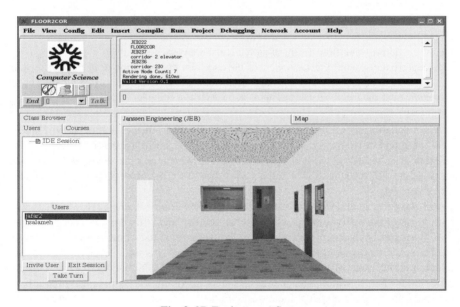

Fig. 2. 3D Environment Scene

A real-time collaborative debugging session begins when one of the participants in the session starts a debugger such as gdb, jdb, or udb[1]. During a collaborative debugging session, developers can take turns controlling the debugger, while other developers watch and discuss the debugging commands and messages (Fig. 3 illustrates a collaborative IDE/Debugging session).

---

[1] UDB is the Unicon language source-level debugger.

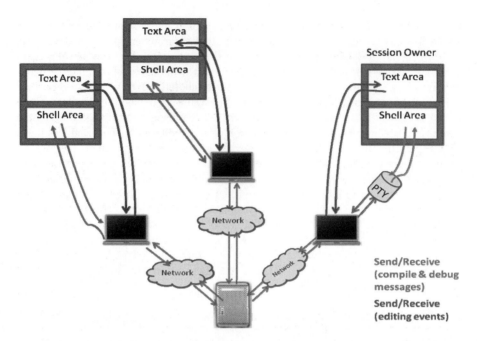

Fig. 3. The information flow design of an ICI collaborative session allows n developers to see and cooperatively control the same view of the code and the execution or debugging session. PTY (pseudo-terminal) denotes a pair of virtual character devices that provide a bidirectional communication channel to an external application such as a compiler or debugger.

ICI's collaborative IDE sessions are conducted within the context of the virtual environment and developers can multi-task or enter and leave a session as needed. The collaborative session is owned by its creator; however, invited developers can take turns requesting session control and act as if they were the owners during the collaborative session.

ICI supports general collaborative software development tasks. However, it was built to serve the specific needs of computer science and software engineering education, particularly distance education. Typical requirements scenarios were computer science teaching environments where an instructor and/or a small team of students are interacting during a software development task. For example, one scenario is a virtual office-hour visit, in which a teacher assists students on their assignments without having to be in the same physical location. Similarly, ICI may be used in a virtual lab to help teaching assistants and tutors in teaching a group of students on remote computers. A third example is a distributed team environment, where team members use a collaborative session to work on a shared task, each from his/her own location [4].

The tools' suitability for software engineering distance education also makes it highly suitable for software engineering teams whose members are spread across multiple sites. ICI presently focuses on interactive collaboration. Although individual scenarios focus on specific collaboration tasks, ICI's requirements were influenced by the expected surrounding context. While an instructor or consultant is collaborating with one student or team, many others may be seeking interactive assistance or sending messages. Providing specific information about individuals' pending requests and queuing options helps all parties.

The rest of this paper discusses the design and implementation of ICI. Section three introduces the design. Section four describes the implementation. Section five compares ICI with the related tools. Section six provides an overview of the planned future work. Section seven ends the paper with some conclusions.

3   Design

ICI's architecture is composed of four major components: 1) a collaborative editor, where developers can share source code editing and navigation, 2) a collaborative shell, where developers can share compilation, program runs, and real-time debugging sessions, and 3) a set of communication tools such as a text and voice chat, provided by the surrounding virtual environment context, and 4) an interface for collaboration control, which allows users to invite other developers, take turns at the IDE controls, and enter and leave the collaborative session.

Users meet in a *collaboration session* that is an interactive work session. The session owner, the person who started the session, is responsible for inviting the other participants. ICI uses a client-server architecture. The clients communicate with a *collaboration server*, a component of the collaborative virtual environment server cluster that implements shared collaborative services. The collaboration server forwards messages to the appropriate clients based on the message type. Section 4.1 shows a list of ICI network protocol messages. Fig. 4 shows the UML diagram for the classes related to the collaborative IDE (ICI). The classes in the diagram named Dialog, EditableTextList and Dispatch are large standard GUI class libraries in the Unicon language; they are subclassed and customized in ICI, which keeps the implementation relatively compact. Similarly, classes Server, LoginSession, Commands, and NSHDialog are not part of ICI, but rather CVE virtual environment classes that provide the context in which ICI executes. The collaborative IDE design did not have to establish communications capabilities or create its own window; instead it needed to interface with an existing infrastructure.

SyntaxETL is a multi-language syntax-coloring collaborative editable text list widget developed for ICI. Its sibling the ShellETL editable text list class adds multiplatform child process execution. The IDE aggregates the source code SyntaxETL widget and the ShellETL execution widget to form the user interface for a collaborative session. The IDESession class coordinates the activity in these widgets with the client's view of the remote users that are participating in the session.

3.1   Real-Time Collaborative Editor

ICI provides a fairly standard programmer's editor, in which a user can edit files privately, and then invite others into a shared session on the fly when consultation is needed. Unlike an ordinary text editor widget, the collaborative editor widget must send and receive network messages for all editing actions to the appropriate session on the collaboration server. The design of the collaborative editor is kept as simple as possible. ICI's editor has two modes: watch and edit. During a collaborative session, only one of the participants may edit; the rest of the participants are in watch mode. In watch mode users can ask questions, provide suggestions, or ask for permission to take a turn at the controls. A participant in watch mode sees and automatically follows any code modification or navigation made by the edit mode user in the collaborative session.

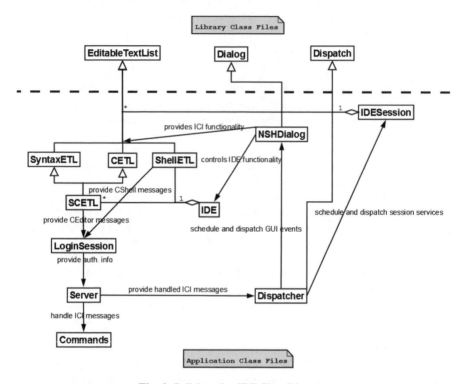

Fig. 4. Collaborative IDE Class Diagram

3.2 Real-Time Collaborative Shell

In order to collaboratively compile, run, and debug a shared program, ICI implements a collaborative shell. The collaborative shell allows developers to see the compilation messages of the target program and to share the inputs and outputs of the running program. A real-time collaborative debugging session can be started by any developer as a private IDE activity and subsequently shared when it is determined that assistance is needed. The collaborative shell uses a simple multiplatform virtual executor facility to interact with the execution or debugger session, and a network protocol to share shell I/O with the rest of the participating developers in the collaborative debugging session.

The virtual executor is a simple two-way I/O channel that funnels bytes from one process to another. On Windows it is implemented as a bi-directional pair of pipes; on UNIX platforms it is a pseudo-tty which behaves like a pair of pipes with the additional property that one end of it looks like a conventional TTY terminal [5].

When a user starts a debugging session, a call to open a virtual executor runs the debugger such as gdb, jdb, or udb appropriate for the shared program. Once the debugging session starts, a prompt appears at the collaborative shell allowing the owner of the session to start entering commands to the debugger, while the other clients can observe the text commands and the debugging messages simultaneously. The collaborative shell uses the same protocol used by the collaborative editor, but with a different set of network protocol messages (see section 4.1).

3.3 Communication, Control, and Activity Awareness

The ICI design perspective is that what happens in between collaboration sessions is just as important as exactly how the shared session works. Much of what happens between sessions is ordinary, non-collaborative IDE work, and the primary emphasis in design was to minimize the transition effort between individual work, collaborative session, and back to individual work, so that this can easily occur dozens of times during the course of a work period. For example, an invitation to a collaboration session might intuitively be delivered via a popup dialog, but the intrusion of such a dialog does not scale well for busy users; better is a Facebook-style visible indicator of pending invitations and the ability to review the queue at one's convenience to see who has been waiting for what, and for how long.

Group awareness is an important factor for a successful collaboration, providing an understanding of other developer activities. ICI provides support for group awareness. In addition to the current collaboration sessions which are temporary in nature, there are persistent groups, modeled after those found in Massively Multiuser Online games (MMOs). Persistent groups (often called "guilds" in games) provide both chat and wiki-style collaboration aids that remain across work sessions. Unlike most MMO "guilds", in ICI one may be a member of as many groups as needed.

4 Implementation

ICI is implemented as a part of the CVE virtual environment. The code for ICI is available as part of the open source CVE virtual environment project (cve.sourceforge.net). CVE is written in Unicon, a very high level object-oriented programming language [5], [6]. Unicon provides a simple interface to the standard internet protocols, TCP and UDP, as well as several higher level communications and messaging protocols [5].

Each set of ICI clients that are working together is associated with a session object on the collaboration server which allows clients to broadcast messages to all members of the group. The shared editor was implemented using an approach similar to the one used by GHT (Group Homework Tool), a "same time different place" groupware tool built to support synchronous, collaborative coding among novice programmers [7]. In ICI, insertions and deletions are executed locally on the client before they are sent to the server. The other clients then apply the modifications to the text.

The concept of collaborative IDE session appears in both the server and the client. The server manages the sessions using a table that contains all the needed information about each collaborative IDE session: the owner, current edit mode user, file, and list of users in the session. On the client, there is another session table that contains all the information for this client about its sessions (session id, owner of the session, reference to the user interface component for the session, file, and list of users in the session).

4.1 Network Protocol

The ICI network protocol messages are strings consisting of a message name followed by arguments which are often a data payload or a list of users. Messages are divided into three different categories:

General ICI messages

CETLOpen: opens a collaborative IDE session when a user presses the Invite button. It takes a string a string argument consisting of the user_name, filename, and encoded file contents.

CETLAccept: informs the server when a user accepts an invitation to collaborate. The message has the parameters (recipient, index_counter, file_name, slave). It opens a collaborative IDE tab on the accepting client, and changes the background color of the editor and shell widgets on all clients to light yellow as an indication for the collaboration session.

RejectIDE: informs the server after the user rejects an invitation to join a collaboration session.

CETLCompile: shows the compilation and linking messages, notably the error messages, to the collaboration users.

CETLLock: requests a lock for the users of the collaboration other than the owner

CETLLockTransfer: indicates that the user requests a turn at program editing. If the owner close the session, then state will change to unlock, and the background color of the editor and shell widgets will change to white.

Collaborative editor messages

CETLevent, CETLmouse, CETLkey, CETLscrol: The syntaxCETL class generates these CETL messages in response to GUI events coming from the collaborative editor. The server calls the *handle_CETL_Event()* method to forward the event to the participating clients.

Collaborative shell messages

SHLevent, SHLmouse, SHLkey, SHLscrol: the ShellETL class generates these SHL messages for both GUI user input events and output received from the external process (the compiler, debugger, or program being executed), and the server calls the *handle_CETL_Event()* method to forward the event to all the collaborative shell at the other side.

4.2 Source Code

ICI's source code is organized into the following classes.

SyntaxETL: this subclass of the Unicon standard library EditableTextList class provides a multi-language syntax-coloring collaborative editor widget.

CETL: is the main class of the IDE. It is a subclass of the Unicon standard library EditableTextList class, which provides a scrollable editable text area. This class issues "CETL events" to send the changes through the network to all the collaborating clients.

ShellETL: this subclass of the Unicon standard library EditableTextList class executes a simple command shell within a collaborative editable textlist widget, in order to fulfill the requirements of the compiling and debugging procedures.

IDESession: is a class responsible of managing the collaborative IDE session (create new sessions; receive events from collaborative IDE, etc.)

NSHDialog: is a class which has methods related to the GUI (buttons, trees, etc.) that is used by the virtual environment and by the collaborative IDE. In the collaborative IDE this class is used to build the GUI which will make it simpler for the user.

Server: is a collaborative virtual environment server class which has methods for managing the virtual environment. The collaborative IDE uses this class as the manager for the collaborative IDE sessions. It creates a session entry there when a user invites another user. Also it adds another user into the users list when additional users are invited into the session. Once a user exits the session or logs out from the virtual environment, they are removed from the users list.

N3Dispatcher: is a subclass of Dispatcher and it sends messages from the server to the client and vice versa. The collaborative IDE uses this class to synchronize different events between the clients and the server. This is done by sending different types of messages between them (invite user to the session, remove user from a session and fire event in the editor which is currently in a collaborative IDE session). Fig. 5 depicts event transmission during a typical collaborative session in which a GUI operation is sent to the server and forwarded to other participating clients.

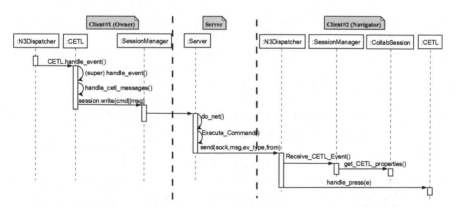

Fig. 5. GUI Events are Transmitted to the Server and Forwarded to other Clients

4.3 Technical Challenges

The major benefit of collaborative editing is to reduce software development task completion time. To achieve this benefit, an IDE such as ICI must overcome both the technical challenges of maintaining consistency together with good performance, as well as the social challenges of supporting group activities.

One challenge of using a real-time collaborative editor comes from network latency. Simultaneous editing by multiple users will either be slow or else potentially inconsistent edits must be reconciled. The challenge is to figure out exactly how to make the changes appear in the document, which were created in remote versions of the document that never existed locally, and make sure that no conflict occurs with the client's local document edits.

ICI dodges the communication lag issue by having just one version of the program and allows the clients to edit the program one at a time. For the scenarios envisioned in ICI's requirements, collaborative editing does not include concurrently editing the same file by multiple users. Concurrency control was added to the editor widget by implementing a way to lock the developers who do not have permission. File editing happens using a simple algorithm such as:

1. Request the 'edit document' token from the server
2. Wait for the server response and loading the file
3. Give the right to edit to the file owner. Tell the server about every edit on the document.
4. If the other client wants to edit, they must ask the owner to give authorization. The reason why authorization is required is to avoid having inconsistent versions of the document.
5. Once authorized; the client may edit the document
6. Inform the server of changes to the document, by sending edit event messages for each change.

5 Related Work

The major difference between ICI and almost all the related work cited in this section is that ICI is part of a 3D multi-user virtual environment. Having said that, a multitude of collaborative text editors have been developed that constitute related work. This section does not cite them all but instead highlights various existing systems and research that provide interactive collaboration for multiple phases of program development. Eclipse (www.eclipse.org) and NetBeans (www.netbeans.org) integrate revision control systems, which constitute non-interactive collaboration tools in the IDE [8]. ICI presently focuses on the harder, orthogonal challenge of interactive collaboration.

Langton et al. [7] presented GHT (Group Homework Tool), a tool developed for programmers coding in synchronous collaborative manner. This tool helps teaching assistants and tutors in teaching a group of students on remote computers. GHT has several components which include: synchronous code editor, HTML frames for an assignment definition and resource page, chat and shared Whiteboard. The GHT collaborative editor performs insertions and deletions (modifications on the text) locally only after they performed remotely on other clients. This way of handling modifications slows down (less than 0.3 seconds under good network conditions) the process of editing the text. The GHT collaborative editor also can be used to send chat messages between developers and this helps in discussing specific lines in the code at the same window. Also, it can be used by teaching assistants to help students with their programming assignments. The GHT collaborative editor has two modes: watch and edit mode. Edit mode means only one student at a specific moment can modify the code and the other students can only ask questions or provide suggestions. Watch mode means the editor just scrolls automatically to show which part is viewed by the other student. Of all the tools in this section, ICI is the most similar to GHT, but avoids the latency issues mentioned above and emphasizes collaboration on execution and debugging tasks, not just code editing.

Booch and Brown [9] point out that a rich collaborative development environment arises from the collection of many seemingly simple collaborative components to support coordination, collaboration, and community building. They also state that IDEs equipped with team-centric features are a step up from those merely augmented with some collaborative support. According to Booch and Brown, merging these features into the IDE's file viewer reduces friction, by saving the developer from having to go outside the IDE and manually dig for such information.

A notable working example of a collaborative IDE is Jazz, a research project at IBM that adds a set of collaboration features for the Eclipse IDE [10], [11], [12]. The objective is to help developing the collaboration within the group. Jazz provides a facility similar to an IM buddy list to monitor who is online and whether they are coding or not. Developers can initiate chats, or use different communication methods such as screen sharing and VoIP telephony. Jazz also provides some awareness features [10].

A number of other Eclipse-based projects focus on the integration of collaborative features into IDEs. GILD is an example of a project that provides cognitive support for novice programmers and support for instructor activities [8], [13]. CodeBeamer is another example of a commercial product that has plug-ins for integrating collaborative capabilities into IDEs such as chatting, messaging, project management, and shared data [14]. Another example is Sangam; a plug-in for the Eclipse platform that features a shared editor and chat for pair programming [15].

In another work Cubranic and Storey [1] introduce a collaborative IDEs that can be used to help students in computer programming courses to work as a team on the programming assignments. This study evaluates how usable and effective is the collaborative IDE. One problem with this collaborative IDE is it does not save what each participant added into the code for monitoring what updates happened. This feature is common in revision control systems and non-interactive document collaboration tools such as Microsoft Word, but is absent from many of the interactive collaborative text editors including ICI.

Ozzie and O'Kelly [16] and Roseman and Greenberg [17] introduce a set of tools (e.g. text editors, chat and whiteboard) which are needed in generic collaborative interactive environments. Hupfer et al. [18] describe a collaborative IDE which can be used as a communication medium between team members, enabling them to ask questions or discuss an issue with other team members. Also, it can be used to help development teams in working together in a project. The system can show if a team member is online or active in his/her machine or not.

ICI provides similar capabilities, with a greater focus on interactive problem solving during development tasks such as debugging. Also, ICI allows teaching assistants and tutors to help students in their homework's on remote computers. ICI developers can see from the tree view of the ICI session users not just who is online or active, but more specifically: users participating in specific ICI sessions, and the owner of those sessions.

6 Conclusions

Collaboration plays a crucial role in software development. For this reason, continuing to improve the collaborative tools available inside integrated development environments is of great potential benefit. Collaborative tools can be used alongside a non-collaborative IDE, but integration reduces the time required during the development process. ICI is a 2D collaborative IDE that lives within a multiuser 3D virtual environment. The merger of these two forms of collaboration tools makes both more interesting. The 3D virtual environment benefits from having real-world purposes for the social networking and game-like interaction that it provides. The IDE benefits from the awareness support and communication context provided by the virtual environment.

ICI does not yet blur the lines sufficiently to test whether value can be added by further virtualization of IDE activities; for example, whether the use of software development tools and artifacts such as source and object files will benefit from 3D representation, or whether 3D views of IDE activities will provide awareness or other benefits sufficient to overcome the loss of resolution incurred in embedding 2D data in a 3D scene.

A lot of work remains in order to evaluate and tune ICI for differing environments and turn it into a production tool. At present, ICI is a good tool to facilitate the education of novice programmers. It helps students taking introductory computer programming to improve their programming skills, and to improve the student-instructor and student-tutor interaction.

Research in collaborative IDE systems can be categorized into: 1) application level, where the focus is coordination management between remote software components; and 2) human level, where collaborative widgets are studied in detail. This work has primarily focused on the application level, although some usability, group awareness, access control, work area management, and similar features are also performed. One major contribution of this work is the support for a flexible collaborative IDE that integrates responsive, and real-time collaborative editing and debugging. This research has been focused on the core technical components to support such tools.

7 Future Work

ICI currently has many limitations. Although successful as a real-time software development collaboration tool, its turn-based explicit control is not appropriate for all collaboration scenarios. Future work will include a "traffic-light" control mode in which editing permission switches between users automatically. For example, a user in edit mode has a green light until another user attempts to edit, after which edit mode user's light switches to yellow, and control transfers after the edit mode user is idle for an appropriate time. It would also be useful to mark user changes by different colors.

Another area for near-term future research is to extend ICI to provide better support for collaborative sessions that span multiple files in a project. Part of ICI that is still under construction is a collaborative UML drawing UML tool.

To test the effectiveness of the supported features, user studies are needed, focusing on evaluating the usability and efficacy of collaboration across multiple interactive sessions. Alpha test sessions suggest that code sharing, editing, compiling, and debugging using the ICI collaborative IDE was easy for all the participating users, but nevertheless suggest several additional areas for future work. Test sessions demonstrated that developers need all the integrated features; such as shared editor, shared shell, and text and voice chat in order to succeed in collaboration. The test session results did not show any need to further reduce the lag time in the group editor. They do point out that further extensions to the group awareness are needed, so that users have different colors or icons, and revealed a need to add an indication to the changes made by the user, for example underlining the changes.

Acknowledgments

This project was supported in part by the National Science Foundation Advanced Technological Education (ATE) grant number DUE-0402572.

References

1. Cubranic, D., Storey, M.-A.: Collaboration support for novice team programming. In: Proceedings of ACM GROUP 2005, November, pp. 136–139 (2005)
2. Sarma, A.: A survey of collaborative tools in software development. Technical report, University of California Irvine, Institute of Software Research (2005)
3. Kubo, M.M., Tori, R., Kirner, C.: Interaction in Collaborative Educational Virtual Environments. CyberPsychology & Behavior 5(5), 399–408 (2002)
4. Bouras, C., Tsiatsos, T.: Educational Virtual Environments: Design Rationale and Architecture. Multimedia Tools and Applications 29, 153–173 (2006)
5. Jeffery, C., Mohamed, S., Parlett, R., Pereda, R.: Unicon book Programming with Unicon (1999-2003), http://unicon.org/book/ub.pdf
6. Jeffery, C., Jeffery, S.: An IVIB Primer (February 21, Unicon Technical Report #6b (2006), http://www.cs.nmsu.edu/~jeffery/unicon/utr/utr6b.pdf
7. Langton, J., Hickey, T., Alterman, R.: Integrating tools and resources: a case study in building educational groupware for collaborative programming. Journal of Computing Sciences in Colleges 19(5), 140–153 (2004)
8. Cheng, L.-T., De Souza, R.B.C., Hupfer, S., Patterson, J., Ross, S.: Building Collaboration into IDEs. ACM Queue 1(9), 40–50 (2003)
9. Booch, G., Brown, A.: Collaborative Development Environments. In: Advances in Computers, vol. 59. Academic Press, London (2003)
10. Cheng, L.-T., et al.: Jazz: a collaborative application development environment. In: OOPSLA Companion, pp. 102–103 (2003)
11. Cheng, L.-T., Hupfer, S., Ross, S., Patterson, J.: Jazzing up Eclipse with Collaborative Tools. In: Proceedings of the OOPSLA Eclipse Technology eXchange Workshop, Anaheim, CA, October 2003, pp. 45–49 (2003)
12. Ellis, C.A., Gibbs, S.J.: Concurrency Control in Groupware Systems. In: ACM SIG- MOD 1989 proceedings, Portland Oregon (1989)

13. Storey, M.-A., Michaud, J., Mindel, M., et al.: Improving the Usability of Eclipse for Novice Programmers. In: OOPSLA Workshop: Eclipse Technology Exchange Anaheim CA, pp. 35–39 (October 2003),
    http://gild.cs.uvic.ca/docs/publications/oopsla.pdf
14. CodeBeamer, http://www.intland.com, http://www.intland.com
15. Ho, C., Raha, S., Gehringer, E., Williams, L.: Sangam: A Distributed Pair Programming Plug-in for Eclipse. In: Eclipse Technology Exchange (Workshop) at the Object-Oriented Programming, Systems, Languages, and Applications, OOPSLA (2004)
16. Ozzie, R., O'Kelly, P.: Communication, Collaboration, and Technology: Back to the Future. White paper, Groove Networks (2003)
17. Roseman, M., Greenberg, S.: Network Places for Collaboration. In: Proc. ACM 1996 Conference on Computer Supported Cooperative Work., Boston, MA, pp. 325–333 (1996)
18. Hupfer, S., Cheng, L.T., Ross, S., Patterson, J.: Introducing Collaboration into an Application Development Environment. In: Proceedings of the ACM Conference on Computer Supported Cooperative Work, pp. 21–24 (2004)

Evaluating a Mobile Emergency Response System

Cláudio Sapateiro[1], Pedro Antunes[2], Gustavo Zurita [3], Rodrigo Vogt[3],
and Nelson Baloian[4]

[1] Systems and Informatics Department, Superior School of Technology,
Polytechnic Institute of Setúbal, Portugal
csapateiro@est.ips.pt
[2] Department of Informatics, Faculty of Sciences,
University of Lisbon, Portugal
paa@di.fc.ul.pt
[3] Management Control and Information Systems Department, Business School
Universidad de Chile
gnzurita@fen.uchile.cl, rodrigovogt@gmail.com
[4] Computer Science Department, Engineering School
Universidad de Chile
nbaloian@dcc.uchile.cl

Abstract. Existing information systems often lack support to crisis and emergency situations. In such scenarios, the involved actors often engage in ad hoc collaborations necessary to understand and respond to the emerging events. We propose a collaboration model and a prototype aiming to improve the consistency and effectiveness of emergent work activities. Our approach defends the requirement to construct *shared situation awareness* (SA). To support SA, we developed a collaborative artifact named *situation matrixes* (SM), which relates different *situation dimensions* (SD) of the crisis/emergency scenario. A method was also developed to construct and evaluate concrete SM and SD. This method was applied in two organizations' IT service desk teams, which often have to deal with emergency situations. The target organizations found our approach very relevant in organizing their response to emergencies.

1 Introduction

Information Systems (IS) development has been traditionally approached by focusing on predefined work models, most of them conceived with efficiency concerns. Nevertheless, many unknown variables, both external (e.g., market dynamics, natural disasters) and internal (e.g., latent problems, emergent work processes or the lack of flexibility in work structures), are among the factors that may lead to the lack of support of existing IS when facing unplanned/ unpredicted/unstructured events. Such situations may often scale to crises, defined in [1] as a series of unexpected events causing uncertainty of action, or emergencies, when time-pressure is also present.

In non-routine or unique emergency situations, the use of anticipated protocols may be quite difficult or even impossible [2]. In order to adapt to a specific situation, the involved participants rely heavily on their experience, and strategic decisions must be made

R.O. Briggs et al. (Eds.): CRIWG 2008, LNCS 5411, pp. 121–134, 2008.

often lacking full insight about the situation. Information shortage, as well as information overload, may lead to an unbalanced response (e.g., overloading some personnel, prioritizing less urgent actions, lack of awareness of mutually exclusive tasks).

Developing IS to support such unstructured scenarios raises several challenges, considering that work processes under such conditions are characterized by: having no best structure or sequence; often being distributed; dynamically evolving; unpredictable actors' roles; and unpredictable contexts [3]. These characteristics challenge the traditional IS assumptions regarding predictability and analyzability.

Our approach to IS support to emergency situations emphasizes the collaborative dimension of the emergency response rather that the more traditional command & control model [4]. The proposed collaboration model is grounded in several principles of resilience engineering. Resilience engineering is characterized as a comprehensive endeavor towards increased resistance and flexibility when dealing with the unexpected [5]. Resilience engineering should be regarded as an important and innovative approach to IS development, at least because the traditional IS approaches have revealed many limitations regarding emergency scenarios.

The main organizational failures addressing emergency situations, pointed out in [6], may be rooted in a lack of collective awareness of the ongoing situation. Our research contributes to the development of *shared situation awareness* (SA) as a mean to improve the emergency response. Our approach to SA relies upon a set of shared artifacts that may be collaboratively updated on a contingency basis. Considering that in many emergency scenarios the involved actors may need to operate in distributed locations, the approach is also based on mobile devices (tablet PCs and PDAs).

The prototype was developed on top of a pen-based application framework developed at the University of Chile. Besides handling all communication and collaboration issues, this framework provides a very rich collection of predefined pen-based gestures supporting the creation and manipulation of visual objects.

Aiming to evaluate our approach in real settings, we conducted experiments with two IT service desk teams operating in two different organizations. These teams often face situations classified as emergencies; for instance, if a network link or a server is down, it may compromise the organization's work. In a number of organizations, these situations are overcome without IS support.

One fundamental constraint of this research was the adoption of an adequate evaluation method. Groupware evaluation has raised many methodological concerns, since the adopted strategies may differ in: product maturity (design, prototype, finished product), time span (hours, weeks, months, years), setting (laboratory, work context), type of people involved (domain experts, final users, developers), and type of research (quantitative, qualitative) [7]. The scope of the evaluation process may also target different dimensions, ranging from the technical dimension (e.g., interoperability, connectivity) to the organizational dimension (e.g., effects on tasks performance, processes structure) [8, 9]. Concerning our objectives, several dimensions could have been considered:

1. Evaluate the collaboration model, including its capability to address emergency situations and incorporate the resilience engineering principles.
2. Evaluate the situation awareness hypothesis, aiming to improve performance in emergency response scenarios, thus focusing on the shared artifacts.

3. Evaluate the prototype usability.
4. Evaluate the technological constraints and its implications to performance (e.g., mobile ad hoc network - MANET issues).

Of course these dimensions are highly interdependent, thus increasing the difficulties accomplishing a comprehensive evaluation. Considering these difficulties, we established the reasonable goal to only evaluate the first two dimensions.

In the next section we present some research contributing to this work. Section 3 describes our conceptual approach. The prototype is briefly described in section 4. Sections 5 and 6 present the details of the evaluation process and the obtained results. We conclude the paper by making some remarks and pointing some future work directions.

2 Related Work

We may find in the research literature several projects addressing how to bring IS operations back to model behavior after deviations caused by unpredicted events [10-12]. The problem addressed by this paper moves the research beyond this perspective towards the support to emergent work structures in emergency situations, adopting a perspective where work models do not serve to prescribe work processes but rather as informational artifacts [13, 14] helping getting the work done.

Several definitions for SA may be found in the research literature typically referring SA as an understanding of the situation elements (people, objects, etc.) and dynamics (interactions, events, etc.) One of the most established models organizes SA in three levels [15]:

1. Perception produces Level-1 SA: the most basic level of SA, providing awareness of the multiple situational elements (objects, events, people, systems, environmental factors) and their current states (locations, conditions, modes, actions).
2. Comprehension produces Level-2 SA: an understanding of the overall meaning of the perceived elements.
3. Projection produces Level-3 SA: awareness of the likely evolution of the situation and possible/probable future states and events.

The recent research on *team shared awareness* highlights that teams need to detect cues, remember, reason, plan, solve problems, acquire knowledge, and make decisions as an integrated and coordinated unit [16]. The research on SA in the Computer Supported Cooperative Work (CSCW) field has developed a functional perspective of SA [17-20]. In our research we emphasize the organizational perspective, considering the orchestration of activities necessary to construct, manage and use SA. In this regard, the team members should not only be able to monitor and analyze SA, but also anticipate the SA needs of their colleagues. Hence, [21] define team SA as SA plus the mutual adjustment of one and another's minds as they interact as a team in a specific context of action.

We also adopted the phenomenological perspective of *contexts of action*, traditionally used in social sciences, which regards SA as evolving dynamically as actions unfold [22]. From an organizational perspective, this means that situated decision making models such as the *garbage can* [23] are more applicable to our context than traditional *rational choice* models [24].

Regarding the support to mobility, several collaborative solutions have already been proposed [25-29]. Although these proposals have shown useful to support specific collaborative activities, they were not designed to address emergency management. Their reuse capability is therefore relatively small.

3   Conceptual Approach

As stated in [6], resilience is a function of the organization's awareness. IS should thus focus on providing SA as a mechanism for efficiently sharing and coordinating actions in emergency contexts.

SA implies an understanding of the entire operating environment and should be built by taking advantage on the experience of the involved participants. In our approach, we aim to facilitate the externalization of the user's experience and tacit knowledge, enhancing the individual contributions to the overall understanding of the situation (supporting the externalization knowledge flow referred by [30]). This deference to expertise is a fundamental resilience principle and is trained in programs like Crew Resource Management [31, 32] adopted by aviation and firefighter organizations.

Considering the Swiss-Cheese Accident Model [33], accidents occur when several organizational defense layers are transposed. In our model we address the emergency situation by collaboratively constructing layers of defense. Involved actors should be able to align and correlate different *situation dimensions* (SD) of the unfolding events and actions. We consider as samples of SD: involved actors, necessary actions, resources allocation, goals, etc. For a given application domain, an initial set of relevant dimensions may be adopted and later on dynamically redefined, as the unplanned situation unfolds.

The existing SD are correlated in an artifact named *situation matrixes* (SM), expressing existing relations among different dimension of the situation. Samples of SM are Actions-Actors, Actor-Allocated Resources, Goals-Actions, etc. Despite a possible starting set, SM may be dynamically defined. Our specific implementation of the SM was inspired by the perspective proposed in [34], which uses several types of matrixes to visualize qualitative data, for instance: concept cluster matrixes, empirical matrixes, and temporal or event driven matrixes.

The SD correlations are specified in the SM as circles, using different sizes and/or colors to express the perceived strengths of such relations. Several alternatives may be considered to express the semantic meaning of such correlations, but in our approach we leave the concrete semantics to be defined by the application domain experts. Figure 1 illustrates the proposed collaboration model and SM artifacts.

The SM artifacts accomplish several goals: support action planning and status reporting; and by providing a shared integrated representation (kind of real-time dashboard), implement a monitor/feedback mechanism. As the situation evolves, the SD may include more items (e.g., more actors involved, more actions proposed), and new SD may be created and related in existing or new SM.

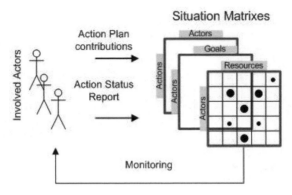

Fig. 1. Collaboration Model and SM artifacts

4   Developed Prototype

As stated earlier, mobility may constitute a requirement in emergency management. The developed prototype operates in Tablet PCs and PDAs (see figure 2). The system is a full peer-to-peer application. This means that every user runs exactly the same application and shares data using the ad-hoc network. Using multicast messages, the application automatically finds other partners and establishes a reliable TCP link with them for transmitting data.

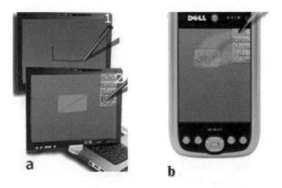

Fig. 2. Prototype a. Tablet PC b. PDA

A key concern while developing the prototype was requiring a minimal overhead to operate the SM. SM are easily created by drawing an half rectangle (figure 2a(1)). The SM may be populated with SD as shown in figure 2b. To specify the contents of the matrix, it should be "expanded" by a double clicking on the rectangle. To create a new column, the user has to double click on the label of the columns (Figure 3a). After this, the user enters the header text for the column as shown in figure 3b. A similar procedure is used for editing rows (figure 3c).

Fig. 3. Prototype a-b. Column creation c. Row creation

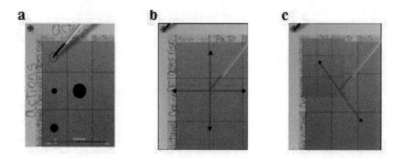

Fig. 4. a. Correlations editing; b. Navigation: Scrolling; c. Navigation: Zooming

Figure 4a shows a user marking a relationship between SD items. This relationship is expressed with a dot of a certain dimension, with bigger dots meaning more importance. Figures 4b-c illustrates the navigation capabilities (scrolling and zooming) through the SM artifact.

5 Evaluation

We have considered several alternatives to evaluate the collaboration model. Typical evaluation strategies include computer simulations, field methods and usability inspections. Although field methods allow capturing more realistic data, they could be difficult to settle in our case for several reasons: time investment, scenario setting, associated costs and prototype maturity.

The computer simulations allow, to some extent, to overcome some of these problems. We may find in the literature different approaches to computer simulations in our research context, from fully automated agent–based simulations [35] to hybrid approaches including humans in the loop [36]. Fully automated agent–based simulations rely heavily on modeling (situation constrains, information flows, actors behaviors, etc.) A combination of computer simulations with humans in the loop may be accomplished with game playing in virtual scenarios. But despite the validity of these options, they all rely to some extent in pre-defined situations. Our work focuses on supporting human behavior in non-predicted scenarios, emerging in real time and from the involved actor's experience, which does not seem adequate to the computer simulation approach.

Usability inspection techniques are much less costly than field methods and they can often be used earlier and more frequently in the development cycle. However, since these techniques are not used in the actual work context, some researchers state that it is unclear whether the usability information they provide is valid for real-world contexts. In [37], the authors discuss that it is possible to integrate usability inspection techniques with work scenarios, jointly constructed by domain experts, and that these techniques may lead to results comparable to the ones obtained from field studies. We based our evaluation method in the combination of the inspection technique with the scenario based approach [38-40].

Our evaluation method consisted in four steps. We started by conducting a set of individual semi-structured interviews to IT service desk team members to present the problem and understand its relevance in the application domain. We also jointly analyzed a set of consequence scenarios aiming to understand which were considering realistic emergency situations and actual work practices. These interviews were audio recorded for future reference and analysis.

In the second evaluation step we administrated a questionnaire to each team member to identify the key requirements of collaboration support in emergency situations. The third evaluation step concerned the realization of a workshop (also filmed for future reference) with all team members, where we presented the collaboration model and a paper prototype. The paper prototype allowed focusing the evaluation on the model, discarding interference of possible usability and technological issues.

Table 1. Evaluation Methodology

Step	Technique	Goals
1.	Semi-structured interviews (audio recorded)	• Introduce the support of unstructured activities problem. • Perceive the relevance of such problem in the IT service desk application domain. • Perceive actual emergency situations and work practices.
2.	Questionnaire 1	• Rate the set of proposed requirements to address unstructured work activities
3.	Workshop (filmed)	• Introduce the collaboration model and prototype. • Discuss its usage in a real scenario. • Collect possible SD and SM
4.	Questionnaire 2	• Evaluate the perceived effectiveness of the implementation of the collaboration model and prototype.

Once all participants were familiarized with this approach, we presented the prototype in more detail and discussed its usage. Finally, a second questionnaire was administrated to evaluate the perceived implementation of the discussed requirements; this constituted the fourth step of our evaluation. Table 1 outlines the various steps of the evaluation method and clarifies the respective goals.

Conducted interviews were structured around the topics summarized in table 2.

Table 2. Interviews structure

Interviews - Discussed Topics
1. Which situations may be described as emergencies
2. Current preventive practices
3. Current diagnosis practices
4. Current registration practices
5. Current recovery formal procedures
6. Current recovery informal procedures
7. Current communication schemas
8. Existing performance metrics
9. Priority near future improvements (address current identified vulnerabilities)

Table 3. Requirements under evaluation

Nº	Requirements	Influence Area
1.	Communication support through shared artifacts	Groupware Collaboration Heuristics
2.	Transitions between individual and team work	
3.	Coordination support	
4.	Facilitate in finding collaborators	
5.	Facilitate in establish context	
6.	Facilitate situation (specific issues) monitoring	
7.	Minimal overhead work demand	
8.	Mobile end device availability	
9.	Assist situation understanding	Situation Awareness
10.	Perceived who is involved	
11.	Assist situation size up	
12.	Assists (overall) situation representation	
13.	Knowledge externalization support	Knowledge Management
14.	Knowledge transfer support	
15.	Incident handling documentation	
16.	Improvement in diagnosis time	Performance
17.	Improvement in recovery time	
18.	Number of coupled incidents simultaneously attended	

Our evaluation method received several influences from different evaluation methodologies. From the groupware studies, we considered the heuristics proposed by the mechanics of collaboration [19, 37], which were developed to evaluate shared workspaces. Since our claims consider externalization of tacit knowledge and evaluation of team performance, we also considered the works from [41] and [42]. Finally, we also considered the situation awareness evaluation techniques proposed by [16]. The Table 3 summarizes the considered requirements for evaluation.

5.1 Conducted Experiments

In this section we present the outcomes of the experiments conducted in the two IT service desks. The experiments involved two teams of IT support in two different organizations. The first team was constituted by three senior and two junior members. The second team had the chief, one senior and one junior member.

We present bellow a brief summary of the main topics discussed in the interviews. Regarding the critical incidents, the most serious cases reported were related with server failures (in which the more frequent problem is the disk failure) and connectivity losses in some network segments (that may be due to switches' firmware problems) compromising a wide variety of services. It was also reported that more untypical problems may occur and lead to emergency situations, "[…] like a flood in the basement where some of the equipment is situated […]" The existing preventive practices rely heavily in monitoring the active network elements trough a control panel fed by SNMP messages, where alerts are displayed and emailed to the technicians. Also, several equipments are under SLA agreements with suppliers and a spare stock exists. Actual diagnosis and recovery practices rely heavily in the field experience of each team member and the fact that they all know the intervention domains of each one (e.g., some team members address Linux and others Windows problems).

The collaboration is essentially supported by meetings, phone calls and chat tools. Despite the existence of a trouble ticket software, it is only used (sometimes) for an incident opening and some (few) occasional post mortem annotations to close it. The reported main concerns regard documenting the intervention process, to facilitate future interventions and knowledge transfer. Considering these teams rely heavily upon experience, the junior members are often less performing. A number of other vulnerabilities were identified that could lead to critical situations; for instance, not all equipments have a spare stock or SLA coverage, and overcoming this situations is done by ad hoc measures and temporary workarounds that, once more, are highly informal and experience dependent. Also, the possible abandon of the team by a senior member may dramatically decrease the capacity to handle some incidents due to knowledge and collaboration losses.

In the second evaluation step, the IT service desk members answered to the first questionnaire, rating the relevance of several requirements to support unstructured work activities. The ratings were done in the scale: 1 - Not perceived as important, 2 - Less important, 3 - Important and 4 - Very important.

The questionnaire results yield that requirement 2 was not perceived as important. Requirements 12, 13 and 15 were rated from Less Important to Very Important. And all other requirements were rated either Important or Very Important. A more detailed analysis of the results in conjunction with the recorded interviews yield the following considerations: Knowledge transfer and incident documentation revealed Very Important to the team leaders; situation representation and knowledge externalization support revealed Important to the junior technicians.

Table 4, provides a description of the scenario collaboratively constructed in the workshops.

Table 4. Workshop scenario description

Scenario
"From several rooms, were reported the lost of network connectivity. Some technicians were notified by email, while others received several complaints by phone. The senior technician that received some of this complaints suspects from the central switch located on the main building." How the proposed approach may help in coordinating, diagnosis and recovery actions?

From the discussions that took place in both workshops, the highly informal and unstructured work practices were obvious to both teams. The courses of action vary according to the involved actors and some discussions took place on the more efficient ways to address this problem. A set of SD and respective SM were drafted in the paper prototypes. Figure 5 shows the paper prototypes used in the workshop sessions and the PDA prototype being operated.

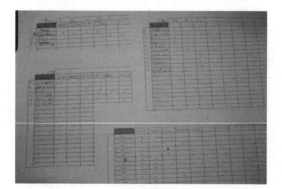

Fig. 5. Prototype a. paper prototype b. PDA prototype

Finally, the results from the last questionnaire confirmed that the proposed approach was perceived as aligned with the requirements that were considered relevant. But some further considerations are worth made: SM should be easily reused and a global representation of the situation (e.g., with all existing SD and which of them correlate) would be much appreciated. Regarding the implementation, some notes about navigating the existing SM were made to ease the use of correlations.

6   Discussion

It was possible to confirm in our experiments that, when facing emergency scenarios, the formalized procedures either do not exist or do not apply to the particular situations. The technicians' experience may dictate the set of actions necessary to inspect or recover some components, to involve specific actors with specific knowledge, etc. But many of these issues rely tacitly and distributed on the team members, which constitutes an additional difficulty when coordinating their actions. At the end of each workshop both teams reported that these sessions revealed to them what they were already suspicious about: the individuals' tacit knowledge and experience strongly conditions the team's efficiency. The issue was not completely new and they were trying to address it by compiling a set of major guidelines to externalize and optimize the use of such knowledge. But due to the lack of time for this task, an interesting feature of the prototype would be to generate such knowledge from the correlations expressed in the SM.

Additionally, since the actions needed to overcome emergency situations may include several dislocations to different physical spaces/buildings, communication and mobility constitute key requirements to maintain shared SA among the distributed team.

As a result of the workshop sessions, a set of specific SD was proposed: Equipments, Actors, Locations, Actions and Activities, which should be correlated in the following suggested SM:

1. Actions-Steps, detailing operational activities (e.g., check router X, reboot switch Y).
2. Actors-Steps, defining responsibilities.
3. Equipment-Actors, expressing the persons responsible for the equipment (e.g., who is empowered to activate a supplier warranty, who is habilitated to inspect a Linux server or a specific service).
4. Equipments-Locations, allowing team members (mostly junior) to know the equipment locations (e.g., main gateway of building C6 is located in room 6.3.0.1).

Finally, regarding the evaluation method, some considerations are also worth made. The first interview revealed crucial to establish a common ground for a richer problem discussion. The paper prototype revealed a good choice to support the discussions about emergency scenarios. Since it did not constrain users regarding usability issues, it focused the discussions on: 1) the SD and SM necessary to address the emergency scenarios; 2) the semantic meanings of the elicited SM relations; and 3) the collaboration model to operate both SD and SM as shared artifacts.

7 Future Work

Besides addressing the various suggestions emerging from the evaluation process, we are also considering studying the timeliness of the situation awareness elements. Timeliness (recent, evolving, outdated, etc.) may be fundamental to further develop SA, since outdated information may considerably degrade SA. But the dependence on explicit user declarations constitutes an overhead work that should be, whenever possible, avoided. We are studying a pulling strategy to handle timeliness: 1) when users input information, a deadline is also introduced (e.g., valid for the next 15 min) and when this expires users are prompted to report information validity; 2) if no deadline is introduced, then the specified correlation will incrementally became more visually transparent as time goes by.

We are also exploring the integration of our approach with the IT Infrastructure Library (ITIL) framework in order to support other organizational levels involved in the different phases of the emergency life cycle management. To accommodate the required service levels and promote the IT infrastructure and business processes alignment, ITIL defines five processes: Incident Management, Problem Management, Configuration Management, Change Management, and Release Management. These processes are related with each other (e.g. incident management may fire a request for change – RFC handled under change management process responsibility) and share a set of ITIL objects (e.g. incidents, problems, RFCs). Our approach to SA regarding the collaborative editing of shared artifacts encompassing relations among situation entities could be extended to expose the relations among ITIL objects and processes tracking both functional and hierarchical escalation.

Acknowledgements. This paper was partially supported by: the Portuguese Foundation for Science and Technology, Project FCT (PTDC/EIA/67589/2006) and Fondecyt 1085010.

References

1. ESSAY. Enhanced Safety through Situation Awareness Integration in training. In: European Community ESSAY project, Contract No GRD1-1999-10450 (2000)
2. Bruinsma, G., Hoog, R.d.: Exploring protocols for multidisciplinary disaster response using adaptative workflow simulation. In: ISCRAM (2006)
3. Markus, M.L., Majchrzak, A., Gasser, L.: A design theory for systems that support emergent knowledge processes. In: MIS Quaterly (2002)
4. Trainor, J.E.: Searching for a system: Multi-organizational coordination in the september 11th world trade center and rescue response. In: Sociology, Delware (2004)
5. Hollnagel, E., Woods, D.D.: Resilience Engineering Precepts. A. Publishing (2006)
6. McManus, S., et al.: Resilience Management: A framework for Assessing and Improving the resilience of organizations (2007), http://www.resorgs.org.nz/
7. Herskovic, V., et al.: Evaluation Methods for Groupware Systems. In: Haake, J.M., Ochoa, S.F., Cechich, A. (eds.) CRIWG 2007. LNCS, vol. 4715, pp. 328–336. Springer, Heidelberg (2007)
8. Gauducheau, N., Soulier, E., Lewkowicz, M.: Design and evaluation of activity model-based groupware: methedological issues. In: 14th International Workshops on Enabling TEchnologies: Infrastructure for collaborative enterprise (WETICE). IEEE Computer Society, Los Alamitos (2005)
9. Vyhmeister, R., Mondelo, P.R., Novella, M.: Towards a Model for Assessing Workers Risks Resulting from the Implementation of Information and Communication Systems and Technologies. Wiley InterScience, Hoboken (2006)
10. Dourish, P., et al.: Freeflow: Mediating between representation and action in workflow systems. In: CSCW, USA (1996)
11. Bernstein, A.: How can cooperative work tools support dynamic group processes? Bridging the specifity frontier. In: CSCW (2000)
12. Mourão, H., Antunes, P.: Supporting effective unexpected exceptions handling in workflow management systems. In: SAC, Seoul, korea (2007)
13. Suchman, L.: Plans and Situated Actions: The problem of human-machine communication. Cambridge University Press, Cambridge (1987)

14. Gasson, S.: A social action model of situated information systems design. The Data Base for Advances in Information Systems 30(2) (1999)
15. Endsley, M.: Toward a theory of situation awareness in dynamic systems. Human Factors 37(1), 32–64 (1995)
16. Salmon, P., et al.: Situation Awareness measurment: A review of applicability for C4i enviorments (2004)
17. Storey, M.-A.D., Cubranic, D., German, D.: On the Use of Visualization to Support Awareness of Human Activities in Software Development: A Survey and a Framework (2004)
18. Neale, D.C., Carroll, J.M., Rosson, M.B.: Evaluating Computer-Supported cooperative work: Models and frameworks. In: CSCW (2004)
19. Gutwin, C., Greenberg, S.: A descriptive framework of workspace awareness for real time groupware. In: CSCW (2002)
20. Bolstad, C.A., Endsley, M.R.: Shared displays and team performance. In: Human Performance, Situation Awareness and Automation (2000)
21. Shu, Y., Futura, K.: An inference method of team situation awareness based on mutual awareness. Cognition, Technology & Work 7, 272–287 (2005)
22. Borges, M.R.S., et al.: Groupware system design the context concept. In: Shen, W.-m., Lin, Z., Barthès, J.-P.A., Li, T.-Q. (eds.) CSCWD 2004. LNCS, vol. 3168, pp. 45–54. Springer, Heidelberg (2005)
23. Cohen, M.D., March, J.G., Olsen, J.P.: A Garbage Can Model of Organizational Choice Administrative Science Quarterly 17(1), 1–25 (1972)
24. Flin, R.: Sittin in the Hot Seat, Leaders and Teams fo Critical Incident Management. John Willey & Sons, Chichester (1996)
25. André, P., Antunes, P.: SaGISC: A Geo-Collaborative System. In: de Vreede, G.-J., Guerrero, L.A., Marín Raventós, G. (eds.) CRIWG 2004. LNCS, vol. 3198, pp. 175–191. Springer, Heidelberg (2004)
26. Guerrero, L.A., Fuller, D.: A Pattern System for the Development of Collaborative Applications. Group Decision and Negotiation 43(7), 457–467 (2001)
27. Muñoz, M.A., et al.: Context-aware mobile communication in hospitals. IEEE Computer 36(9), 38–46 (2003)
28. Zurita, G., Baloian, N.: Handheld Electronic Meeting Support. In: Fukś, H., Lukosch, S., Salgado, A.C. (eds.) CRIWG 2005. LNCS, vol. 3706, pp. 341–350. Springer, Heidelberg (2005)
29. Neyem, A., Ochoa, S.F., Pino, J.A.: Designing Mobile Shared Workspaces for Loosely Coupled Workgroups. In: Groupware: Design, Implementation, and Use, pp. 173–190. Springer, Heidelberg (2007)
30. Nonaka, I., Takeuchi, H.: The knowledge-creating company. Oxford University Press, Oxford (1995)
31. Helmereich, R.L., Merrit, A.C., Wilhelm, J.A.: The evolution of Crew Resource management Training in commercial Aviation. International Journal of Aviation Psychology 9(1), 19–32 (1999)
32. Tippet, J.: Crew Resource Management Manual - A positive change for the fire service. I.A.o.F. Chiefs (2002), http://www.iafc.org/
33. Reason, J.T.: Managing the risks of organizational accidents. Ashgate, Aldershot (1997)
34. Miles, M.B., Huberman, A.M.: Qualitative data analysis. Sage Publications, Thousand Oaks (1994)
35. Johnson, C.W.: Using evacuation simulations to ensure the safety and security of the 2012 olimpic venues. Safety science 46(2), 302–322 (2008)

36. McGrath, D., McGrath, S.P.: Simulation and Network-Centric Emergency Response. In: Interservice/Industru Training, Simulation and Education (I/ITSEC) (2005)
37. Steves, M.P., et al.: A comparison of usage evaluation and inspection methods for assessing groupware usability (2001)
38. Carroll, J.M.: Making Use: scenario-based design of human-computer interactions. MIT Press, Cambridge (2000)
39. Haynes, S., Purao, S., Skattebo, A.: Situating evaluation in scenarios of use. In: CSCW. ACM, New York (2004)
40. Stiermerling, O., Cremers, A.: The use of cooperation scenarios in the design and evaluation of a CSCW system. IEEE Transaction on Software Engineering (1999)
41. Vizcaino, A., et al.: Evaluating collaborative applications froma knowledge management approach. In: 14th IEEE International Workshops on Enabling Technologies: Infrastructure for collaborative entreprise WETIC 2005 (2005)
42. Baeza-Yates, R., Pino, J.: Towards formal evaluation of collaborative work and its application to information retrieval. Information Research 11(4), 271 (2006)

Maturity Levels of Information Technologies in Emergency Response Organizations

Raphael S. Santos[1], Marcos R.S. Borges[1], José Orlando Gomes[1,2],
and Jose H. Canós[3]

[1] Graduate Program in Informatics - IM & NCE
Federal University of Rio de Janeiro, Brazil
[2] Industrial Engineering Department
raphaels@posgrad.nce.ufrj.br,
{mborges,joseorlando}@nce.ufrj.br
[3] Dept. of Computer Science
Technical University of Valencia, Spain
jhcanos@dsic.upv.es

Abstract. In emergency response organizations with very limited resources, information technologies are not adequately explored. In such organizations, the simple adoption of new information technologies is not productive, as their efficient use depends on many other interrelated technologies. This work describes a model to help understanding these interrelationships. The model allows the cooperative evaluation of an organization through different perspectives. The model also helps the performing of the evaluation from different perspectives, making it suitable to collaborative evaluation. Using the model, an organization can measure its maturity level and guide the investment in emergency response capabilities. The information technology dimension of the model has been applied to the firefight organization in Brazil.

Keywords: Emergency organizations, collaborative assessment, maturity models.

1 Introduction

The emergency domain is gaining greater evidence in the most varied sectors of society. Complex emergencies affect big areas putting in risk an increasing number of people and properties. This complexity makes the interaction between the various organizations involved an essential requirement [18]. The systematic and organized management of emergencies can reduce their consequences.

Emergency management can be divided into four stages that cover the full course of an emergency: mitigation, preparation, response, and recovery [8]. Of these, the response phase is possibly the most complex. It has a high degree of dynamism and uncertainty, demanding speed in the actions realized and not tolerating faults. The dynamic and uncertainty nature of the emergency response prevent a complete definition of actions, the time they will take place, the resources needed and the performers

R.O. Briggs et al. (Eds.): CRIWG 2008, LNCS 5411, pp. 135–150, 2008.
© Springer-Verlag Berlin Heidelberg 2008

[17]. The complexity may be augmented as the professional involved often execute their tasks without the necessary information [2].

The problem of lacking information may be mitigated, or even solved, through information technologies. These technologies fulfill a fundamental role in emergency responses, aiding decision making as well as action execution. However, information technologies are not adequately exploited by emergency response organizations, especially those with limited resources. This situation is observed principally in developing countries, where these organizations often don't have adequate resources. In these organizations, just adopting information technologies is not productive. Understanding the relationship with other technologies and resources may better direct investments.

In this work, we describe a model and a method to assist in understanding this relationship and to guide the investments relative to the response activities of emergency organizations. In this way, we hope it will be possible to increase these organizations' response capacity. However, such complex evaluation requires specialists from different backgrounds and expertise, who should integrate their views to portray the organization's technology maturity. The model and the method were designed in such way to make them appropriate to be used by groups of experts working collaboratively.

The proposed model uses some concepts from maturity models used to assess organizations in various domains [7] [13]. The model presented here consists of levels composed of several variables. These levels are organized according to the complexity of assessing the organization relative to those variables. The higher levels are decomposed until a level is reached whose variables can be easily measured through analysis or observation.

Once applied, the model assesses the emergency organization's response capacity along different dimensions, determining its maturity in the response activities. After the evaluation the model allows the organization to see its knowledge about response actions. From this visualization the organization identifies its positive and negative points, and can thereby plan possible improvements to increase its response capacity.

The work is divided as follows. Section 2 presents a revision of the main concepts used in the model: Emergency Response, Information Technology relative to Emergency Response, and Maturity Models. Section 3 presents a method for the construction of the model. Section 4 presents an application of the model built, and the results of this application, in Rio de Janeiro (Brazil) state's Fire Department. Section 5 concludes the paper.

2   Basic Concepts

Emergency responses are short-term activities, designed to reduce the effects of an emergency. They begin when a dangerous situation requiring immediate action happens and end when that situation is resolved [3]. They are complex and do not tolerate failures, as these may have serious consequences.

The main objectives of emergency responses are saving lives, stabilizing the incident, and preserving property and infrastructure [1]. To reach them, it is often necessary that one or more teams, from one or more organizations (fire departments, police, medical organizations, civil defense, public agencies, etc.), interact satisfactorily.

These teams must possess an adequate level of preparation as they always operate under the pressures characteristic of this phase, can't fail, must act rapidly, all often without the information necessary for this [5].

Thus, the success of a response operation depends on the collaboration and coordination among the teams (and their members) involved. According to Oomes [10], the problems pertaining to collaboration and coordination during response operations can often be resolved by dividing their command. The author states that command, whenever possible should be divided in three well defined and specified parts: strategic, tactical, and operational. This division is able to reduce the possibility of decision making conflicts as well as the information and task overloads on the professionals involved in the response operations. This problem is an interesting topic in CSCW studies, but it is not addressed in this paper.

2.1 Information Technologies for Emergency Response

Emergency responses require that decisions be made quickly and precisely, as their activities rarely admit delays. However, their complexity makes it so that this is not always feasible. This poses a problem, as wrong or late decisions during emergency responses can cause loss of life and property.

Information can generate the knowledge necessary to facilitate the understanding of emergency situations and making decisions about them. This facilitation can happen as long as the information has quality. In other words: information must be constantly updated, disseminated in the right amount, and directed to the correct individuals [14] [11].

To make the information available correctly during emergency responses is not an easy task due to the large amount of information, which may exist during this phase [12]. Information technologies can make this task easier. In the response operations these technologies can assist in capturing, representing, and in disseminating the information to the professionals involved. Besides this, they can help in connecting the different organizations involved in these operations [20] [15].

Just as they are able to assist the response operations, information technologies can obstruct them if they are not applied correctly. Information overload, underperformance, poor usability, and inappropriateness for some risk situations can be cited as problems involving information technologies in emergency responses.

Capturing information for emergency responses can assist in understanding emergency situations, and so assist in the response activities. Information technologies can help the organization's professionals in the capture process, making it more effective. This is possible as these technologies have high processing power and allow several capture methods to be used [6].

For the captured information to be used it is necessary to store it. The information technologies allow the storage of large amounts of information. Without their support, the information remains within each individual's tacit knowledge, and may result in the loss of important information.

Dissemination is another way for professionals to obtain the information necessary for the response operations. Effective dissemination transmits good quality information to all individuals involved in the response [14]. The support of information technologies to this activity is important as they are able to rapidly disseminate a large amount of information, as well as filter it.

Supporting decision making is another use of information technologies in response operations [14]. The complexity of these operations often imposes a large cognitive load on the professionals involved, and can hamper decision making. The information technologies can assist in filtering the information, so that only that which is useful to response operations is made available, allowing those responsible for making the decisions to focus only on the decision process.

As examples of information technology in emergency responses, the ones that most stand out are information systems. They stand out principally due to their capacity to process information. Examples of information systems for emergency responses are:

- ETOILE [4] used to train professionals;
- WIPER [16], a system focused on decision making;
- IMI [19], a system that seeks to integrate the diverse organizations that participate in emergency responses;
- MIKoBOS [9], a system for information exchange during emergency response operations.

3   Assessing Emergency Organizations' Response Capability

The emergency response activities assessment can facilitate the organization's knowledge about its performance in these activities. Through the analysis of response activities, it is believed the organization can identify and develop improvements so that its response actions are more effective [15].

The main objective of an organization's evaluation is to provide it with knowledge about itself, so that it can, among other things, plan improvements to its activities. This is important since clients' demand for service quality, reliability and consistency is ever greater, and to remain competitive, the organization has to adapt to their needs.

However, it is observed that organizations often don't use assessment methods, and when they do, the methods used are often inefficient and superficial. This is understandable as these methods are generally limited to issues related to the organization's infrastructure, and often are not adequate for the reality of the organization's response operations. In other words, these methods focus more on the final result of the response actions, than how the outcome was achieved.

In an attempt to overcome the problems perceived in existing methods, a model whose objective is to allow the assessment of emergency organization's response activities in a structured way was considered. The model makes it so the organization better understands its actions in response activities, through the visualization of its knowledge about these activities. This fact can facilitate the identification and implementation of possible improvements to the organization's response activities.

The use of maturity models is an interesting way to assess organizations. These models are divided in increasing maturity levels, and allow the organization, in addition to the evaluation, to plan how to reach more mature levels. This is possible as they position an organization relative to its performance of a given task, and so allow possible improvements to be identified.

The model allows the organization to be assessed from the point of view of different dimensions related to emergency response activities. These dimensions can be considered jointly or separately. Thus, the model induces the assessment to be made by independent specialists, each one in charge of a specific dimension. But, at the same time, the model also induces the specialists to combine their assessments into a single framework, and, for this, they will need to cooperate. Examples of these dimensions are communication, collaboration, coordination, and information technologies.

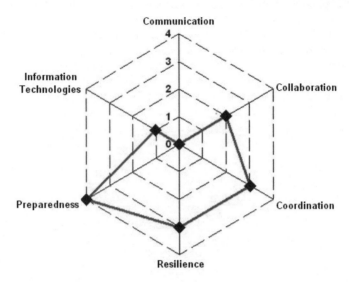

Fig. 1. Radar graph example

The model is divided into maturity levels. Each of these levels has characteristics that must be complied with for an organization to be allocated to it. Thus, once situated at some maturity level, the organization comes to know the points to prioritize in order to improve its response capacity and achieve a higher maturity level.

As a means of visualizing the model's maturity level structures, radar graphs will be used. These graphs were selected as they afford a multidimensional view of the model. Radar graphs allow the visualization of each dimension separately or grouped into a higher-level dimension. An example of these graphs is presented in Figure 1.

As response activities vary among domains, there is a need to build a model for each domain [15]. Thus, emergency organizations from a given domain can only be assessed using that domain's model. To ease this task, a method for building the model will be described. This method, among other things, may induce organizations to build its own assessment model.

3.1   The Method for Building the Model

The first step of the method described here is the definition of the component structure that will constitute the model for the domain in question. This structure must be defined in such a way that it can faithfully portray the response activities carried out by the emergency organizations belonging to the domain.

The starting point for the description is the definition of the dimensions related to the domain's response activities. However, analyzing the organization relative to each component is not trivial. To ease this analysis, our method proposes their decomposition. The components are divided into sub-components whose analysis is less complex. The decomposition should go on until components are obtained whose assessment is possible through direct observation and/or measurement of the organization's variables.

Each of the model's components is a level of abstraction. A level of abstraction may have "m" elements for which the assessment of an organization's response capacity is similarly complex. The levels of abstraction are organized hierarchically so that the greater ones, those with more complex organization assessment, are at the top of the hierarchy. An illustration of this hierarchy is presented in Figure 2

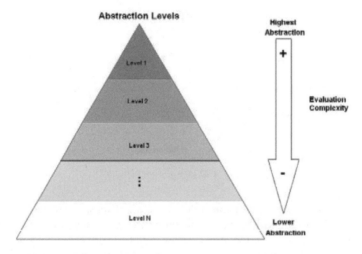

Fig. 2. Illustrative example of the abstraction levels hierarchy [15]

As can be seen in Figure 2, in the proposed model there may be "n" levels of abstraction. The number of levels will depend on the domain for which the component structure is being developed. There should be as many levels as necessary to reach a level of abstraction in which the organization's assessment can be obtained.

Each element of the levels of abstraction is divided into **maturity levels**. It is important to better clarify the difference between **levels of abstraction** and **maturity levels**. Levels of abstraction are elements that will guide the assessment of emergency organizations' response capacity, or be it, they determine what should be observed to assess the organizations. On the other hand, maturity levels correspond to the division of the elements of one given level of abstraction.

Returning to the definition of the component structure, it should be developed by experts in the domain for which the model will be built. It is interesting that more than one specialist create the definition, as different points of view can bring the component structure closer to the reality of emergency response organizations' operations in the given domain.

3.1.1 The Model's Maturity Levels

The model to be built is based on the maturity model concepts [13]. The second step in the method consists of adjusting the model to these concepts. In other words, the second step consists of the definition of the maturity levels of the elements of each abstraction level of the model. For each element of the levels of abstraction, the maturity of an emergency organization is assessed, relative to it.

The maturity levels are organized hierarchically from the most immature to the most mature. Each maturity level has a set of characteristics that an organization must fulfill to be deemed at that level. Once situated at a maturity level relative to an element of the levels of abstraction, the organization comes to know its negative and positive points regarding its emergency response operations. Additionally, the organization can identify procedures that will allow it to overcome its negative points.

As the organization overcomes its negative points, it increases its maturity level in the model's hierarchy, or be it, increases its response capacity. Ideally, the organization manages to make itself mature (reach the highest maturity level) in all of the model's components, gradually and evenly.

Figure 3 shows how an emergency organization's maturity is defined in relation to the different abstraction levels. First, the maturity of the organization relative to the elements of the least level of abstraction is obtained. After this classification, heuristics are applied to the maturity (ratings) obtained, giving rise to the organization's maturity relative to the elements of the abstraction level immediately greater than the lesser. The application of heuristics is repeated until the organization's maturity relative to the elements of the highest level of abstraction is reached.

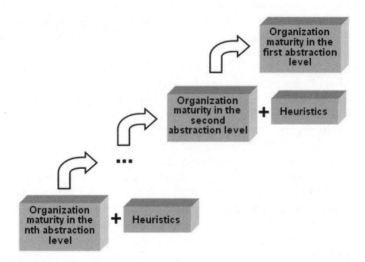

Fig. 3. Obtaining the maturity levels of the model's components [15]

3.1.2 Model Verification

The fourth and last step of the method proposed here is the verification of the model built. No matter how perfect the model built, little details may not represent emergency response operations faithfully. Verification is necessary for these details to be fixed. Domain specialists are interviewed so that an analysis of the model's items is done. This analysis happens through comparisons with the response operations of emergency organizations in the domain, from the interviewees point-of-view.

During the professionals' analysis, any consideration about the model constructed can be made, and if possible, registered. The pertinent considerations must be implemented so that the resultant model is apt to be applied in assessing emergency organizations' response capacity, producing results compatible with the organization's performance in response operations.

3.2 Building a Model

Now that the method for building a model has been presented, we'll show a model built with this method. The domain chosen for this was fire-fighting organizations. A detailed analysis of this domain was undertaken first, so that the components (levels of abstraction) of the model could be defined in a way compatible with the domain organization's operating reality. This analysis consisted of interviews with domain specialists and studies undertaken about the domain.

This analysis initially found the need for three levels of abstraction for the model. This conclusion was arrived at from observations, and from reports found in the literature and obtained in the interviews. In an attempt to facilitate the understanding of these levels of abstraction, we decided to call them as follows: **indicators** (first abstraction level), **criteria** (second abstraction level), and **aspects** (third abstraction level). The indicators are divided into criteria, which are in turn divided into aspects.

The indicators correspond to the dimensions present the domain emergency response operations. Some indicators were identified for the fire-fighting organization domain, such as communication, collaboration, coordination, information technologies, information management, resilience, preparedness, etc. In addition to these, other indicators may be identified in future domain analyses, due to different knowledge and points of view among the professionals responsible for these analyses.

To finalize the models construction we chose one of the indicators to structure the other levels of abstraction. The indicator chosen for this was the one relative to **information technologies**. This indicator was chosen as it is related to one of the most important elements present in response activities: information.

The information technologies must service a number of requirements to be effectively used. The requirements are those relative to information capturing, storage, dissemination, and quality. However, information technologies are often used without a comprehensive analysis of these requirements. Thus, information technologies often do not satisfy the needs for which they're employed, which may indicate wasted resources, a critical issue in places where investments are scarce.

The model allows the visualization of how the organization's information technologies perform relative to their requirements, which may facilitate the use of these technologies. This visualization became possible, as the model was developed to assign priority to the requirements related to information technologies. After the study in the fire-fighting organization domain, and taking into account the concepts related to information technologies in the responses, the criteria that compose the selected indicator were defined. These criteria don't take into account specific technologies, but rather the functionalities they are able to provide. The criteria defined for this indicator were:

- Information capture;
- Information storage;
- Information dissemination;
- Information quality;
- Inter-organizational information;
- Previous information;

Information Capture

This criterion seeks to assess how the organization's information technologies are used to capture information useful for emergency responses. It is expected it will facilitate the organization's planning of measures seeking to improve the use of information technologies to capture information for emergency responses. The aspects verified for this criterion are:

- The policies and strategies to stimulate the use of information technologies to capture information for emergency responses;
- The techniques employed to capture information for emergency responses;
- Information filtering while capturing;

Information Storage

This criterion seeks to verify how the organization uses technologies to store information captured for emergency responses. It is believed that it will be possible to assist the organization in planning improvements related to the storage of its information, and to a certain degree, foresee what its information dissemination may be like. The aspects verified for this criterion are:

- The bases for the storage of the information captured for emergency responses;
- The standardization of the information captured for emergency responses.

Information Dissemination

This criterion seeks to verify how the organization uses information technologies in the dissemination of information during emergency responses. It is important that during the response activities, the right information be transmitted to the right people, at the right time. Thus, it may be possible to assist the organization in planning improvements related to the dissemination of its information. The aspects verified for this criterion are:

- The importance of the disseminated information to the emergency responses;
- The reach of the dissemination of information for emergency responses;
- The availability of the disseminated information to the emergency responses;
- The filtering of information during dissemination;
- The use of disseminated information in emergency responses;

Information Quality
This criterion seeks to verify the quality of emergency response related information which is treated by the information technologies used by the organization. We considered quality information that is relevant for a given emergency response. This criterion enables an organization to improve the quality of its response related information, and consequently the outcome of its response activities. The aspects verified for this criterion are:

• The relevance of the information to emergency responses;
• The accuracy of the information for emergency responses;

Inter-Organizational Information
This criterion seeks to verify how the organization's information technologies deal with inter-organizational information during emergency responses. The importance of this criterion is justified as this type of information circulates during most emergency response activities. The aspects verified for this criterion are:

• The policies adopted to promote getting and exchanging inter-organizational information during emergency responses with information technology support;
• The strategies used to promote getting and exchanging inter-organizational information during emergency responses with information technology support;
• The tactics used to promote getting and exchanging inter-organizational information during emergency responses with information technology support;
• The technological infrastructure for getting and exchanging inter-organizational information during emergency responses;
• The ability for inter-organizational communication.

Previous Information
Previously formalized information includes consolidated concepts and information extracted from previous events, deemed important for future emergencies. This criterion seeks to verify how this type of information is treated by the organization's information technologies, for using during emergency responses. As this type of information is related to future events, we believe that with this criterion, the organization will be able to better prepare its professionals for possible emergency responses. The aspects verified for this criterion are:

• The formalization of previous information for emergency responses;
• The use of previous information in emergency responses;
• The dissemination of previous information for emergency responses;
• The previous inter-organizational information for emergency responses.

3.3   Verification of the Model Constructed

Once the model was built, it was necessary to check with specialists if it was consistent with the response operations in the domain. Non-domain specialists, who as such were unfamiliar with the domain, generated the first version of the model. Three specialists with different knowledge of response operations in the domain participated in the verification. They were asked to analyze the division of the model's components and the maturity level structure, indicating and suggesting changes and improvements so that the modeled indicator became as consistent as possible with the reality of the response operations in the domain. The resultant model, after this verification session, according to the specialists, faithfully portrays the response activities in the domain, and can, therefore, be used in the assessment.

4 Applying the Model

For this application, an actual organization was selected: the Rio de Janeiro Fire Department (CBMERJ). To instantiate the model for an organization a set of steps was followed which included the study of the target organization, the definition of the participants in the assessment process, and the assessment itself.

Initially, a study of the organization was done, conducted mostly through interviews with the organization's professionals with the intent of adapting the model's application to the organization's characteristics. This study permitted the identification and the understanding of the organization's structure, which contributed to its assessment within the framework established by the model.

The assessment was conducted through interview sessions supported by a tool. There was a separate session for each of the three interviewees. Before each session, a presentation showed how the assessment would be conducted and the expected results. The model's levels structure concepts were presented, as well as the use of the tool. This training sought to facilitate the assessment task.

A tool was conceived to facilitate data collection and graphically represent the organization's "picture". From the data supplied by the interviewees, the tool automatically generates graphs representing the organization's maturity regarding its emergency response activities. The tool was implemented in an Excel spreadsheet using VBA (Visual Basic for Applications).

The user must first select one of the predefined heuristics, or define a new one. The two predefined heuristics are: simple sum, and weighted sum of the maturity levels. After this selection, the tool applies the calculations relative to the selected heuristic to construct the organization's maturity graphs.

To arrive at conclusions regarding the assessment, it was necessary to compile the data collected. The tool synthesized this information automatically into graphs. With them, it became easy to identify which were the organization's positive and negative points regarding its emergency response activities. Once these were exposed, it was possible to identify probable improvements to the organization's response capability. This identification also enabled a comparison between the evaluated units'.

Next we'll present the data collected and compiled for each of the two units (a third unit was analyzed, but its data is not presented, due to lack of space. The compilation was obtained using the weighted sum of maturity levels heuristic. The weighting of each of the model's components was accomplished by consensus among the interviewees. This was interesting in that it enabled a comparison of the three units' assessments, besides reducing discrepancies between results.

Unit A

The first unit evaluated serves a relatively small area where the frequency of occurrences is low. The total number of occurrences in the unit's target area also is not high. The unit has on average 100 fire-fighters, divided into shifts, performing the various duties to assure success of the emergency responses.

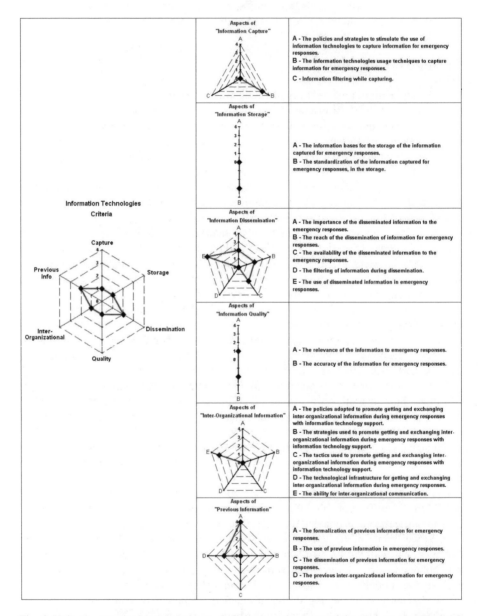

Fig. 4. Unit A's maturity levels relative to the criteria and aspects of the "Information Technology" indicator

The assessment showed that the unit is in the first maturity level relative to information technologies. This means that its use of information technologies is still *ad hoc*. This was not a uniform result, as in some criteria relative to the utilization of these technologies the unit showed satisfactory responses while in others it showed problems. The reasons the unit is at the first maturity level relative to information technologies can be seen in Figure 4. The figure shows graphs that indicate the unit's maturity levels relative to the criteria and the aspects that make up the "Information Technology" indicator.

By analyzing the data in Figure 4, it can be seen that the unit makes best use of information technology in dissemination and in treating previous information. In contrast, its use of information technologies is not very effective relative to the capture and storage of emergency response activity information, and shows little concern with the quality of response information and with inter-unit information.

Although the organization has in place relatively effective measures for information dissemination during responses, the amount is not high because the Unit has not systematized capture and storage procedures. Besides, the response information lacks quality, and may impair the response activities.

The interviewee considered the assessment method to be simple and very useful for the unit. According to him, the method got close to the reality of the Unit's response operations, providing a better visualization of the knowledge related to the use of information technology in the response activities and facilitating the identification of positive and negative points of the organization in these activities.

On the other hand, the interviewee criticized the way in which the assessment was conducted. According to him, other professionals should have participated in the assessment. Only one professional participating in the assessment may skew the results. Also, the interviewee concluded that the assessment should be carried out during a longer period, and not in only one interview.

Unit B

The second unit evaluated serves an area larger than the first. This area also presents a larger number of events. This may be caused by the area's high demographic density, which may affect emergency-causing factors such as in-traffic, types of housing, commercial and industrial installations, etc. Consequently the manpower of the Unit is greater, as is its infrastructure, both installation and equipment wise.

This Unit, like the first, is at the first maturity level relative to the "Information Technologies" indicator. However, the reasons for this classification are different. The graphs present in Figure 5 show the Unit's information technologies are used efficiently for the storage of emergency response activity information. They also indicate that the unit uses its information technologies to capture and disseminate quality information during its emergency response operations.

The breakdown of each indicator in less-complex components, according to the interviewee, allowed the identification of factors he did not imagine were able to influence the organization's response performance. Examples the interviewee cited include the systematization of the use of technologies to treat inter-organizational information, and the use of information technologies to support the formalization of previous information. As a drawback the interviewee mentioned the need to consult professionals more familiar with certain response activities.

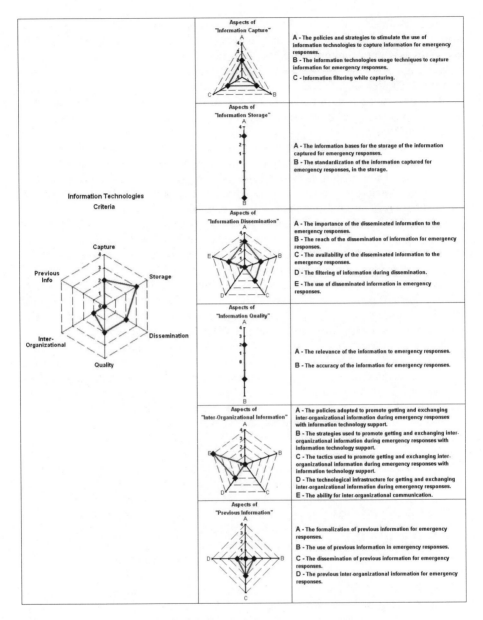

Fig. 5. Unit B's maturity levels relative to the criteria and aspects of the "Information Technology" indicator

5 Conclusions

The solution presented here began from an important issue: an emergency organization's knowledge of how it operates in the course of its response activities. After studying several assessment methods used in emergency responses it was seen that it is difficult for the organization to satisfactorily understand how it has been operating in emergency responses through them. From there, a reference model based on the maturity model concepts was proposed to evaluate the response capacity of emergency organizations.

In spite of being based on the maturity model concepts, the model proposed does not have their characteristic rigidity. This method makes the model dynamic, and can adapt it to evaluate organizations of any emergency related domain, in addition to allowing the concern of any dimension involved in possible emergencies. The method also allows the model to be extended to other phases of emergency management.

Evaluation through the model positions the organization within its emergency response activities performance spectrum, showing it a portrait of that performance. This portrait has the organization's positive and negative points regarding response activities, which may more effectively guide the organization's focus in the search for improvements to these activities.

At this point we can say that there is some indication that our hypotheses hold in the development of a model for the fire-fighting domain, and its application in the Fire-Fighting Corps of Rio de Janeiro, Brazil. The model was developed with only one dimension: the **information technologies**. In it some issues related to these technologies in this specific domain's response activities were raised and detailed

Although specialists did not develop most of the model, the assessment participants praised its proximity to the organization's response operations reality. This proximity produced results coherent with the current use of information technologies by the organization in emergency responses. The participants praised the model and reported the importance of an assessment that translates the organization's reality. According to them, the results translated clearly the organization's real needs regarding information technologies for emergency responses.

Acknowledgements

Marcos R. S. Borges and Jose Orlando Gomes were partially supported by grants No. 305900/2005-6 and 484981/2006-4, respectively, from CNPq (Brazil). The cooperation between the Brazilian and the Spanish research groups was partially sponsored by the CAPES/MECD Cooperation Program.

References

1. Bigley, G.A., Roberts, K.H.: The incident command system: high reliability organizing for complex and volatile task environments. Academy of Management Journal 44(6), 1281–1299 (2001)
2. Carver, L., Turoff, M.: Human-Computer Interaction: The Human and Computer as a Team in Emergency Management Information Systems. Communications of the ACM 50(3), 33–38 (2007)
3. Diniz, V.B., Borges, M.R.S., Gomes, J.O., Canós, J.H.: Knowledge Management Support for Collaborative Emergency Response. In: Proc. of the 9th International Conference on Computer Supported Cooperative Work in Design, pp. 1188–1193. IEEE Press, London (2005)

4. Dörner, R., Grimm, P., Seiler, C.: ETOILE – An Environment for Team. Organizational and Individual Learning, CG Topics 13(3), 5–6 (2001)
5. Federal Emergency Management Agency (FEMA): Emergency Management Guide for Business and Industry: A Step-by-step Approach to Emergency Planning, Response and Recovery for Companies of All Sizes (1998)
6. Gu, Q., Mendonça, D.: Patterns of group information seeking in a simulated emergency response environment. In: Proceedings of the 2nd International ISCRAM Conference, Brussels, Belgium (2005)
7. King, W., Teo, T.: Integration between Business Planning and Information Systems Planning: Validating a Stage Hypothesis. Decision Sciences 28(2), 279–307 (1997)
8. Lindel, M.K., Prater, C., Perry, R.W.: Emergency Management. John Wiley & Sons, New York (2007)
9. Meissner, A., Wang, Z., Putz, W., Grimmer, J.: MIKoBOS: A mobile information and communication system for emergency response. In: Proceedings of the 3rd International ISCRAM Conference, Newark, New Jersey (2006)
10. Oomes, A.H.J., Neef, R.M.: Scaling-up support for emergency response organizations. In: Proceedings of the 2nd International ISCRAM Conference, Brussels, Belgium (2005)
11. Palen, L., Hiltz, R., Liu, S.: Online forums supporting grassroots participation in emergency preparedness and response. Comm. of the ACM 50(3), 54–58 (2007)
12. Paton, D., Flin, R.: Disaster stress: an emergency management perspective. Disaster Prevention and Management 8(4), 261–267 (1999)
13. Paulk, M.C., Weber, C., Curtis, B., Chrissis, M.: The Capability Maturity Model: guidelines for improving the software process. Addison-Wesley, Reading (1995)
14. Quarantelli, E.L.: Problematical aspects of the information / communication revolution for disaster planning and research: ten non-technical issues and questions. Disaster Prevention and Management 6(2), 94–106 (1997)
15. Santos, R.S.: A reference model for the assessment of the organization capability in emergency responses. M.Sc. Dissertation, Graduate Program in Informatics, Federal University of Rio de Janeiro, Brazil (2007) (in Portuguese)
16. Schoenharl, T., Szabo, G., Madey, G., Barabasi, A.L.: WIPER: A multi-agent system for emergency response. In: Proceedings of the 3rd International ISCRAM Conference, Newark, New Jersey (2006)
17. Turoff, M.: Past and Future Emergency Response Information Systems. Communications of the ACM 45(4), 29–33 (2002)
18. Turoff, M., Chumer, M., Hiltz, R., Clasher, R., Alles, M., Vasarhelyi, M., Kogan, A.: Assuring Homeland Security: Continuous Monitoring, Control and Assurance of Emergency Preparedness. Journal of Information Technology Theory and Application (JITTA) 6(3), 1–24 (2004)
19. Van der Lee, M.D.E., Van Vugt, M.: IMI – An information system for effective multidisciplinary incident management. In: Proceedings of the 1st International ISCRAM Conference, Brussels, Belgium (2004)
20. Zimmerman, R., Restrepo, C.E.: Information technology (IT) and critical infrastructure interdependencies for emergency response. In: Proceedings of the 3rd International ISCRAM Conference, Newark, New Jersey (2006)

The Semantic Architecture Tool (SemAT) for Collaborative Enterprise Architecture Development

Frank Fuchs-Kittowski and Daniel Faust

Fraunhofer Institute for Software and Systems Engineering ISST
Mollstr. 1, 10178 Berlin, Germany
{frank.fuchs-kittowski,daniel.faust}@isst.fraunhofer.de

Abstract. In this paper the semantic architecture tool (SemAT) for collaborative EA development is presented. This includes the concept of a semantic wiki-like collaboration tool for collaborative EA management and an EA ontology as a formal representation of the EA. In addition, the prototypical implementation of the semantic collaboration environment is described. Finally, the benefits of the approach are discussed.

1 Introduction

Enterprise architecture (EA) management is widely accepted as an essential instrument for ensuring an enterprise's agility, consistency, compliance, and efficiency, and is especially used as a basis for a continually aligned steering of IT and business (IT business alignment) [1], [2], [3]. EA management is the field of managing whole EA's as well as the artifacts that constitute EA's. While an EA model represents an enterprise's as-is or to-be architecture [4], [5], an EA framework provides meta-model(s) for EA description and method(s) for EA design, development, use, and evolution [6], [7].

Nevertheless, the design and evolution of an EA is still a challenging and complex task [5], [8]. It is a cost intensive and time-consuming process, especially in large scale enterprises with numerous, spatially distributed locations. It consists of many participants responsible for different kinds of information for different parts of the EA using different methods and tools for information gathering. Different business functions (data owners) provide information required for the divergent needs of various stakeholders with different interests. Structuring such a process is difficult due to the involvement of many stakeholders from different business functions and cultures, thus resulting in increased communication and coordination efforts for all involved.

Our goal is the design of a collaboration environment to support the collaboration of all individuals involved in the process of EA design and evolution. In particular, it must be possible for a large and spatially located group of individuals to gather information about EA collaboratively and with minimal effort. Furthermore, information gathering for EA management must be possible without having to plan and structure this process in advance. Our approach to achieve this goal of a participative EA management is a semantic, wiki-like collaboration environment. This

R.O. Briggs et al. (Eds.): CRIWG 2008, LNCS 5411, pp. 151–163, 2008.

is a solution based on the concept of semantic web and the paradigm of Web-2.0, e.g. user-generated content, participation, collective intelligence. This semantic collaboration tool allows the combination of formal, semantic structuring of EA information (in an EA ontology) with informal, participative processes of gathering this EA information (supported by a wiki-like collaboration environment).

This paper presents the concept of a semantic collaboration tool for collaborative EA management. An EA management application scenario is presented to further characterize the problem and to derive requirements on a collaboration environment supporting collaborative gathering and maintenance of EA information (section 2). Based on this scenario, the approach is presented (section 3) to include the concept of a semantic, wiki-like collaboration tool for collaborative EA management (section 3.1) and an EA ontology as a formal representation of EA (section 3.2). Additionally, the prototypical implementation of the semantic collaboration environment (section 4) is described and the benefits of the approach are discussed (section 5).

2   EA Application Scenario

A large number of methods for EA management have been developed by academia and practitioners (e.g. [6], [8], [9], [10], [11], [12], [13]). These methods usually distinguish between the following EA management processes as a life cycle model of the EA: (a) strategic architecture visioning and definition, (b) EA development, (c) EA use, (d) EA maintenance. Almost all of these methods pay little attention to specifying information gathering procedures for EA model data in detail – especially during EA development and use [14].

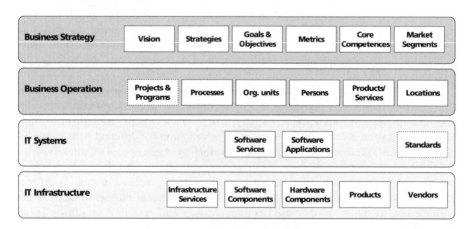

Fig. 1. Simple 4-layered EA

An EA is comprised of a large number of artifacts. Since creating EA models is expensive and without intrinsic value, it is desirable to only create EA models that support good decision making [15]. Therefore, an EA needs to include only those artifacts (and relationships) that promote well-defined analyses for a certain EA application scenario (e.g. IT consolidation). As an example, following current divisions of EA in layers and artifacts in EA frameworks (e.g. [6], [8]), scientific literature (e.g. [16], [17], [18], [19], [20]), or meta-models of commercial EA tools (e.g. E2AF [21], MEAF [22]), we decompose EA into four EA layers and propose a simple 4-layered EA with core artifacts (figure 1):

- Business strategy (goals, markets, products/services, etc.),
- Business operation (processes, organizational structure, etc.),
- IT systems (software service, software application, software components, etc.), and
- IT infrastructure (infrastructure services, hardware component, products, etc.).

In practice, the different (sets of) EA artifacts are developed and maintained by different individuals. Very often, EA coexists with other, more specialized or detailed artifacts that cover a subset of these artifacts [17], [23]. Therefore, EA should integrate these existing artifacts as well as the used modeling techniques and tools [24] and useful interfaces have to be specified and established [17]. Interfaces could be established, for example, to an ERP system (products and services), process modeling tool (business process models), or hardware inventory (hardware data).

Moreover, an EA has to represent not only isolated artifacts but also the entire complexity of an enterprise with all its relationships and dependencies. Supporting EA-relevant decisions requires knowledge of multiple artifacts of the EA. For example, "IT planning's" main task is to combine the business process requirements with appropriate IT systems within a limited budget. This need-driven planning has to follow an integrative approach where the perspectives of (a) business structures and processes, (b) IT systems and infrastructures, and (c) finances have to be combined. To achieve this, an EA has to gather and document not only the isolated artifacts but also the relationships between them.

As most of the required EA information in heterogeneous and spatially distributed environments is owned and maintained by different individuals, the integration process becomes very important. This process has to ensure that various information from different business departments, functions, domains and individuals have to be interlocked. Only if the information is integrated, inconsistencies and cost-intensive incorrect planning can be avoided. The need for integration ties the separate information gathering processes of different individuals much closer together. The situation of multiple participants and stakeholders results in diverse requirements. Besides the modeling of isolated and shared artifacts, this includes communication and documentation among the various participants and stakeholders involved in the EA development process. In addition, different tools and description formats may be used.

The foremost method of information gathering for EA models in practice is to task external consultants with collecting and modeling the EA information in enterprises. This approach is expensive since it requires experienced consultants, time consuming since the required data is not on hand, frequently incomplete since efforts and expenses to document the entire EA are too high. Compared to this method - which focuses on EA information gathering by only a small group of "outsiders" - our approach focuses on involvement of the data owners and other individuals such as stakeholders. Therefore, the technical support of the collaborative process of a large group of individuals is necessary.

Existing EA methods and most of the commercial EA tools (e.g. planningIT by alfabet, Metis by Troux Technologies) focus on a process-oriented EA management approach[1]. Characteristics of the process-oriented approach are: (a) the detailed, formal description of a process necessary to gather EA information, (b) the specification of roles to execute, manage and control all process activities, (c) the mapping of roles to process activities by means of responsibility charting (i.e. by specification of responsibility, accountability, etc. for each process activity). But for a large number of participants the process-oriented approach is inadequate. It is hardly possible to structure the entire process in advance if a lot of individuals from different functions, domains and cultures are involved. In addition, fixed roles do not allow for participative flexibility.

In order to address the challenge of EA information gathering by a large group of individuals, we propose a participative approach. It does not focus on the definition of processes, roles and responsibilities, and gathering EA information by a small group of external staff but rather on internal stakeholders (data owners), their interaction and the resulting processes of change. In the participative approach the responsibility for gathering EA information is delegated to the data owners and other individuals such as stakeholders. The participative modeling process involves a large group of individuals (stakeholders) that identify, document, and consolidate their different knowledge and interests concerning the problem modeled - as the required knowledge mainly exists in their minds, opinions, intentions, work routines, experiences, etc..

Participative EA modeling methods (for example, [25], [26], [27], [28]) lead to improved quality as well as an increased consensus and acceptance of the business decisions. An empirical study [29] shows that participative EA modeling can successfully support both, business development objectives and quality assurance objectives. It also facilitates maintaining and sharing knowledge about the business as well as organizational learning (for example, [30], [31]). The full and positive effects of participative EA modeling heavily depend on the ability of its users to manage situational factors which characterize, influence and constrain development situations where EA modeling is used [32].

The goal of a participative EA development is the involvement of data owners and stakeholders. Hence, the integration of a large group of individuals as well as a tool support of the collaborative process of this large group – providing an integration space for the participants - is necessary.

The proposed semantic collaboration tool for participative, collaborative EA management (especially EA information gathering and development) enables - compared to existing (semantic) wikis (e.g. SemanticMediaWiki [33], IkeWiki [34], Kaukolu [35], Sweet wiki [36]) where textual content is addressed - the semantic representation and structuring of EA information (ontology) and - compared to existing commercial EA tools which promote a process-oriented approach to EA management - a participative process of EA design and evolution and - compared to existing ontology editing tools (e.g. Protégé) which do not focus on a collaborative process and require expertise in formalisms and ontology engineering [37] - a collaborative, fine-granular participative and community-oriented EA ontology information gathering and development process by a large number of non-experts.

---

[1] See [16] for a detailed description of existing commercial EA tools.

3 Concept and Ontology of a Semantic, Wiki-Like Collaboration Environment

To resolve the before-mentioned problems, the presented concept will support collaborative EA design and development more effectively. To reach this goal a participative approach is proposed which will be achieved by a semantic web 2.0-like collaboration tool. Web 2.0 tools, like wikis and weblogs, support the generation, gathering and exchange of knowledge in large, spatially distributed groups. However, they do not offer possibilities for the (semantic) representation of EA information and are, therefore, inadequate for EA design and evolution. Semantic web tools support the classification, structuring, and representation of large, unstructured stocks of information but do not support collaborative EA design and evolution effectively.

Hence, our solution is a semantic web 2.0-like collaboration tool that combines the advantages of the two innovative technologies. This solution provides an integration space to bridge the gap between the participants in EA information gathering since it (a) provides means for collaborative information gathering and communication regarding a certain subject and (b) it provides a formal foundation for handling and integrating EA information. This solution supports the collaboration of a large number of individuals based on a semantic representation (ontology) of the EA. The following sections describe the main components of the concept:

- A semantic (web 2.0-oriented) collaboration tool to support EA information gathering and modeling (section 3.1).
- A semantic EA (ontology) to enable an integrated, formal, semantic representation and structuring of EA information (section 3.2).

3.1 Concept of a Semantic, Wiki-Like Collaboration Environment for EA Management

The semantic, wiki-like collaboration environment organizes the collaboration of a large number of spatially located individuals based on the semantic representation (ontology) of the EA. The concept is composed of three main components: (a) community features, (b) EA information repository, and (c) interfaces to external systems (figure 2).

A core component is the EA information repository. It enables the manipulation and management of semantic EA information (ontology). Furthermore, it serves as an integration space for all participants. All individuals involved in the EA design and development work on only one common, shared model of the EA (instead of each individual modeling his/her own model). This way EA development is not restricted to one partition of the EA (i.e. a subset of artifacts) which is relevant to only one domain. With the aid of the semantic collaboration tool they are able to model and document the EA with a standardized notation (ontology). It is essential that all individuals involved use the same tool to access the same repository and the same EA (ontology) data. This integration of the information gathering and modeling for the different artifacts of an EA (e.g. organizational goals, processes, software applications, etc.) enables a continuous, integrated design and development of the EA (bird's eye view).

Another core component are the community features. There are mechanisms for community support which enable an informal, participative process of EA information gathering and modeling and ensure a high degree of quality and up-to-dateness of the EA information. We propose the concept of annotating these community features to each ontology element. Community features include mechanisms for feedback, review, discussion, rating, and negotiation. These are annotated to each item of the model data (ontology). This way communication about the process and the results of information gathering is focused on smaller items which enables efficient collaboration in a very large group.

As a third component, the semantic collaboration tool provides interfaces (adapters) for import and export of EA information from or to external systems. Thereby, the formal representation or notation of the EA (ontology) supports the seamless import/export. On one hand, EA ontology data is imported from a holistic EA model or uses data from existing specialized models to keep modeling efforts to a minimum. On the other hand, EA ontology data can be exported (e.g. for analysis or visualization) to enable reuse.

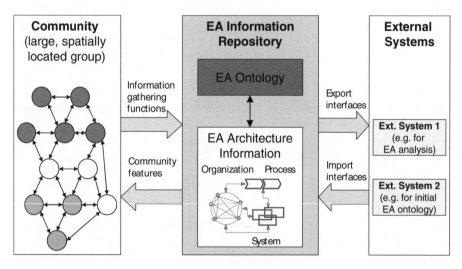

Fig. 2. Components of a semantic, wiki-like collaboration environment for EA management

The functionality of the semantic collaboration tool allows users to modify the EA, create relations between EA items, and annotate communication to certain parts of the EA ontology. The aim is to minimize efforts for information gathering and maintenance and, therefore, to minimize communication and process costs. Most importantly, this functionality provides possibilities for fine-granular and intuitive gathering and editing of the EA ontology, e.g. EA classes and EA instances as well as their properties and relationships.

3.2 Concept of an Ontology Representing the EA (EA Ontology)

Ontology has been introduced as a key concept in informatics to facilitate and encompass access to domain knowledge in different application domains. Gruber has defined ontology as an "explicit specification of a conceptualization" [38]. Very often existing standards for ontology representation such as the Resource Description

Framework (RDF) and the Web Ontology Language (OWL) are adopted. RDF is a simple graph-like format for describing metadata about resources [39] that are described using a Uniform Resource Identifier (URI). OWL is defined on top of RDF/RDFS and provides a standard ontology vocabulary for describing ontologies based on description logics [40].

A first attempt to represent an EA as ontology is the Enterprise Ontology (EO) [41]. The purpose of EO includes: (1) to guarantee smooth communication between participants to facilitate sharing the unified understanding about the enterprise model by providing necessary and sufficient vocabulary, (2) to provide an infrastructure that is stable but at the same time adaptable to the change of understanding about and requirements to the enterprise model, (3) to augment interoperability of various application programs of an enterprise model by using EO as a mutual language (Interlingua) for information exchange. Typical terms (concepts) contained in this EO are: activity (e.g. activity specification, sub-activity, event, plan, process specification, capability, skill, and resource), organization (e.g. person, machine, corporation, partner, and organizational unit), strategy (e.g. purpose, objective, vision, mission, and goal), marketing (e.g. sale, vendor, customer, product, market, and competitor) and time (for example, time line, and time interval). However, these concepts only cover the business (upper) layer of an EA.

For the EA application scenario, a conceptual structure - an EA ontology (see figure 3) - specifying concepts and their relations might be useful to predetermine the initial structure of the information space (repository). Contrary to (semantic) wikis supporting the emergence of wiki content or knowledge structures (ontologies) out of existing wiki content [33], [34], [35], [36], this alternative proposal results in a semantic, wiki-like collaboration tool based on a predefined knowledge structure that supports the emergence of the knowledge structure (EA ontology) or the emergence of ontology data (EA information) out of the knowledge structure (EA ontology). Thus, the EA ontology provides a frame for its own development and for adding further ontology data (EA information) within a limited application scenario, e.g. business continuity planning (BCP), portfolio management (applications, partners, projects), evaluation of strategic options, IT service management.

The main purpose of the EA ontology is to supply the initial structure to the EA information space (repository) of the semantic, wiki-like collaboration environment enabling the gathering and documentation of EA information by the different groups involved (e.g. detailed process structures by business, detailed application architectures by architects, detailed software architectures and data flows by developers). Thereby, the EA ontology serves as a shared mental model to facilitate communication among the participants by providing a shared understanding of the EA under development and enforcing conceptual integrity. Instead of determining the collaborative process, only the goal ("the big picture") is roughly described by the EA ontology. This way, the EA ontology serves as a shared mental model of the EA to be developed, supports communication and provides orientation in a self-organized, participative collaboration process, and ensures convergence of the collaboration process towards a result to be achieved. In addition, the EA ontology also provides means for formalization. A more formalized EA allows validation and verification of the EA or other tool-supported analyses based on a formally defined semantic as well as a seamless import and export of external resources.

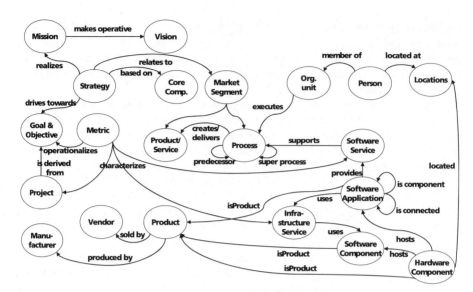

Fig. 3. Enterprise Architecture Ontology

For this purpose the ontology has to include central layers and artifacts (concepts) of an EA (see figure 1). It consists of the relevant classes (persons, locations, products, etc.), their attributes (name, org. unit, position, location, room, etc. for a person), the associations between classes (persons and hardware components are located at one location, persons are members of organizational units, products are sold by vendors, etc.), and an additional glossary collecting specialized terms and their meanings.

In an EA every single artifact (e.g. process model, software architecture) can be decomposed into detailed models. Accordingly, the EA ontology decomposes artifacts into a hierarchy of artifacts (class hierarchy). The prototype implementation in the next section includes an example EA ontology (see figure 4). It shows a class hierarchy. Here, for example, the class "IT Infrastructure" contains the subclasses "Hardware Component" and "Software Component", and "Computer" and "Network Devices" are subclasses of "Hardware Component".

But, additionally, the EA ontology focuses on relationships between artifacts across all layers. For example, "Hardware Component" hosts "Software Application" which provides "Software Service" to support "Process" which is executed by an "Organizational Unit". This provides an integrated information model connecting the different layers of an EA and knowledge across multiple artifacts or layers of the EA that is required to support EA relevant decisions. Because of these relationships across layers and artifacts, EA ontology can better address business driven questions, e.g.: Which company products and services are affected if an application is outsourced? Which processes are affected if a system platform fails? Therefore, EA addresses strategic business issues and diminishes the gap between business and IT perspectives. Each concept shown in figure 3 includes a class hierarchy of concepts with arbitrary relations to other concepts in the same or other class hierarchies. The prototype implementation in the next section shows an example for the concept "IT Infrastructure" with its direct sub concepts (figure 4). Other information displayed for a concept/class are its instances, object properties, and data type properties.

4 Prototype Implementation

This section presents the prototype implementation of the approach to a semantic, wiki-like collaboration tool for collaborative EA design. This implementation is based on the open source, semantic web application development framework pOWL [42] and is incorporated in the integrated knowledge and collaboration environment "WiKo" (German: Wissens- und Kooperationsplattform) [43] which also provides the relevant community functionality. WiKo incorporates collaborative knowledge work in an integrated collaboration environment with organizational knowledge work in an integrated organizational knowledge environment.

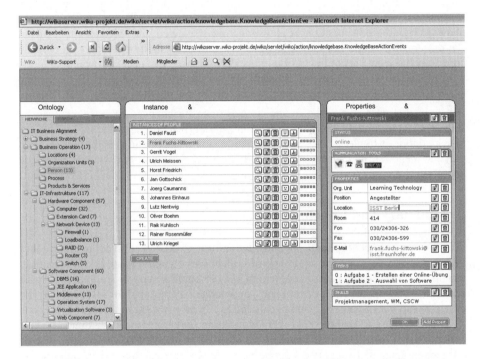

Fig. 4. Prototype implementation of the semantic wiki-like collaboration environment

The web-based user interface is consistently subdivided into three areas: (1) "Ontology" (orientation: navigation & search), "Instance & Function" (ontology content and functions), "Properties and Community" (instance properties and community features). Furthermore, a toolbar represents the menu for central WiKo functionality (figure 4).

Under the heading "Ontology" (on the left) orientation and finding EA information is supported by means of "navigation" and "search". The user gets an overview of ontologies available in the system. The class hierarchy of a selected ontology is presented. Via a context menu to each class of the class hierarchy a new subclass or instance can be created, all existing instances can be listed, and the class description can be viewed, edited or deleted.

Under "Instance and Function" (center) the ontology data and functionality is presented according to the current navigation and search context. Upon selection of a class (or ontology) (on the left), the center area shows an overview of the instances. For each instance functionalities are available that are shown as a button on the right-hand side of the instance name. These functionalities enable the user to present the instance properties (⌖), edit (✐) and delete instances (🗑). In addition, each instance is provided with community functionality to support convergence in the collaborative EA development process. There are community functionalities for rating (▪▪▪▪▪), discussion (⌗) and negotiation (✎).

Under "Properties and Community" (on the right) for each instance selected under "Instance & Function" (center) its available instance property functions (present, edit, delete) and its available community functionalities (rate, discuss, negotiate) are shown. Once the edit function of an instance was selected, the screen switches from the presentation mode (as shown in figure 4) to the edit mode to display an editable form (on the right) according to the class definition of the instance which supports editing all existing properties as well as adding new properties. Even in the presentation mode single properties can be edited by clicking on the "edit" button (✐). In this case, the value of the property can be changed or resources can be linked. If a community function is selected, the corresponding data and actions are provided. For example, in case of a negotiation, a new negotiation can be started or an existing negotiation can be participated in, e.g., vote for one of the available negotiation alternatives.

5 Summary and Conclusion

We presented the concept and implementation of a novel semantic collaboration tool that supports a participative approach of collaborative gathering of EA information. Popular web 2.0 concepts for participation and community support are integrated with semantic web methods for semantic knowledge representation resulting in the effective support of gathering and documenting EA information by a large group.

This approach to collaborative EA information gathering and development has – among others – the following benefits: (a) It builds on the general advantages of web 2.0 technologies that support collaborative settings of a large number of individuals, thus providing a scalable means for the development of an EA. (b) It integrates different groups of individuals involved in the information gathering and EA development process. The (ontology) repository provides an integration space. EA information and their relationships previously gathered and maintained separately are now collected, integrated, and managed in a central location. (c) The formal representation (ontology) of the EA supports focusing on the relationships between EA artifacts, a better searching and browsing of the EA elements, ensures semantic consistency, enhanced analysis of the EA and the import/export of EA information from/to external resources.

The underlying technologies and concepts of web 2.0 are still in an early stage of maturity. Therefore, statements still cannot be made about the acceptance of the approach. Currently, an implementation and evaluation of our approach is realized in cooperation with a German logistics service provider.

References

1. Bucher, T., Fischer, R., Kurpjuweit, S., Winter, R.: Analysis and Application Scenarios of Enterprise Architecture: An Exploratory Study. In: Proceedings of the EDOC Workshop on Trends in Enterprise Architecture Research TEAR 2006, Tenth IEEE International EDOC Conference EDOC 2006, p. 28. IEEE Computer Society, Los Alamitos (2006)
2. Ross, J.W., Weill, P., Robertson, D.C.: Enterprise Architecture as Strategy: Creating a Foundation for Business Execution. Harvard Business School Press, Boston (2006)
3. Wagter, R., von den Berg, M., Luijpers, J.: Dynamic Enterprise Architecture: How to Make It Work. John Wiley & Sons, Hoboken (2005)
4. Rood, M.A.: Enterprise Architecture: Definition, Content, and Utility. In: Proceedings of the IEEE Third Workshop on Enabling Technologies Infrastructure for Collaborative Enterprises, pp. 106–111. IEEE Computer Society, Morgantown (1994)
5. Lankhorst, M.: Enterprise Architecture at Work: Modeling, Communication and Analysis. Springer, Berlin (2005)
6. Open Group: TOGAF - The Open Group Architecture Framework (Version 8.1 "Enterprise Edition"). The Open Group, San Francisco, CA, USA (2003)
7. Schekkerman, J.: How to Survive in the Jungle of Enterprise Architecture Frameworks: Creating or Choosing an Enterprise Architecture Framework, 2nd edn. Trafford Publishing, Victoria (2004)
8. CIO (Chief Information Officer) Council: A Practical Guide to Federal Enterprise Architecture. Technical report. Federal CIO Council, Springfield, VA, USA (2001)
9. Aziz, S., Obitz, T., Modi, R., Sarkar, S.: Enterprise Architecture: A Governance Framework - Part I: Embedding Architecture into the Organization. Technical report. Infosys Technologies Ltd., Bangalore, India (2005)
10. Aziz, S., Obitz, T., Modi, R., Sarkar, S.: Enterprise Architecture: A Governance Framework - Part II: Making Enterprise Architecture Work within the Organization. Technical report. Infosys Technologies Ltd., Bangalore, India (2006)
11. Bittler, R.S., Kreizmann, G.: Gartner Enterprise Architecture Process: Evolution 2005. Gartner Inc., Stamford, CT, USA (2005)
12. DoDAF Working Goup: DoD Architecture Framework - Volume I: Definitions and Guidelines (Version 1.5), 23, The United States Department of Defense, April 2007, http://www.defenselink.mil/cio-nii/docs/DoDAF_Volume_I.pdf
13. IFIP–IFAC: GERAM: Generalised Enterprise Reference Architecture and Methodology, Version 1.6.3. IFIP–IFAC Task Force on Architectures for Enterprise Integration (1999), http://www.cit.gu.edu.au/~bernus/taskforce/geram/versions/geram1-6-3
14. Jonkers, H., Lankhorst, M., ter Doest, H., Arbab, F., Bosma, H., Wieringa, R.J.: Enterprise Architecture: Management tool and blueprint for the organisation. Information Systems Frontier 8, 63–66 (2006)
15. Narman, P., Johnson, P., Nordstrom, L.: Enterprise Architecture: A Framework Supporting System Quality Analysis. In: Proceedings of the 11th IEEE International Enterprise Distributed Object Computing Conference (EDOC 2007), pp. 130–141. IEEE Computer Society, Washington (2007)
16. Sebis: Enterprise Architecture Management Tool Survey 2005. Report. Technical University Munich, Chair for Informatics 19, Munich, Germany (2005)

17. Winter, R., Fischer, R.: Essential Layers, Artifacts, and Dependencies of Enterprise Architecture. In: Proceedings, EDOC Workshop on Trends in Enterprise Architecture Research (TEAR 2006) within The Tenth IEEE International EDOC Conference (EDOC 2006), p. 30. IEEE Computer Society, Washington (2006)
18. Bucher, T., Fischer, R., Kurpjuweit, S., Winter, R.: Analysis and Application Scenarios of Enterprise Architecture: An Exploratory Study. In: Proceedings of the 10th IEE International Distributed Object Computing Conference Workshop (EDOCW), p. 28. IEEE Computer Society, Los Alamitos (2006)
19. Kaisler, S.H., Amour, F., Valivullah, M.: Enterprise Architecting: Critical Problems. In: Proceedings of the 38h Hawaii International Conference on Systems Sciences (HICSS-38 2005). IEEE Computer Society, Big Island (2005)
20. Janssen, M., Hjort-Madsen, K.: Analyzing Enterprise Architecture in National Governments: The cases of Denmark and the Netherlands. In: Proceedings of the 40th Hawaii International Conference on Systems Sciences (HICSS-40 2007). IEEE Computer Society, Waikoloa (2007)
21. Schekkerman, J.: Extended Enterprise Architecture Framework (E2AF): Essentials Guide v1.5. Institute for Enterprise Architecture Developments, Amersfoort, Netherlands (2006)
22. TrouxMetis: Metis Enterprise: Metamodeling and Modeling with Metis Enterprise Architecture Framework (MEAF), v5.2. Troux Technologies Inc., Austin, TX, USA (2006)
23. Bernard, S.A.: An Introduction to Enterprise Architecture, 2nd edn. Authorhouse, Bloomington (2005)
24. Lankhorst, M.: Enterprise architecture modelling: the issue of integration. Advanced Engineering Informatics 18(4), 205–216 (2004)
25. Bubenko, J.A.j., Persson, A., Stirna, J.: User Guide of the Knowledge Management Approach Using Enterprise Knowledge Patterns. Deliverable D3, IST Programme project Hypermedia and Pattern Based Knowledge Management for Smart Organisations, project no. IST-2000-28401, Department of Computer and Systems Sciences, Royal Institute of Technology, Stockholm, Sweden (2001)
26. Bubenko, J.A.j., Stirna, J., Brash, D.: EKD User Guide. Technical report, Department of Computer and Systems Sciences, Royal Institute of Technology. Stockholm, Sweden (1997)
27. F3: F3 Reference Manual. Deliverable of ESPRIT III Project 6612, F3-Consortium, Stockholm, Sweden (1994)
28. Loucopoulos, P., Kavakli, V., Prekas, N., Rolland, C., Grosz, G., Nurcan, S.: Using the EKD Approach: The Modelling Component. Technical report, UMIST, Manchester, UK (1997)
29. Persson, A., Stirna, J.: An explorative study into the influence of business goals on the practical use of Enterprise Modelling methods and tools. In: Proceedings of the 10th International Conference on Information Systems Development (ISD 2001), pp. 465–468. Kluwer, London (2001)
30. Stirna, J., Mikelsons, J., Kalnins, J.R., Kapenieks, A., Kazakovs, M., Vanaga, I., Sinka, A., Persson, A., Kaindl, H.: Trial Application in the Riga City Council. Technical report, deliverable D6, IST Programme project Hypermedia and Pattern Based Knowledge Management for Smart Organisations, project no. IST-2000-28401, Riga City Council, Riga, Latvia (2002)
31. Persson, A., Stirna, J., Dulle, H., Hatzenbichler, G., Strutz, G.: Introducing a Pattern Based Knowledge Management Approach: the Verbundplan Case. In: Proceedings of the 4th International Workshop on Theory and Applications of Knowledge Management (TAKMA 2003) in cooperation with 14th International Workshop on Database and Expert Systems Applications (DEXA 2003). IEEE Computer Society, Prague (2003)

32. Persson, A.: Investigating the influence of situation factors on Participative Enterprise Modelling: making a case for a qualitative research approach. In: Proceedings of Conference on Advanced Information Systems Engineering (CAiSE 2000), Stockholm, Sweden (2000),
    http://page.mi.fu-berlin.de/ hinze/caise00_DC/papers.html
33. Völkel, M., Krötzsch, M., Vrandecic, D., Haller, H., Studer, R.: Semantic Wikipedia. In: Proceedings of the 15th international conference on World Wide Web, WWW 2006, pp. 585–594. ACM Press, New York (2006)
34. Schaffert, S.: IkeWiki: A Semantic Wiki for Collaborative Knowledge Management. Technical report, Salzburg Research, Salzburg, Austria (2006),
    http://ikewiki.salzburgresearch.at
35. Kiesel, M.: Kaukolu: Hub of the Semantic Corporate Intranet. In: Völkl, M., Schaffert, S. (eds.) Proceedings of First Workshop on Semantic Wikis From Wiki to Semantic at the 3rd Annual European Semantic Web Conference (ESWC), CEUR Workshop Proceedings, pp. 31–42 (2006), http://CEUR-WS.org/Vol-206/
36. Buffa, M., Crova, G., Gandon, F., Lecompte, C., Passeron, J.: SweetWiki: Semantic Web Enabled Technologies in Wiki. In: Völkl, M., Schaffert, S. (eds.) First Workshop on Semantic Wikis 'From Wiki to Semantic' at the 3rd Annual European Semantic Web Conference (ESWC), CEUR Workshop Proceedings, vol. 206, pp. 74–88 (2006),
    http://CEUR-WS.org/Vol-206/
37. Noy, N., Grosso, W., Musen, M.: Knowledge-Acquisition Interfaces for Domain Experts: An Empirical Evaluation of Protégé-2000. In: Proceedings of the Twelfth International Conference on Software Engineering and Knowledge Engineering (SEKE 2000). Knowledge Systems Institute, Chicago (2000)
38. Gruber, T.R.: A translation approach to portable ontology specifications. Knowledge Acquisition 5, 199–220 (1993)
39. W3C: Resource Description Framework (RDF) (2004)
40. W3C: Web Ontology Language (OWL) (2004)
41. Uschold, M., King, M., Moralee, S., Zorgios, Y.: The Enterprise Ontology. The Knowledge Engineering Review (Special Issue on Putting Ontologies to Use) 13(2) (1998),
    http://www.aiai.ed.ac.uk/project/enterprise/enterprise/ontology.html
42. Auer, S.: Powl: A Web Based Platform for Collaborative Semantic Web Development. In: Proceedings of 1st Workshop Scripting for the Semantic Web (SFSW 2005), Hersonissos, Greece (2005), http://www.ceur-ws.org/vol-135/paper9.pdf
43. Fuchs-Kittowski, F., Faust, D., Loroff, C., Reuter, P.: WiKo: Eine integrierte Wissens- und Kooperations-Plattform. I-Com 4(1), 12–19 (2005)

Towards a Virtual Environment for Regulated Interaction Using the Social Theatres Model

Ana Guerra[1], Hugo Paredes[1], Benjamim Fonseca[2], and F. Mário Martins[3]

[1] UTAD, Quinta de Prados, Apartado 1013, 5001-801 Vila Real, Portugal
anaguerra166@gmail.com, hparedes@utad.pt
[2] UTAD/CITAB, Quinta de Prados, Apartado 1013, 5001-801 Vila Real, Portugal
benjaf@utad.pt
[3] Universidade do Minho, DI/CCTC, Campus de Gualtar, 4710 Braga, Portugal
fmm@di.uminho.pt

Abstract. The last decade brought about several virtual communities spread all over the world composed of thousands of people with different ages and social, cultural and physical characteristics. These communities enable users to communicate and share information, often with the aim of achieving a common goal. Due to the vast diversity of users there's often a need to control the activities that occur inside the virtual environment to avoid inappropriate behaviors. However, this control can cause a feeling of digital surveillance, but an appropriate design of the interface can help minimize its impact, becoming an important success factor for the community. For this purpose we analyzed a selection of virtual environments, conducted a survey on users' preferences and analyzed the corresponding results. From these activities we defined a set of requirements to build a 3D interface for a regulated virtual environment.

Keywords: Interaction Environments, Social Interaction, User interfaces, Virtual Environments.

1 Introduction

The development of the Web empowered the emergence of social interaction environments where users, spread all over the world, interact with each other, represented either by a real or a virtual identity. These virtual environments comprise a growing number of heterogeneous users that establish different kinds of relationships, according to the community goals.

An important success factor for virtual environments is user interface, since it exposes the virtual environment functionalities and mediates the communication among users.

Depending on how the functionalities are exposed to the user and how he/she is informed on what is allowed to do, the user will feel more or less power over his/her actions. The lack of action independence can leave the user with a feeling of digital surveillance.

In this context, it is our aim to define and build a 3D interface to regulated social interaction environments, based on the Social Theatres model [1], taking into account

R.O. Briggs et al. (Eds.): CRIWG 2008, LNCS 5411, pp. 164–170, 2008.

the characteristics and preferences of the target audience. For this purpose we analyzed the features of a selection of virtual environments and conducted a survey to evaluate users preferences concerning virtual environments interfaces. From this analysis we defined a set of requirements that helped us to design a 3D interface for the regulated virtual environment.

This paper is organized as follows: first, it summarizes the main aspects of social interaction and the Social Theatres model; then it presents the analysis of selected virtual environments and the results of a survey on users' preferences; based on these studies we summarize the interface requirements and present the modelation of the 3D virtual environment; finally, we present some preliminary conclusions and define future directions.

2   Related Issues

Social interaction is defined as "the acts, the actions and the practices mutually oriented between two or more people"[2]. The author also refers that these actions or behaviors take into account the experiences or mutual intentions of the intervenient.

Sztompka, referred in [3], defines four types of social interaction: accidental, repeated, regular or regulated. Accidental social interaction occurs when the interaction was not planned, with low probabilities of being repeated. Moreover, the participants weren't previously aware of each other's existence. Repeated social interaction occurs when participants know each other beforehand. Regular interaction is similar but referring to interactions that are at least somewhat often, whereas repeated interaction takes place occasionally. Finally, regulated social interaction occurs when the interaction follows predefined rules that set the way users interact with each other. Summarizing, social interaction occurs when there is a relationship among two or more persons that are aware of each other's presence and act in conjunction to achieve some objective.

Each interaction's participant will act according also to the actions of the other participant. To facilitate this, a virtual environment must have immersive capabilities that allows the user to abstract from its surrounding and gives him or her presence awareness (of himself and the others). A high presence awareness feeling makes the user behave like if he or she is really inside the virtual environment [4]. Co-presence, also designated as shared presence or social presence, refers to the feeling of presence of other users inside the same environment, increasing the feeling of belonging to a community. Slater in [5] says that the feeling of being together, even by application-generated avatars, makes the user more participative.

As referred above, there are several types of social interaction, though not all adequate for a virtual community. A community with regulated interaction, previous knowledge of the users and assignment of specific roles, can achieve its goals more quickly, when compared with a community with accidental interaction.

Social Theatres is a model for coordination and regulation of social interaction in virtual environments. This model proposes social interaction in a virtual environment to be regulated based on the concept of theatre, with each user assuming a role following a well defined interaction workflow, regulated by a set of rules [1].

A study referred in [3] shows that in regulated environments users (particularly the most experienced ones) feel more digital surveillance as if the environment was not regulated. In this way, it's important that the interface help to minimize this feeling of surveillance to allow users to concentrate in interaction and the surrounding environment.

3 Virtual Environments Features

In [6] Pirkola e Mannien conducted a study in which they compared some social interaction environments, analyzing features like scalability, avatar characteristics, environment realism, communication among users or graphical user interface. Examples of the environments included in this study are Internet Relay Chat (IRC), WorldsAway and Quake.

Meanwhile, over the last years, several new environments appeared with advanced interfaces, more realistic graphics and the capability to accommodate an increasing number of users, allowing them to communicate in different ways. So, a new analysis is needed to characterize these new environments in several aspects, like environment realism, avatar features, available functionalities and immersive capabilities. For this purpose, we selected the virtual environments that we considered more important, according to their vast communities spread all over the world [7], the graphical quality of their interfaces and the number of functionalities provided. We were particularly interested in aspects like the integration in the community and age control. It was also analyzed the graphical aspect, namely aspects related with colors and menus. Also very important is the set of functionalities provided, focusing in communication among users, avatar assignment and customization, environment visualization, ways to move inside and between environments, and object behaviors.

Table 1. Main features observed in virtual environments

	Second life	**Active Worlds**	**There**	**Kaneva**	**Moove**
Avatar personalization	Yes	No	Yes	Yes	Yes
Navigation in the environment	Keyboard arrows, 2D menu	Keyboard arrows	Keyboard arrows	Keyboard arrows	Keyboard arrows
Navigation between environments	Walking, teleport and 2D menu	Walking, teleport and 2D menu	Walking, teleport and 2D menu	2D menu	Web Page
Chat field	Yes	Yes	No	Yes	Yes
Voice communication	Yes	Yes	Yes	No	Yes
Instructions to the user within the 3D environment	Yes	Yes	Yes	Yes	No
Import external objects to the environment	Yes	No	No	No	Yes
Avatar and object behaviors	Pre-defined, Programmable	Pre-defined	Pre-defined	Pre-defined	Pre-defined, Programmable
Avatar movements in the environment	Flying, running, walking, programmed by the user and predefined animations	Walking, running, and predefined animations	Walking, running, and predefined animations	Walking, dancing, and predefined animations	Walking and predefined animations
Sound	Yes	Yes	Yes	Yes	No

Despite its several differences, these environments also have several similarities. For example, all of them require that users must be registered and make control over their age. Another common characteristic is the need to install a specific client application or a browser plugin. The environments analyzed provide their functionalities through a 2D menu, enabling some customizations (like changing the user's perspective) and interaction with the objects present in the scenario. Finally, all the environments have private spaces and allow avatars to communicate using predefined gestures.

Table 1 summarizes the main features observed for the environments analyzed.

4 Survey on User Preferences

Beyond the study of the main virtual environments available nowadays we wanted to know the features most valuable for users, particularly those for 3D interfaces. For this purpose we conducted a survey on users preferences that involved a total of 192 respondents, divided into three groups with different characteristics, for the sake of heterogeneity. Two of these groups were composed of more experienced users: one group composed of undergraduate students from 3 programmes of the Computer Science area at University of Trás-os-Montes e Alto Douro (UTAD); and the other of undergraduate students from the Informatics programme (Computer Science area) at Instituto Politécnico de Leiria (IPL). A third group was composed of students living in an University Residence, studying distinct programmes at UTAD, with heterogeneous technological backgrounds.

The questions in the survey were divided in groups. The first inquired the user about its experience with computers and interaction environments. The results collected showed that the majority of users consider themselves equally experienced in both subjects, despite many of the students from the Computer Science area consider they are more experienced with computers in general than with interaction environments.

The second group of questions tried to define the user's preferences regarding graphical interfaces in general. Mainly, we observed that the users from Computer Sciences exhibited a preference for Web based 3D interfaces, while the others prefer Web based text chat.

The next group inquired about aspects of 3D visualization (graphical quality, immersion capabilities, waiting times, intuitive use) and functionalities provided to the user. Mainly, the students from the Computer Science area considered all the features very important and the other students considered it just important (these students only considered very important the waiting time).

The users were also inquired on their preferences regarding the environment representation: real, imaginary or hybrid environment. The majority of them preferred the hybrid representation. Another question was about the possibility of changing the appearance of the environment (colors, lighting) and the majority considered it important.

Another question was about user's representation in the environment: human representation with user's self characteristics; human representation with different characteristics; an animal; an object. Figure 1 depicts the results obtained and shows that the answers are not much different from group to group. The most chosen representation was the first one, human with user's self characteristics. However, a human representation with different characteristics was also chosen by many people.

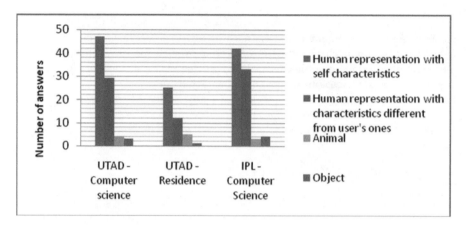

Fig. 1. User representation in the virtual environment

The way users want to control the movement inside the environment was also questioned and 4 possibilities were provided: command line, buttons, mouse and keyboard arrows. The results were similar in all the 3 groups and the majority preferred to control the movements with the mouse, being the second most chosen option the keyboard arrows.

An issue observed before the survey was that there are several ways to visualize the environment: first person, in which the user sees the environment just as in the real world; and top, front, back and side views, that enable the user to see the environment and the own avatar. We included in the survey a question regarding the view of the environment and we gave 3 options: first person, top and side views. The most chosen was the first person view, as can be seen in Fig.2, followed by a still considerable number of preferences by the top view.

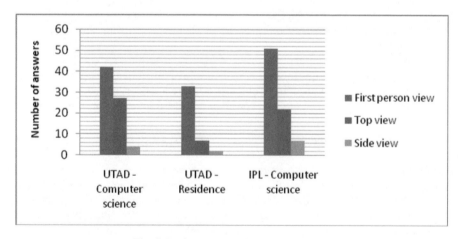

Fig. 2. Preferences on visualization mode

Regarding the way users prefer to receive instructions we provided 3 options: in 3D objects inside the environment, in an external 2D menu or preference for trying without any instructions. The preference of the majority relies on the first option and this result strengthens the notion of need for realism, with users giving particular importance to the immersive capabilities.

The last question in the survey aimed to analyze the user's preferences with respect to the functionalities available in the 3D environment. For this purpose we defined a set of functionalities that users must classify according to their importance for the success of the interface: the possibility of changing the visualization mode and the avatar characteristics, the avatar movement, the ability to interact with objects in the scenario and with other users, and teleporting capability. Generally, everybody considered all functionalities important, but while less experienced users classified all of them equally, more experienced users differentiated many of them and attributed higher values.

By analyzing the overall results, we can observe that generally the answers are very similar independently of the level of user's experience. Some results, like those related with the visualization mode and the user's representation, showed that a great number of persons use a virtual environment to communicate with other people, transposing his/her reality to it, assuming the own real life characteristics. However, a significant part of the users find in these environments an escape from their real lives, assuming one or more parallel identities.

5   Implementation

The analysis of existing environments and the results obtained in the survey enabled us to specify the main features of the virtual environment for regulated social interaction. Generally, it must be simple, attractive and intuitive, in order to reduce learning times.

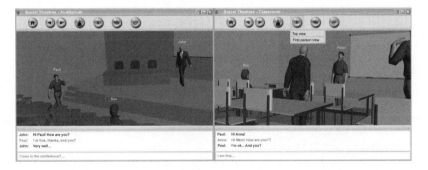

Fig. 3. 3D interfaces for the auditorium and the classroom

The 3D interface must represent a hybrid or a real environment, with high quality graphics. It also must be highly immersive, with the aid of resources like sound, notion of free space and depth. The user must be represented by a human-like avatar that has the ability to change its appearance and to move around (using the mouse and keyboard arrows). It is also important to interact with other avatars and objects present in the scenario. The operating instructions must be provided in objects present in the environment and there must be a 2D menu providing functionalities like changing the scene visualization and some visual aspects of the scenario (lighting, for example), the customization of the avatar.

Based on these requirements, some 3D environments were modeled according to two case studies presented in the Social Theatres model [3]: a class and a conference. For this purpose, we modeled the outside of a building where the activities occur, a lounge that serves as the entry to the building, a classroom and an auditorium. Figure 3 shows the 3D interfaces for the auditorium and the classroom.

6 Final Remarks

The user interface is an important element in a virtual environment, regarding communication and collaboration between users, as well as the interaction with the application. Its construction must take into account several technological and social factors and it requires an adequate research on the environment aim and user's profiles, preferences and needs.

For this purpose we defined a set of requirements based on the analysis of the main existing virtual environments and in the user's preferences collected through a survey in which participated 192 university students.

These requirements enabled us to start the construction of the environments for two case studies, a class and a conference. These environments include several 3D spaces (building, lounge, classroom and auditorium) and some auxiliary artifacts like the functionalities menu. nowadays, there's still some work to be done, namely the insertion and control of the avatars, as well as the integration with the Social Theatres Software Architecture.

References

1. Paredes, H., Martins, F.M.: Social Theatres: A Web-based Regulated Social Interaction Environment. In: Groupware: Design, Implementation, and Use, pp. 87–94. Springer, Heidelberg (2007)
2. Rummel, R.J.: The conflict helix, in Understanding Conflict and War. Sage Publications, Thousand Oaks (1976)
3. Silva, H.A.P.G.d.: Uma Arquitectura de Software Dinâmica para a Criação de Ambientes de Interacção Social Regulada na Web. PhD thesis. Departamento de Informática, p. 317. Braga, Universidade do Minho
4. Schroeder, R.: Social Interaction in Virtual Environments: Key Issues, Common Themes, and a Framework for Research. In: Schroeder, R. (ed.) The Social Life of Avatars, pp. 1–18. Springer, London (2002)
5. Mel Slater, A.S.: Meeting People Virtually: Experiments in Shared Virtual Environments. In: Schroeder, R. (ed.) The Social Life of Avatars, pp. 146–171. Springer, London (2002)
6. Pirkola, T.M.a.J.: Comparative classification of multi-user virtual worlds (1997), http://citeseer.ist.psu.edu/manninen97comparative.html
7. Monica, P.R.L.: Life beyond Second Life. CNNMoney.com (2007)

Addressing Co-text Loss with Multiple Visualizations for Chat Messages

Torsten Holmer[1], Stephan Lukosch[2], and Verena Kunz[3]

[1] Upper Austria University of Applied Sciences, School of Informatics, Communications, and Media, 4232 Hagenberg, Austria
`torsten.holmer@fh-hagenberg.at`
[2] Delft University of Technology, Faculty of Technology, Policy and Management, Systems Engineering Department, Jaffalaan 5, 2628BX Delft, The Netherlands
`s.g.lukosch@tudelft.nl`
[3] FernUniversität in Hagen, Department of Mathematics and Computer Science, 58084 Hagen, Germany
`verena.kunz@fernuni-hagen.de`

Abstract. In this article we address the problem of co-text-loss in chat communication, identify requirements for a solution, discuss related work and present a new approach for addressing co-text loss in text-based chats. We report about first experiences with our solution and give an outlook on future work directions. The core idea of our solution MuViChat (multiple-visualization chat) is to support multiple visualizations of referenced chat transcripts in which users can choose their preferred view. By enabling the recording and replaying of chat discussions and an extensible modular architecture we are supporting evaluation and further integration of advanced visualization concepts.

Keywords: chat tool, multiple visualizations, threading, chat transcript.

1 Introduction

Text-based conferencing systems, also known as chat systems, support written communication. They are widely accepted by users as they have low requirements concerning computing power or network bandwidth. Recently, text chats are often mixed with audio-chats. This leads to new problems and questions that have to be considered when communicating with each other, e.g. what has to be written and what has to be said or how is it possible to link textual information with audio information. However, this article focuses on text-based chat systems and users that use a text-based chat system to communicate with each other. Hence, we refer to text-based chat systems when talking about chat systems.

As chat systems enable computer-mediated interaction and communication, chat systems are part of the interdisciplinary research domain of computer-supported cooperative work (CSCW) and cooperative learning (CSCL). Additionally, chat systems are a research subject for communication science as well as linguistics. Therefore,

R.O. Briggs et al. (Eds.): CRIWG 2008, LNCS 5411, pp. 171–182, 2008.

there are several studies that deal with the linguistical particularities of chat communication and the special role of chat communication in the field between speech and writing (cf. [1], [2]).

There are several advantages of using chat systems for communicating with each other. Chat systems do not depend on a specific location. From a cost perspective, chat systems are a cheap form for communicating with each other. As users are not aware of what the communication partners are doing, it is not considered as unpolite to deal with other things in addition to chat with each other. The latter includes the participation in several chat-based communications. For a lot of users, communicating via chat systems is attractive because of the humorous and playful speech. Finally, compared to oral communication, most chat systems allow users to refer to previous communications and check the own or the communication partner's statements [3].

However, there are also disadvantages for communicating via chat systems. Smith et al. [4] identify five major problems for communication via chat systems:

(P1) Lack of links between people and what they say
(P2) No visibility of listening-in-progress
(P3) Lack of visibility of turns-in-progress
(P4) Lack of control over turn positioning (co-text loss)
(P5) Lack of useful recordings and social context

To what degree these problems occur or influence a chat-based communication is highly dependent on the characteristics and offered functionalities of the used chat system, P1, P2, and P3 mainly influence the communication dynamics. P4 and P5 mainly influence the presentation of the chat-based communication.

Current chat systems do not sufficiently address the co-text loss, i.e. P4. In this article, we will therefore especially tackle this problem by introducing a new chat system that offers different visualizations to overcome co-text loss. We first identify the requirements for such a chat system. For that purpose, we take a closer look on an example scenario and define our understanding of co-text loss. Based on the requirements, we review the state-of-the-art. Then, we present our chat-system MuViChat (multiple-visualization chat) which fulfills the identified requirements to tackle the co-text loss problem. Before concluding and giving an outlook on future work directions, we report on experiences and a preliminary evaluation of MuViChat.

2   Requirements Analysis

Standard chat systems visualize chat messages in the order in which they arrive at the server. As all chatters can simultaneously write and send messages all have the right of speech. Hence, the server defines the order of the messages as they appear in the chat window [1]. As result, the shown conversation is not linear, i.e. a chat message may not refer to the previous one but to another which has been posted earlier. This results in parallel intertwined conversation threads which may even discuss different topics. Due to these parallel threads a chat participant might not know to which previous chat messages a message refers and thus might not be able to follow and understand the conversation. Pimentel et al. [5] describe such a situation as co-text-loss. To illustrate this table 1 gives a prominent example for co-text-loss.

Table 1. Example for co-text loss [6]

Green:	Did you see that new Mel Gibson movie – I think it is called "Payback"?
Blue:	I saw the academy awards last night. Did you watch it?
Blue:	yep.
Blue:	It was very violent, but funny.
Green:	You saw it? You liked it?
Green:	How did it end up – who won?
Yellow:	I heard it was good.
Blue:	It was OK. At least Titanic didn't win everything.
Green:	I guess you can only be king of the world once.

Linguistics considers "co-text" as text which has been written immediately before and after a message and which is helpful or necessary to understand a message. According to Pimentel et al. [5], users which are looking for the co-text cannot follow the "conversation rhythm". Pimentel et al. [5] distinguish four different actions once a user detects co-text loss:

1. The chat participant searches for the co-text in the previous messages. If the co-text is quickly identified the conversation continues.
2. If the chat participant cannot quickly identify the co-text and continues the search, this takes time and causes the loss of "conversation rhythm" while the conversation continues.
3. If the chat participant stops searching and does not declare the co-text-loss, the participant might not understand the conversation anymore.
4. If the participant states co-text loss another participant may step in and help to understand the conversation. When the help of this participant is successful, the other participant may declare his/her understanding of the conversation.

Especially, when a lot of users participate in a chat, these users need a high degree of concentration and a high reaction rate, to disentangle the different conversation threads, avoid co-text loss, and to follow the conversation. Identifying the co-text loss in such a setting gets even more complicated when there is a huge time gap between the messages which belong together. When it is unclear which messages belong together, co-text loss can lead to ambiguities. Chat messages which simply consist of an answer like "yes" or "no" even increase the possibility of co-text loss.

Co-text loss is often also considered as "intention confusion" [6] when the recipient of a chat message is unclear or as "thread confusion" when it is unclear to which conversation thread a message belongs. Herring [3] considers co-text loss as "interactional coherence" while Pimentel et al. also call it "chat confusion" [7]. All agree that co-text loss occurs when it is not possible to identify the message to which another refers. A chat system should therefore help users to identify the messages which belong together and thereby allow users to follow the conversation without co-text loss.

For that purpose, users must be able to reference another chat message when posting a new chat message. This leads to the following requirement:

(R1) A chat system has to support users to reference a chat message when posting a new chat message.

The chat messages and the references between chat messages have to be visualized so that users can easily identify the messages which belong together. The visualization has the main goal to reduce to probability of co-text loss. The visualization has to take care that the relation between a chat message and the user which has posted the chat message is still obvious (cf. P1). Users should be able to choose between different visualization forms so that they can select the visualization form which suits them best in a certain context (discussing with one person vs. in a big group or arguing vs. brainstorming). By applying different visualizations in different contexts user studies can show which visualization overcomes the problem of co-text less most efficiently. This leads to the following requirement a chat system has to address when tackling the co-text loss problem:

(R2) A chat system has to offer multiple visualization forms for displaying chat messages and the references between chat messages.

In some cases, co-text loss might occur even when the requirement R1 and R2 are met by a chat system. To resolve the co-text loss subsequently to a chat conversation, users must be able to store the conversation and to replay the conversation. When replaying a conversation users should as well have the possibility to choose between the different available visualization forms. By offering such a replay mode, the chat system also addresses the problem of unfeasible chat protocols (cf. P5). For storing a chat conversation, a chat system should use a format which can also be read by the users and which can easily be used to import chat conversations which have been stored otherwise. This broadens the usability of the chat system as it then can be used to comprehend other chat conversations. Summing up, this leads to the following requirement.

(R3) A chat system has to allow users to store as well as replay chat conversations.

3   Related Work

In order to improve the interface of chat systems a lot of tools have been developed but most of them neglected the problem of co-text loss. In the following we will only mention those approaches which have addressed the requirements (R1)-(R3).

The systems ThreadedChat [4], HyperDialog [5,7] and ThreadChat [8] provide referencing functionality and visualize the chat transcript as an indented tree. New contributions are integrated into the tree structure and can appear in different areas of the screen especially when participants are discussing in parallel threads. The structure is clear and the co-text of a message is easily identifiable but in the process of communication some problems are likely to occur: if a message is not referenced by accident, it will be sorted into a new thread and messages can be missed because new messages appear in areas which are not visible on the screen.

These problems are tackled in systems like KOLUMBUS [9] and ConcertChat [10] by showing the transcript in linear order but providing a reference indicator at each message which shows a pointer to the co-text when selected. Thereby it is clear where new messages will appear and how the co-text of a message could be inferred. The drawback of these approaches is that they do not provide means for getting a visual overview of the discussion and its structure.

The system factChat [11] allows users to put the messages on a two-dimensional surface in proximity to their co-text. In addition it is possible to create references to messages which are not visible anymore on the screen (older messages fade out and seem to disappear in the background). Although is possible to review the transcript by using a timeline it is not possible to get a visual representation of the entire structure.

In summary there are many different approaches which address partially some problems related to co-text loss but none of them addresses all of the identified requirements, esp. multiple visualizations (R2).

4 Approach

Our approach aims at solving co-text loss in text-based chats. Therefore, it has in the first case to offer standard text chat functionality. Then, it has to offer functionality to include references between different chat messages and support a visualization of these references. In the following, we will present our approach and how it fulfils the identified requirements. Our approach is based on XMPP and therefore our chat client can be used with huge variety of different chat servers which already support this standard.

4.1 Referencing Chat Messages (R1)

At the content level, a chat message can refer to none previous messages, exactly one, or even multiple ones. A chat message does not refer to another chat message if the author starts a completely new topic. When a chat message references exactly one message, it is often an answer or comment to a previously posted question. Multiple references may exist, when a user, e.g., summarizes previous chat messages. The latter occurs quite rarely and would make the visualization and the comprehension of the visualized references more difficult. Like other chat systems that support references [4, 8, 9, 11], we will also focus on supporting references between two different chat messages only. ConcertChat [10] is a chat system which allows users to refer to multiple messages but up to now there are no reported experiences if and how users used this feature.

Another important question concerning the interaction with the chat system concerns whether users are allowed to post new messages without referencing another chat message. The experiences with the chat system HyperDialog have shown [5,7] that especially missing and wrong references were the reason for not being able to overcome the co-text loss by referencing functionality. Additionally, forcing users to reference other chat messages confines the conversation possibilities. Due to this, we do not force users to reference chat messages. Instead, we allow user to reference a chat message before or while creating a new chat message by selecting the referenced message from the displayed chat messages.

Experiences with the chat system KOLUMBUS [9] have shown that referencing is used if the message cannot be placed directly after the referred message. Users are willing to use the functionality if the value is directly visible. In most of our provided vizualisations the benefit of referencing is very clear and thereby should lead to increased usage of the referencing functionality.

4.2   Multiple Visualization Forms for Chat Messages and the References between Chat Messages (R2)

We designed our chat system so that users can select between different forms of visualizing chat messages as well as the references between chat messages. This allows users independently from each other to choose their favourite visualization alternative. Furthermore, all visualization alternatives offer further configuration possibilities to tailor the visualization even more to the users' preferences. The following list contains all of our visualization alternatives:

1. Classical visualization
2. List view with highlighting
3. Simple tree view
4. Tree view with highlighting time by fading colors
5. Tree view with highlighting time by layout
6. Sequential tree view

Our classical visualization alternative corresponds to the visualization of all major chat clients. We choose to integrate this variant for future experiments and evaluation because this variant allows us to compare the new forms of visualization with the standard case. Nevertheless, users can configure the font size, the font colour, and the background colour.

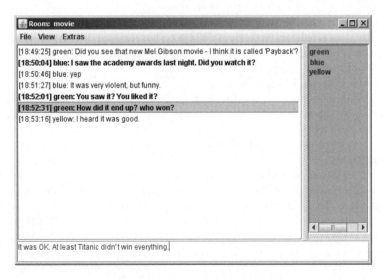

Fig. 1. List view with highlighting

The list with highlighting represents a first variant of the classical view. Here, chat messages are displayed in a list and are chronologically ordered. The newest chat message is highlighted in bold font. If this message references another message, this message is displayed in bold font as well. If this message references another one, this also highlighted etc. By this all referenced messages are highlighted on a path from the most actual message to the root message (cf. Figure 1).

Apart from these two textual visualization alternatives our chat systems offers different tree-based visualizations. All of these tree-based visualizations have in common that

- chat messages are shown as colored rectangles,
- references are shown as lines between these rectangles,
- each user is represented by a different color,
- chat messages can be selected by clicking on them with the left mouse button,
- selected chat messages are highlighted, and
- the size of the chat window can be changed arbitrarily.

Users can choose whether the view is automatically focussed on the most recent chat message. As the tree visualizations require more visualization space than the list-based alternatives, we included a bird's view which shows the complete chat conversation and highlights the part that is currently shown in the chat window. In total we support four different tree-based visualizations:

1. **Simple tree view**: This tree-based visualization ignores the time factor and simply orders the chat messages which reference each other in a tree (cf. Figure 2).
2. **Tree view with highlighting time by fading colors**: This visualization alternative considers the time at which a message was created by fading the color of the rectangle for older messages while new messages are displayed using a bright color (cf. Figure 3a).

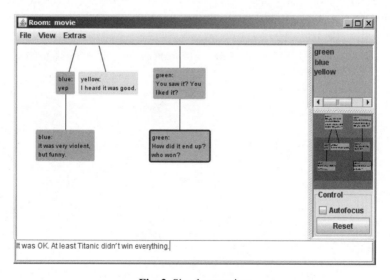

Fig. 2. Simple tree view

3. **Tree view with highlighting time by layout**: Apart from fading the color, time can also be considered by reflecting it in the distance between two messages. This variant increases the distance between two messages the more time has passed between them. In this visualization alternative, the y-axis corresponds to the time (cf. Figure 3b). This view is especially useful to identify nearly simultaneous messages as well as pauses in the communication flow.

4. **Sequential tree view**: When linearly mapping the time to the y-axis, a lot of visualization space can be wasted. The sequential tree view considers this observation and uses the y-axis to visualize the order of the messages. The distance between two messages can be configured by the user (cf. Figure 3c).

Providing different visualizations gives users the flexibility to explore them and to find out which suits them best for a certain task. We expect that the simple tree view is especially useful in situations where users want to see the structure of the discussion and to look for, e.g., the most complex threads or the longest linear sequence. If the discussion is still ongoing, users can see the most actual messages and current threads by visualizing time by lighter colors (Figure 3a). If users want to look at parts of the discussion in which the frequency of interaction has been very intense or in which have been pauses in the discussion, they can use the view with highlighting time by layout (Figure 3b). This allows users to understand the dynamics of the ongoing discussion. In our opinion, the sequential tree view (Figure 3c) is the most convenient one for actively participating in discussions because it shows the most actual messages at the bottom of the screen and makes it easier to follow multiple parallel discussions.

The list view with highlighting (Figure 1) can be used as "power user mode" because it has the same look & feel as traditional chat interfaces but users can reference and can see referenced messages inside their classical list view. This mode is very similar to other tools like ConcertChat [10] and KOLUMBUS [9] and allows direct comparisons with other visualizations instead of comparing the tools.

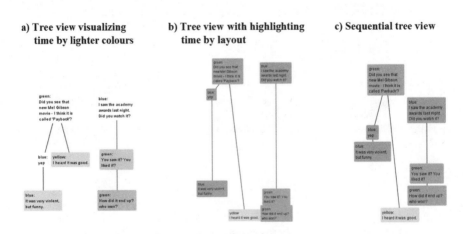

a) Tree view visualizing time by lighter colours   **b) Tree view with highlighting time by layout**   **c) Sequential tree view**

Fig. 3. Overview about the tree-based visualizations

4.3 Persistence and Replay of Chat Conversations (R3)

To reconsider chat conversations, users are able to store chat conservations in the XML-based format TreeML[1]. Thus, a conversation stored by our chat system can be read by users and as TreeML is based on XML with a publicly available DTD it is also simple to implement import filters for other formats. Once a conversation has been stored in this format, it can be imported via our chat system and replayed. Thereby, we implement the REPLAY pattern [12]. A chat conversation replay does not need any input fields or a user list. Instead the user interface for the replay contains a control panel for the replay (cf. Figure 4). In this control panel, users have the possibility

- to start the replay,
- to pause the replay,
- to stop the replay, and
- to vary the speed of replay between real time and a faster replay.

Fig. 4. Replay control panel

5 Experiences

In order to gain experience with the new interface and the supported interaction we conducted some informal experiments with students at the university campus in Hagenberg, Austria. By using MuViChat in 90-minute discussions with 18 active participants we could show that the concept worked for this group size (see Fig. 5). The referencing concept was immediately understood and used correctly. Participants tried the different visualizations and developed their own strategy to use them for their purpose. In order to browse the discussion, participants used the simple tree visualization (cf. Figure 2) or the tree variant shown in Figure 3a because in this visualization the distance between references is the smallest and therefore the under-standing of the conversation is simplified the most. For participating actively in the conversation and following ongoing threads the participants used the sequential tree (cf. Figure 3c). These preliminary results show that participants embraced the differ-ent visualizations and used them for their own purposes.

A further pilot study used the replay function for testing co-text loss while reading chat transcripts. In a 30-minute session the same transcript was given to groups of three individuals which had to read them and answer questions in between regarding references between messages. While with the tree visualization all participants could answers all questions correctly, using the classical list visualization participants had problems to identify all references and needed more time to identify them.

---

[1] TreeML is an XML-based format that can be used to store tree-based structures. It was devel-oped for the IEEE InfoVis Contest 2003. The document type definition is available at: http://www.nomencurator.org/InfoVis2003/download/treeml.dtd

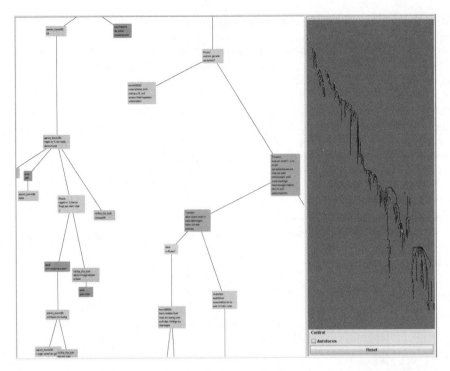

Fig. 5. Tree view with highlighting time by layout of a chat transcript with 18 participants

Although the identification of references seems to be better supported in tree visualizations the pilot study showed some drawbacks. Similar to the study of Smith et al. [4] the participants sometimes ignored messages because they appeared in different locations or the auto-focus function focused the view area on newest messages faster than participants were able to read the current message. These problems will be addressed in future studies and should be addressed in the next extensions of MuViChat. Therefore our next steps will go into two directions of research: further development of new features and more user studies.

An important feature which is missing in most chat tools is typing awareness, which indicates the typing activity of other users before their messages are posted into the chat transcript (cf. P3). Regarding the graphical structure of MuviChat we are thinking about an ACTIVITY INDICATOR [12] which appears when a user is referencing a message and thereby shows that there will be a reply to that message in the future. This helps to get an overview about the concurring activities in the chat room.

In the current implementation messages which were not referenced by accident could not be linked afterwards. In order to give better support for structuring this feature should be implemented.

Up to now the spatial layout of threads is the same for all participants although threads in which a user is actively participating should be more important than other threads. Bringing active threads to the center of the screen and letting older threads disappear into the background are personalized views which could help to stay focused on the subjective most interesting parts of a chat discussion.

All these ideas have to be tested in user studies in order to evaluate their role in referenced chat communication. Currently we are setting up a lab in order to investigate the effects of synchronous communication and cooperation. MuViChat will be one of the first tools we will use for studying effects of group size and awareness support.

6 Conclusions

MuViChat is both an innovative chat tool and an environment for experimenting and evaluating new variants of chat visualizations. By using widespread standards (Java and Jabber) MuViChat can be used on every java-capable computer with internet connection. Storing the log files in XML-format is useful for visualizing them in many different ways as well as using them for visualizations which we will be integrated in the future. Thereby it is possible to make controlled experiments between different visualizations with least effort in order to check their usefulness and readability. Because of the open source license and the modular architecture of MuViChat interested researchers can develop their own visualizations and integrate them in order to share their ideas.

In the future we will use MuViChat for conducting experiments regarding the usefulness of different visualizations for reading chat transcripts as well as for supporting different kinds of discussions. These experiments will clarify under which conditions the provided functions are useful, in which way successful discussions can profit from the functionality and which new visualizations have to be developed to overcome the identified deficits and dangers of co-text loss.

References

1. Garcia, A., Jacobs, J.: The Eyes of the Beholder: Understanding the Turn-Taking System in Quasi-Synchronous Computer-Mediated Communication. Research on Language and Social Interaction 32(4), 337–367 (1999)
2. Werry, C.: Linguistic and Interactional Features of Internet Relay Chat. In: Herring, S. (ed.) Computer-Mediated Communication: Linguistic, Social and Cross-Cultural Perspectives, pp. 47–63. John Benjamins, Amsterdam (1996)
3. Herring, S.: Interactional Coherence in CMC. Journal of Computer-Mediated Communication 4(4) (1999)
4. Smith, M., Cadiz, J., Burkhalter, B.: Conversation trees and threaded chats. In: Proceedings of the 2000 ACM Conference on Computer supported cooperative work, pp. 97–105. ACM Press, New York (2000)
5. Pimentel, M., Fuks, H., Lucena, C.: Co-text Loss in Textual Chat Tools. In: Blackburn, P., Ghidini, C., Turner, R.M., Giunchiglia, F. (eds.) CONTEXT 2003. LNCS (LNAI), vol. 2680, pp. 483–490. Springer, Heidelberg (2003)
6. Vronay, D., Smith, M., Drucker, S.: Alternative interfaces for chat. In: Proceedings of the 12th Annual ACM symposium on User interface software and technology, pp. 19–26. ACM Press, New York (1999)

7. Pimentel, M., Fuks, H., Lucena, C.: R-U-Typing-2-Me? Evolving a chat tool to increase understanding in learning activities. International Journal of Computer-Supported Collaborative Learning 1(1), 117–142 (2006)
8. Holmer, T., Wessner, M.: Tools for Cooperative Learning in L^3. In: Ehlers, U., Gerteis, W., Holmer, T., Jung, H. (eds.) E-Learning Services in the Crossfire. Pedagogy, Economy and Technology, pp. 138–152. Bertelsmann, Bielefeld (2006)
9. Holmer, T., Kienle, A., Wessner, M.: Explicit Referencing in Learning Chats: Needs and Acceptance. In: Nejdl, W., Tochtermann, K. (eds.) Innovative Approaches for Learning and Knowledge Sharing, pp. 170–184. Springer, Berlin (2006)
10. Mühlpfordt, M., Wessner, M.: Explicit referencing in chat supports collaborative learning. In: Proceedings of the 2005 Conference on Computer Support for Collaborative Learning: Learning 2005: the Next 10 Years!, Taipei, Taiwan, May 30 - June 04, 2005, pp. 460–469 (2005)
11. Harnoncourt, M., Holzhauser, A., Seethaler, U., Meinl, P.: Referenzierbarkeit als Schlüssel zum effizienten Chat. In: Beißwenger, M., Storrer, A. (eds.) Chat-Kommunikation in Beruf, Bildung und Medien: Konzepte - Werkzeuge - Anwendungsfelder, Stuttgart, pp. 161–179 (2005)
12. Schümmer, T., Lukosch, S.: Patterns for Computer-Mediated Interaction. John Wiley & Sons, Ltd., Chichester (2007)

Modifiers for Quality Assurance in Group Facilitation

G.L. Kolfschoten[1], P. Grünbacher[2], and R.O. Briggs[3]

[1] Delft University of Technology, Faculty of Technology, Policy and Management, Department of System Engineering, Jaffalaan 5, 2628BX, Delft, The Netherlands
G.L.Kolfschoten@tudelft.nl
[2] Johannes Kepler University, Linz, Austria
paul.gruenbacher@jku.at
[3] University of Nebraska at Omaha, Department of Business administration, Institute for Collaboration Science, Roskens Hall Room 512B, Omaha, NE 68182 USA
RBriggs@mail.unomaha.edu

Abstract. One of the key tasks of a facilitator is to assure the quality of the knowledge outcome created in the collaborative effort. To manage the quality of the knowledge generated facilitators need to follow along in the communication, and to judge the quality of contributions, decisions and consensus. When facilitators or group members detect quality deficiencies, facilitators have the difficult task of making interventions to support the group in improving the quality of their output, without interrupting the flow of the process. In this paper we will present a framework for quality assessment and a toolbox with flexible interventions that can be added to the process design on the fly, as soon as quality deficiencies are detected. The toolbox exists of a set of conditional adjustment interventions, which can be used to enhance the selected facilitation techniques to guard for quality.

Keywords: Facilitation, Quality Assurance, ThinkLets, Collaboration Engineering, Modifiers.

1 Introduction

Collaboration is a critical skill and competence for organizations in the knowledge economy. However, collaboration is challenging, and groups can benefit from collaboration support (technology and facilitation) to collaborate more effective and efficiently [1]. In order to help groups achieve their goals, facilitators design a sequence of activities, possibly supported by tools such as Group Support Systems. Next facilitators instruct the group in each of these activities, guide discussions, and capture outcomes. Last they evaluate the collaboration process to see if the group achieved their goal. Goal achievement can be expressed in terms of quantity of outcomes, quality of outcomes and satisfaction of the users. One of the key tasks of a facilitator is therefore to assure the quality of the knowledge outcome created in the collaborative effort [2]. Grünbacher et al. [3] indicated that quality assurance can be performed at three phases in the process; during the design of the process through validations of the process design, during the execution of the process and post-process. In this paper we will focus on techniques for quality assurance during the process design and execution.

R.O. Briggs et al. (Eds.): CRIWG 2008, LNCS 5411, pp. 183–195, 2008.

Some quality assurance can be initiated and performed by group members. In this paper we will focus on quality assurance by facilitators. Facilitators are not necessarily domain experts, and therefore quality assurance can be a difficult task. To manage the quality of the knowledge generated facilitators need to follow along in the communication, and judge the quality of contributions, decisions and consensus. The dynamic nature of this process makes it difficult to perform rigorous quality analyses, and yet, this is the phase in which the expertise of the entire group can be used to ensure quality. Facilitators rely on criteria for analyzing quality, based on which they monitor the input and communication of the group members. The required quality of the output can vary largely for each session. While tasks and objectives can be vague during a kick-off meeting, they should be precisely defined in a planning effort. When facilitators or group members detect quality deficiencies, facilitators have the difficult task to make interventions that support the group in improving the quality of their output, without interrupting the flow of the process. In this paper we will present a framework for quality assessment and a toolbox with flexible interventions that can be added to the process design on the fly, as soon as quality deficiencies are detected. The toolbox will exist of a set of conditional adjustment interventions, which can be used to enhance the selected facilitation techniques to guard for quality.

In the remainder of this paper we first present a background on facilitation techniques and quality of knowledge or information, which will result in a framework for the quality assessment and the framework for interventions. Next, we will present a toolbox for quality assurance. Finally, we will discuss implementation and implications, and offer suggestions for future research.

2   Quality Goals of Collaboratively Generated Knowledge

The relative importance of different quality aspects will always depend on the specific goal of the collaborative effort. Wand et al [5, 6] performed an analysis on data quality dimensions and found the following most used dimensions for data quality: Accuracy and precision, Reliability, Timeliness and currency, Completeness, Consistency and Relevance. Another frequently used set of quality dimensions is called the 3C's = correctness, completeness, consistency. Both show some overlap, and we will discuss each of these dimensions to determine which ones are most appropriate to qualify collaborative outcomes. As described in the Goal-Question-Metric approach [4], the specific meaning of a quality aspect should be defined using a set of metrics that are established based on a goal and precise questions to derive performances indicator for this goal.

Correctness shows some overlap with both accuracy and precision. They indicate the quality of the information in terms of truthfulness, or the degree to which data actually reflect the reality they model. Precision means that the information is formulated in sufficient detail to ensure that it is well understood. Accuracy and Correctness can be seen as synonyms of truthfulness from a rationale perspective. Reliability has to do with the probability of errors in the data. When knowledge is created in collaborative settings it has to do with the trust in the expertise, ability or sincerity of the contributors. Reliability can be seen as a sub-dimension of correctness from a more subjective perspective. Timeliness and Currency are less relevant in a collaborative setting when the knowledge is assessed at the moment it is created. However, as for reliability people can question whether the knowledge brought into the meeting is accurate and up to date.

Correctness is an important quality factor in collaborative outcomes. In order for knowledge to be regarded as a group result it should be created in shared understanding. When the knowledge created is formulated correctly, it is more likely that it is understood by the group member and other users of the output of the group effort. When information is incorrect, or incorrectly interpreted it might cause misunderstanding and incorrect decisions. *We define correctness as conforming to fact or common understanding or in accordance with an accepted standard.*

Completeness can be an important factor, but in case of complex, knowledge intensive tasks, completeness is also linked to parsimoniousness; over-completeness can cause information overload which requires increased cognitive effort to converge into meaningful knowledge. In some cases completeness is less relevant and time-consuming. Completeness can be assessed at two levels, completeness of individual contributions and completeness of the resulting group outcome, a set of contributions. *We define completeness as containing all necessary elements. We define parsimoniousness as containing only necessary, non superfluous elements.*

Consistency refers to an absence of contradictions and conflicting information in the data set or with respect to information quality requirements. Consistency is especially important when concepts are to be compared or evaluated. For instance, when group members need to compare the feasibility of solutions they need to be described at the same level of detail. In case problems are to be assessed no solutions should appear in the list of problems. *We define consistency as similar in structure or abstraction level.*

Relevance is also important when the collaborative effort focuses on a specific scope, resources should not be wasted on contributions that are outside this scope, for instance on problems that are beyond the scope of the team to solve. *We define relevance as instrumental to achieving the goal.*

Given the analysis above we will consider the following quality dimensions that can be used to characterize the quality of the outcomes of knowledge intensive collaboration:

- Correctness
- Consistency
- Completeness
- Parsimoniousness
- Relevance

The set of dimensions is not necessarily complete and emphasis in quality dimensions might differ in specific domains. However, this set offers a first basis to identify a set of key facilitation interventions for quality assurance.

3 Facilitation Interventions

While facilitation for long has been regarded an art and skill learned through experience and apprenticeship, recently facilitation skills and methods have become more transferable in the tradition of Collaboration Engineering, an approach to design and deploy collaboration processes in organization [7, 8]. In Collaboration Engineering processes are designed as a sequence of activities with facilitation interventions to guide the group to this process and to structure and focus their effort to goal achievement. Facilitation techniques in Collaboration Engineering are documented as design patterns called thinkLets. The conceptual design of a thinkLet exists of a set of rules [9, 10]. Each rule describes for a role and action that needs to be performed using a capability under some set of constraints to restrict those actions. There are two types of rules; instruction interventions; rules to instruct actions, and adjustment interventions; conditional rules that are only communicated when some discrepancy is found between the intended action or outcome, and the actual action or outcome as a result of the instruction [11]. Besides these rules thinkLets contain a script to convey the rules and a description of their effects and implications. An example of a thinkLet can be found in articles by Vreede et al [10, 12, 13].

ThinkLets are used as building blocks to design a collaboration process. We can divide thinkLets in basic thinkLets and variations. Variations on thinkLets are created though the addition of rules, or by replacing rules with slightly different rules. We call these additional or replacement rules modifiers. One of the key functions of modifiers is to improve the quality of output [14]. Modifiers can be used to create a variation on thinkLets for different patterns of collaboration. Santanen and also Shepherd showed that such variations can create significant effects on the quality of the outcomes of a collaborative effort [15, 16].

Modifiers describe an intervention that has a predictable effect on the pattern of collaboration and the outcome of the collaborative effort described in the thinkLet. A metaphor to understand the modifier concept is a virus. In one sense, a modifier is to a thinkLet as a virus is to a cell. A virus, invokes predictable changes on the way the cell performs. In like manner, modifiers can be applied to thinkLets to create predictable *changes* in the patterns of collaboration the thinkLet invokes. An example of a modifier is the use of anonymity. When brainstorming or voting in a group process, it can be useful to work anonymously. A modifier can thus be used in combination with various thinkLets to create different patterns of collaboration. Using anonymity creates a predictable change in the process; it removes certain barriers to share information and critique, resulting in more open discussion and higher participation [17].

Since quality assurance interventions are used when the quality of the output shows deficiencies, they are likely to exist mainly of adjustment interventions. This means that based on a quality deficiency, a modifier is chosen to assess and resolve the quality deficiency. Quality deficiency can be identified in different ways. The group can judge each contribution on certain criteria, the facilitator can monitor input based on one or more criteria, or the contributions can be organized or compared to check completeness or consistency. The adjustment interventions for quality exist of a detection method, a way to find quality deficiencies, and a way to improve the quality. In many cases focusing attention on the quality deficiency will be sufficient; a group member will improve the quality of the contribution, for instance a group member can edit his contribution to improve its quality or can re-classify a contribution when inconsistently organized. In some cases the specific quality criteria for contributions or modifications need to be further discussed and defined. In some cases consensus should be created about the definition of quality criteria.

Besides making adjustment interventions when quality deficiencies are detected, there are some ways to assure quality directly. These methods involve the direct emphasis or enforcement of the constraints for contribution. For instance, when brainstorming for solutions for a particular problem, it helps to explicitly define the problem and to keep the problem statement visible for reference. When people define risks in terms of cause and effect, a template might be useful to formulate risks accordingly. Also in some case it might be useful to train group member in an exercise to create contributions that meet specific quality requirements.

4 Quality Assurance Framework

Given the quality dimension we identified and the methods to make adjustment interventions and to emphasize contribution constraints, we can now create a framework. This framework will help to give an overview of quality assurance modifiers. In the framework we distinguish methods for prevention of quality deficiencies, for discovery of such deficiencies and for fixing of such deficiencies. The framework offers an overview of modifiers based on the thinkLet book [18]. The thinkLet book offers an overview of best practices of GSS use, captured as design patterns [10]. The thinkLet library has been validated in various ways; the thinkLets have been documented best practices, and were reviewed by professional facilitators [19]. They have been recognized in GSS session transcripts [20], and they have been used in various case studies [21-24]. Based on the thinkLet library we distilled modifiers that can be used for quality assurance. Some of these modifiers were described independently; others are distilled parts of other thinkLets. It is a first step to identify quality modifiers, and can be extended with more techniques for quality assurance.

Table 1. Framework for quality assurance

	Correctness	Consistency	Completeness contribution	Completeness combined contributions	Parsimoniousness	Relevance
Deficiency prevention						
Input template	x	x	x		x	x
Constraint emphasis	x	x	x		x	x
Constraint re-emphasis	x	x	x		x	x
Contribution training	x	x	x		x	x
Definition of quality criteria	x	x	x		x	x

Table 1. (*continued*)

One up	x					x
Deficiency discovery						
Input monitoring	x	x	x	x		x
Quality evaluation	x		x		x	x
Comparison		x		x		x
Step by step checking/Review	x	x	x		x	x
Cluster check		x		x		
Deficiency Fixing						
Expert fixing	x	x	x	x	x	x
Chauffeur fixing	x	x	x	x	x	x
Parallel fixing	x	x	x	x	x	x

5 Quality Assurance Modifiers

We identified modifiers for deficiency prevention, deficiency discovery and deficiency fixing. These are listed in table 1 above. On the left hand side, the quality
modifiers are listed. In the table an 'x' marks the ability to use the quality assurance
technique to evaluate this particular quality criterion. Each modifier is briefly described below. The rule contains various <parameters> indicated between '<>'. These
need to be instantiated with actual criteria, roles or other parameters to use the modifier in a specific context.

5.1 Deficiency Prevention

Input Template
To ensure the quality of a contribution, we can also create an input template. This
template is used to support participants in framing their contributions in a more complete, precise, consistent, parsimonious or relevant way. While this rule can be used as
a guideline or instruction presented by the facilitator, it can also be enforced though
technology restriction, e.g. multiple fields need to be completed in order to add a contribution.

```
<role> make their contributions using the <input tem-
plate>, incomplete templates are not added to the list.
```

Instantiation example: To add a risk, please specify cause and effect using the template. You can only submit your template when it is completed.

Constraint Emphasis
A constraint is a restriction to the activity, for instance a brainstorming topic is a restriction to the scope of the brainstorm. Emphasizing the constraints of activities will
sharpen the input of participants. This is a very straight-forward technique, but in
practice it might be overlooked. In any collaborative activity it is useful to have a
clear description or definition of the constraints for the activity (brainstorm question,
evaluation assignment, etc.) visible throughout the activity.

```
Clearly state the <constraints> to the activity, and
keep them visible during the activity.
```

Implementation example: The scope of the brainstorm is explained and its definition is projected on the wall during the brainstorm.

Constraint Re-emphasis

This technique has been known as 'one minute madness' [18]. It is used to verify if participants understood the quality criterion used. After a first round of contributions, modifications, etc., the facilitator can stop the group and go through the list, verifying if the contributions have the required quality. If not, the facilitator can point out quality deficiencies, and explain again, the quality criteria.

```
<role> explains the <quality criteria> to the group
giving examples based on <quality deficiencies> in the
set of contributions from the group.
```

Instantiation example: We are looking for problems, please refrain from indicating solutions, for instance here it says, we do not have a monthly meeting, but this indicates a desired solution, not the problem that calls for this solution.

Contribution Training

This technique works the same as the contribution template, but first we let participants fill out the template for an example case, then we perform a "clarification of criteria" to bring contributions of insufficient quality to their attention and we train participants to create contributions that meet the quality criteria. Then we start working on the real topic. This technique is a combination of modifiers.

```
<role> make their contributions using the <input tem-
plate>, incomplete templates are not added to the list.
```

```
<role> consider each <contribution> and judge it based
on the <quality criterion/criteria>. If the <contribu-
tion> does not meet the <criterion>, ask group members
to highlight or improve it.
```

```
<role> look at the highlighted <quality deficiencies>,
can we improve or remove them from our list? <role> im-
proves the contribution or <role> removes the contribu-
tion.
```

Instantiation example: Please brainstorm risks with respect to events that happened when you went to work this morning using the input template. We will go through your contributions and refine them to ensure that you formulate precise and consistent risks, than we will start working on the risks involved in this project.

Defining quality criteria

To ensure that participants understand and accept the criteria used for contributions it is useful to define them in discussion with the group. Creating definitions is simplest based on a proposal for the definition, for instance from a dictionary or other independent source. Some quality criteria are relevant for the entire process; others are only relevant for a specific step in the process.

```
<role> proposes a definition for the <quality con-
straint> for contributions, ask the group members if
they understand and agree with the definition.
```

Instantiation example: Let's define a task as an activity, with a specific objective. For instance, "organize a workshop to create the basis for our strategy report". Does everyone agree with this definition and understand how to precisely contribute tasks?

One Up

This technique is used to get contributions that excel on a specific quality criterion. The criterion is used to encourage participants to come up with contributions that excel on this criterion. To increase the effect, participants can be asked to also present the argument that describes why the new contribution is better than the current set on this specific criterion.

```
<role> add a <contribution> that is better than the
<set under consideration> on <criterion>.
```

Instantiation example: Please choose solutions that are cheaper than the ones we already selected for further consideration.

5.2   Deficiency Discovery

Input Monitoring

Input monitoring can be done by the facilitator, a chauffeur or an expert. In a brainstorming activity, unclear, imprecise, inconsistent or incomplete, or irrelevant contributions can be highlighted for improvement during convergence; items that are added or selected to the smaller set can be monitored and refined during organizing; items that are related can be monitored. During evaluation and consensus building monitoring is possible, but it is preferable to use high quality input for these activities.

```
<role> monitor input of the group members, if the <con-
tribution> does not meet <quality criterion> ask <role>
to edit the contribution to improve its quality.
```

Instantiation example:
The facilitator monitors the input of the group members, if the solutions are not relevant to the problem under discussion, ask the contributor to edit the contribution to improve its relevance to solve the problem.

Quality Evaluation

When monitoring is difficult, for instance in large groups, or when the facilitator is busy with other interventions, a quick quality evaluation can be made after the input is generated or collected. In this case participants or people in a specific role are asked to highlight contributions that are not meeting one or more of the quality criteria. For this modifier a domain specific checklist could be used.

```
<role> consider each <contribution> and judge it based
on the <quality criterion>. If the <contribution> does
not meet the <criterion>, ask group members to high-
light or improve it.
```

Instantiation example: please read through the list of problems and mark each problem that is unclear to you, please clarify contributions that you made and that are highlighted.

Comparison

Some quality criteria require comparison of contributions, for instance to judge if contributions are consistent, or to see if the entire set of contribution is complete. For this purpose we need to judge the set of contributions rather than the individual contributions.

```
<role> consider the set of <contributions> to find
<quality deficiency> and highlight it.
```

Instantiation example: Please go trough the list of tasks and find and highlight tasks that can be decomposed in sub-tasks to meet a similar abstraction level as the rest of the list.

Step by Step Checking

Once a list of contributions is created, it can still contain quality deficiencies. These deficiencies can be highlighted using comparison or quality evaluation, and need to be resolved. There are two options to resolve quality deficiencies; when the quality deficiency is caused by a lack of relevance, the contribution can be removed. When the quality deficiency has to do with completeness, consistency, precision or parsimoniousness, the deficiency can be resolved by modifying the contribution.

```
<role> look at the highlighted <quality deficiencies>,
can we improve or remove them from our list? <role> im-
proves the contribution or <role> removes the contribu-
tion.
```

Instantiation example: please look at the highlighted problems, these are unclear. Please either clarify your contribution, or mark it with a cross. These will be removed by the facilitator.

Cluster Check

With a cluster check the completeness of the set of contributions is evaluated, and based on this the relevance and consistency of the set of contributions. The method is to create labels for clusters, and to move the contributions in the best fitting cluster. Based on this the group gets an overview of clusters for which only a few contributions are generated. When the clusters cover the scope of the problem, the group can see if the contributions are relevant to one or more aspects of the scope and the consistency of the contributions within one cluster can be verified.

```
<role> move <contributions> to the <cluster> indicating
part of the <scope> it belongs to.

<role> verify if each <cluster> in the <scope> is suf-
ficiently covered.
```

Instantiation example: Cluster the ideas to the four key focus points, let's see if we covered these focus points sufficiently.

5.3 Deficiency Fixing

Expert Fixing

In this modifier an expert reviews the deficiencies and fixes them based on his/her expertise. In some cases it is important to show the participants the revisions made and to verify acceptance of those revisions.

```
<expert> look at the highlighted <quality deficiencies>
and revise the <contribution> to resolve the quality
deficiency.
```

Instantiation example: The group has highlighted criteria that have a deviating abstraction level in comparison to the overall list. The expert is asked to resolve these issues by either decomposing or generalizing the contributions.

Chauffeur Fixing

In this modifier a chauffeur fixes deficiencies in discussion with the group. Each deficiency is discussed and the revision is recorded by the chauffeur.

```
<chauffeur> record the revision of the <contribution>
to resolve the <quality deficiencies> as suggested by
the group.
```

Instantiation example: The group has highlighted risks that do not meet the contribution template. The group discusses each risks and sharpens it's formulation to fit the template.

Parallel Fixing

In this modifier participants get the right to edit contributions that contain a quality deficiency. It is possible to allow participants to only edit their own contributions with a deficiency.

```
<Participant> look at the highlighted <quality defi-
ciencies> and revise the <contribution> to resolve the
quality deficiency.
```

Instantiation example: the group has highlighted ideas that are outside the scope of the solution space. The participants get a chance to rephrase their idea to better fit the scope. They can edit their ideas to create a better fit, in parallel.

6 Implementation and Implications

The techniques described above can be used in different combinations with a variety of thinkLets. Each of these techniques can be implemented though facilitation. However, facilitation might not be easy, especially when large amounts of contributions are processed or created during the activity. Therefore, support tools should be developed. Tools that enable participants to highlight quality deficiencies, tools that restrict contributions or modifications using a specific template or tools that automatically detect specific qualities of a contribution, or quality deficiencies. In domain specific settings, self-learning tools can be built to detect quality deficiencies.

Quality assurance, especially if quality means clarity, precision and thus some level of shared understanding among the group members of the contributions, will require time and considerable cognitive effort. The more constraints to a contribution, the more cognitive effort it requires to create or modify the contribution. Therefore introducing quality criteria needs to be considered carefully. For instance in a creative process, to many contribution constraints can harm creativity; too much cognitive effort is spend on the careful construction of the contribution, which leaves too little cognitive capacity for the processing of stimuli that trigger creativity [25].

The same problem can occur for the facilitator. When too many quality constraints need to be verified it might be better to split the verification process in several steps to work on various quality constraints. For instance, a group can first rephrase contributions to improve precision, and then check for consistency or relevance. While it would be theoretically possible to check all three quality criteria at once, it might be challenging practically, and splitting the task might even be more efficient, and result in a higher quality result.

It is also very important to consider the need for certain qualities, and more important the need to establish this quality with the group. For instance, it might not be necessary to create precision or consistency with the group. In such a case one group member could modify the contributions to improve their quality and send the result for verification and approval to the other group members. However, when creating a code of ethics for instance, precision completeness might be important to establish as a group as this will also increase support for the code and consistency in its execution. After all, if a code of ethics is interpreted differently, it is of little use in guiding behavior.

7 Conclusions

This paper presents a framework for quality assurance in collaborative activities. Based on the framework we offered an overview of techniques that can be used as a variation on thinkLets to ensure quality during different collaborative activities that create different patterns of collaboration. Furthermore, we presented guidelines for the instantiation and implementation of these quality assurance techniques.

We predict that all quality interventions will increase cognitive effort of the task, but when combined well, and taken into account the additive effect of the cognitive load imposed by each quality constraint, techniques can be designed that make optimal use of the cognitive capacity of the group, creating a high quality outcome given the time and effort availed for the task.

Based on this quality framework we can examine the effect of these techniques on the quality of collaborative results. The framework offers both the interventions to improve quality, and the intended effect on quality (increased precision, consistency, etc.). Also, it would be useful to compare these effects with the usefulness of the results and satisfaction with the results. Last, it would be interesting to measure the effect of these quality interventions on cognitive effort of the task, and in comparable conditions of group and task on the efficiency of the task.

The techniques presented here can be used by facilitators to design and modify their interventions in groups to ensure specific qualities of the collaborative result.

They can also be used by designers of group support systems to create tools that support the group and the facilitator in quality assurance. Last the framework can be used as a research model to study the effect of these quality interventions.

References

1. Fjermestad, J., Hiltz, S.R.: A Descriptive Evaluation of Group Support Systems Case and Field Studies. Journal of Management Information Systems 17, 115–159 (2001)
2. Kolfschoten, G.L., Rouwette, E.: Choice Criteria for Facilitation Techniques: A Preliminary Classification. In: Seifert, S., Weinhardt, C. (eds.) International Conference on Group Decision and Negotiation. Universtatsverlag Karlsruhe, Karlsruhe (2006)
3. Grünbacher, P., Halling, S.B., Kitapci, H., Boehm, B.W.: Integrating Collaborative Processes and Quality Assurance Techniques: Experiences from Requirements Negotiation. Journal of Management Informaiton Systems 20, 9–29 (2004)
4. Basili, V., Caldiera, G., Rombach, H.D.: The Goal Question Metric Approach. Encyclopedia of Software Engineering, pp. 528–532. John Wiley & Sons, Inc, Chichester (1994)
5. Wand, Y., Wang, R.Y.: Anchoring Data Quality Dimensions in Ontological Foundations. Communications Of the ACM 39, 86–95 (1996)
6. Wang, R.Y., Storey, V.C., Firth, C.P.: A Framework for Analysis of Data Quality Research. IEEE Transactions on Knowledge and Data Engineering 7, 623–640 (1995)
7. Briggs, R.O., de Vreede, G.J., Nunamaker Jr., J.F.: Collaboration Engineering with ThinkLets to Pursue Sustained Success with Group Support Systems. Journal of Management Information Systems 19, 31–63 (2003)
8. de Vreede, G.J., Briggs, R.O.: Collaboration Engineering: Designing Repeatable Processes for High-Value Collaborative Tasks. In: Hawaii International Conference on System Science, IEEE Computer Society Press, Los Alamitos (2005)
9. Kolfschoten, G.L., Briggs, R.O., de Vreede, G.J., Jacobs, P.H.M., Appelman, J.H.: Conceptual Foundation of the ThinkLet Concept for Collaboration Engineering. International Journal of Human Computer Science 64, 611–621 (2006)
10. de Vreede, G.J., Briggs, R.O., Kolfschoten, G.L.: ThinkLets: A Pattern Language for Facilitated and Practitioner-Guided Collaboration Processes. International Journal of Computer Applications in Technology 25, 140–154 (2006)
11. Kolfschoten, G.L., van Houten, S.P.A.: Predictable Patterns in Group Settings through the use of Rule Based Facilitation Interventions. In: Kersten, G.E., Rios, J. (eds.) Group Decision and Negotiation conference. Concordia University, Mt Tremblant (2007)
12. de Vreede, G.J.: Collaboration Engineering: Current Directions and Future Opportunities. In: Seifert, S., Weinhardt, C. (eds.) International Conference on Group Decision and Negotiation. Universitatsverlag Karlsruhe, Karlsruhe (2006)
13. de Vreede, G.J., Briggs, R.O.: ThinkLets: Five Examples Of Creating Patterns Of Group Interaction. In: Ackermann, F., de Vreede, G.J. (eds.) Group Decision & Negotiation, La Rochelle, France, pp. 199–208 (2001)
14. Kolfschoten, G.L., Santanen, E.L.: Reconceptualizing Generate ThinkLets: the Role of the Modifier. In: Hawaii International Conference on System Science. IEEE Computer Society Press, Waikoloa (2007)
15. Santanen, E.L., de Vreede, G.J.: Creative Approaches to Measuring Creativity: Comparing the Effectiveness of Four Divergence ThinkLets. In: Hawaiian International Conference on System Sciences. IEEE Computer Society Press, Los Alamitos (2004)

16. Shepherd, M.M., Briggs, R.O., Reinig, B.A., Yen, J., Nunamaker Jr., J.F.: Social Comparison to Improve Electronic Brainstorming: Beyond Anonymity. Journal Of Management Information Systems 12, 155–170 (1996)
17. Valacich, J.S., Jessup, L.M., Dennis, A.R., Nunamaker Jr., J.F.: A Conceptual Framework of Anonymity in Group Support Systems. Group Decision and Negotiation 1, 219–241 (1992)
18. Briggs, R.O., de Vreede, G.J.: ThinkLets, Building Blocks for Concerted Collaboration. Delft University of Technology, Delft (2001)
19. Briggs, R.O., de Vreede, G.J., Nunamaker Jr., J.F., David, T.H.: ThinkLets: Achieving Predictable, Repeatable Patterns of Group Interaction with Group Support Systems. In: Hawaii International Conference on System Sciences, IEEE Computer Society Press, Los Alamitos (2001)
20. Kolfschoten, G.L., Appelman, J.H., Briggs, R.O., de Vreede, G.J.: Recurring Patterns of Facilitation Interventions in GSS Sessions. In: Hawaii International Conference on System Sciences. IEEE Computer Society Press, Los Alamitos (2004)
21. Boehm, B., Grünbacher, P., Briggs, R.O.: Developing Groupware for Requirements Negotiation: Lessons Learned. IEEE Software 18 (2001)
22. Grünbacher, P., Halling, M., Biffl, S., Kitapchi, H., Boehm, B.W.: Integrating Collaborative Processes and Quality Assurance Techniques: Experiences from Requirements Negotiation. Journal of Management Information Systems 20, 9–29 (2004)
23. de Vreede, G.J., Koneri, P.G., Dean, D.L., Fruhling, A.L., Wolcott, P.: Collaborative Software Code Inspection: The Design and Evaluation of a Repeatable Collaborative Process in the Field. International Journal of Cooperative Information Systems 15, 205–228 (2006)
24. Bragge, J., Merisalo-Rantanen, H., Hallikainen, P.: Gathering Innovative End-User Feedback for Continuous Development of Information Systems: A Repeatable and Transferable E-Collaboration Process. IEEE Transactions on Professional Communication 48, 55–67 (2005)
25. Santanen, E.L., de Vreede, G.J., Briggs, R.O.: Causal Relationships in Creative Problem Solving: Comparing Facilitation Interventions for Ideation. Journal of Management Information Systems 20, 167–197 (2004)

Linking to Several Messages for Convergence: A Case Study in the AulaNet Forum

Mariano Pimentel[1], Hugo Fuks[2], and Carlos J.P. Lucena[2]

[1] Department of Applied Informatics
Federal University of State of Rio de Janeiro (UNIRIO)
Av. Pasteur, 458, CCET, Urca, Rio de Janeiro – 22290-240 RJ, Brazil
pimentel@unirio.br
[2] Department of Informatics
Catholic University of Rio de Janeiro (PUC-Rio)
R. Marquês de São Vicente, 225, Gávea, Rio de Janeiro – RJ, 22453-900, Brazil
{hugo,lucena}@inf.puc-rio.br

Abstract. In this article, a piece of research on the development and use of a discussion forum is presented. In an online course, it was identified that one of the problems of the educational use of discussion forums is the high number of messages that go unanswered. To attenuate this problem a convergence mechanism was elaborated: links to messages beyond the replied one. In the case study conducted in two course editions, it was verified that the use of links decreased the number of unanswered messages.

Keywords: LMS, Discussion forum, convergence, links.

1 Introduction

This article presents a case study on the use and development of the discussion forum tools that are available in some learning management systems: Moodle [1], Black-Board [2], AulaNet[3] etc. Research on the development and use of forum tools in the educational context has been on the increase [4].

As presented in Section 2 of this article, forum tools have typically structured the discussion in a tree format, implying in a diverging development of the discussion as the breadth of the tree tends to increase. Diagram tools, on the other hand, have structured the discussion in a graph format that favors the convergence of the discussion but is more difficult to use and not as widely adopted.

As addressed in Section 3, according to a certain discussion dynamics conducted through the forum in a given online course, a discussion is considered good when learners discuss more among themselves than directly reply to the questions posted. In that course, it is considered that a deeper tree level and a low percentage of leaves are indicators of a good discussion. A lower leaf percentage indicates that fewer messages were left unanswered, i.e., more messages were used in the conversation chain.

As discussed in Section 4, one strategy to prevent a high number of unanswered messages proposes the use of Diagram tools, which makes possible having a graph structured discussion. However, learners have difficulties in using such tools. The

R.O. Briggs et al. (Eds.): CRIWG 2008, LNCS 5411, pp. 196–203, 2008.
© Springer-Verlag Berlin Heidelberg 2008

solution proposed in this paper is the use of an enriched hierarchical structure: beyond the father, links to previous messages in the forum can also be established.

According to the case study analyzed in Section 5 links were effectively established decreasing the number of unanswered messages, as it was desirable in the discussion dynamics of that course. The conclusion and future works of this research are presented in Section 6.

2 Conversation Structuring

In a communication tool messages are generally organized in one of the following structures: in a list (linear), tree (hierarchical) or graph (map) [5] – Fig. 1. Although the list is a particular case of a tree, and a tree is a particular case of a graph, none of the structures is always better than the others.

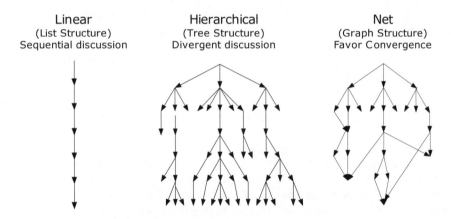

Fig. 1. Models of Discussion Structuring

In the list structure, typically used in e-mail and list discussion tools, explicit relationships among messages are not established. A message can relate to the text of other previous messages, but if there are too many relationships the reader will have difficulty in identifying them and getting a general vision of the discussion. The messages are listed according to the date they were received and can be rearranged according to other parameters such as the sender's name or the message's heading. The linear organization is propitious for communication where the chronological order is more important than the eventual relationships among the contents of messages, as it is the case of memos, bulletins and news.

In the tree structure, typically used in forum tools, participants choose which messages they wish to reply to, thus creating ramifications of the discussion. This discourse structure promotes the topical organization of the discussion: it favors the chaining of messages about a given subject in a single branch and the separation of messages into different branches about different subjects. In a tree the only explicit relationship is between the father message and the children messages. If a message makes a reference to a sister or grandmother message or to a message located in another branch of the tree, this reference will not be explicit in the structure of the discourse. The hierarchical organization is propitious for the visualization of the width and depth of the discussion. A problem with this structure is the difficulty to converge a discussion [6] since the conversation proceeds into divergent lines and the breadth of the tree tends to increase.

The graph structure is used when it is necessary to express relationships that are more complex than message hierarchy. It is useful when seeking convergence of a discussion or negotiation, taking decisions, seeking consensus, or when a high degree of structure in the dialog records is desired, as in the joint construction of semantic webs or in the study of concepts and their relationships. One of the first tools to deal with a discussion using the graph structure was gIBIS [7, 8, 9]. The discussion is structured on the basis of the IBIS - Issue Based Information System [10] - which proposes the categorization of messages into Question, Position and Argumentation. The QuestMap tool [11] is an evolution of gIBIS based on argumentative processes for project decision-taking that presents a larger set of categories. With the potential to express more complex relationships comes also the potential to generate a relationship mess that can hinder the understanding of the discussion. There are frequent reports on the difficulties encountered in using these tools that indicate the need for better training on their adequate use [12]. To try to decrease this problem some tools make use of rules that restrict the number or type of links among messages.

3 Problem: A High Number of Unanswered Messages

The research presented in this article was prompted by the identification of a problem in the use of the forum in an online course: the high number of unanswered messages. The course in question, ITAE (Information Technologies Applied to Education), is a course offered by the Department of Informatics at PUC-Rio that has been conducted totally at a distance through the AulaNet environment since the second semester of 1998 [13, 14]. Amongst the activities carried out in this course learners participate in seminars to discuss issues regarding the course's topics. The seminar is conducted in the Conference service, the AulaNet environment's discussion forum tool. The discussion is initiated through the Conference service with the "Seminar" message after which 3 messages using the "Question" category are chained. After that, learners start the discussion sending messages using the "Argument", "Counter-argument" or "Clarification" categories and establishing chaining with previous messages.

The form of the resulting discussion tree supplies indications on the quality of the discussion [4, 15, 16]. For example, in the tree presented in Fig. 2.b the average level was very low: 2,1. A barren discussion took place in which practically all learners answered the questions presented in the seminar directly, without discussing each other's ideas. The objective of the educational activity is to promote discussion among the learners themselves so that the forum will not become merely a questionnaire to be answered by all learners in the class. It can be observed that in the tree in Fig. 2.b, 75% of the messages were not answered (high percentage of leaves). When sending a new message, practically no learner took into consideration what the other participants had already said about the questions under discussion. Comparatively, the tree in Fig. 2.a presents a much fruitful discussion resulting in a smaller leaf percentile and a higher average depth level.

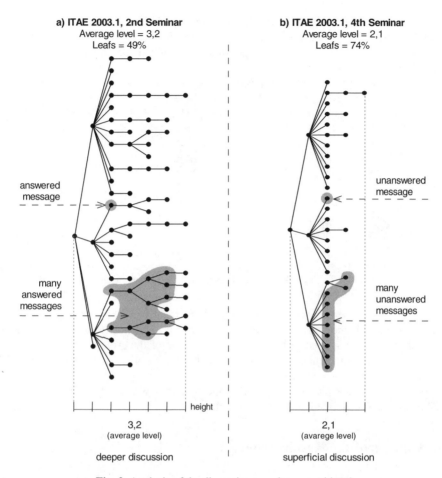

Fig. 2. Analysis of the discussion tree: leaves and level

Throughout several editions of the ITAE course it can be observed that the percentage of leaves is generally between 52% and 62%. In this course, a discussion that results in a tree with a percentage of unanswered messages above 62% is considered inadequate. The average depth level of the tree is inversely moderately correlated with the percentage of leaves: the more unanswered messages, the lower the depth of the discussion, which is also considered inadequate for the educational activity conducted within the course's forums.

4 Solution Proposed: Enriched Hierarchical Structure

To decrease the number of unanswered messages, the solution proposed was the use of an enriched hierarchical structure. Using this structure, as illustrated in Fig. 3, the linking with any other previous message can also be established besides the hierarchical association reply mechanism. This structure does not reproduce the structure used in the Graph Structured tools since the spanning tree [17] is kept as the main structure. The goal is to provide a mechanism that keeps the typical message hierarchy of a discussion forum but also makes it possible to establish the multiple associations typical of the graph structured tools.

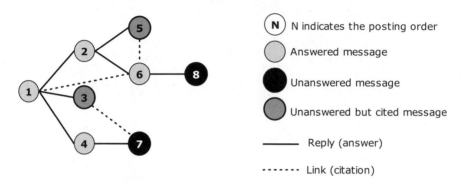

Fig. 3. Enriched Hierarchical Structure

Using the enriched hierarchical structure there is an increase in associations that promotes the reduction in the number of unanswered and not-cited messages. This structure was implemented in the Conference service of the AulaNet [18]. When replying a message the sender may also establish links to other messages resulting in the enriched hierarchical structure.

To investigate the use of this solution the new tool was used in two editions of ITAE course, as described and analyzed in the next section.

5   Analysis of the Results from a Case Study

The hypothesis put forward is that linking mechanism will be used by the forum's participants and thus few messages will remain unanswered and not-cited. The tool implemented was used in the ITAE 2006.1 and 2006.2 editions (first and second semesters of 2006).

As shown in Fig.4, In the ITAE 2006.1 edition, 9 learners participated and sent a total of 251 messages during the 8 seminars of the course. In the 4 first seminars the tool with hierarchical structure was used and 52% of the messages were left unanswered. In the 4 last seminars, the enriched hierarchical structure was introduced and the learners used it in 58% of the messages, resulting in only 30% of unanswered and not-cited messages. Without the linking mechanism 53% of leaf-messages would have occurred in the 4 last seminars. In effect, its use reduced the number of unanswered and not-cited messages.

Similar results occurred in the ITAE 2006.2 edition; the use of the enriched hierarchical structure also decreased the percentage of unanswered and not-cited messages: from 56% it dropped to 34%.

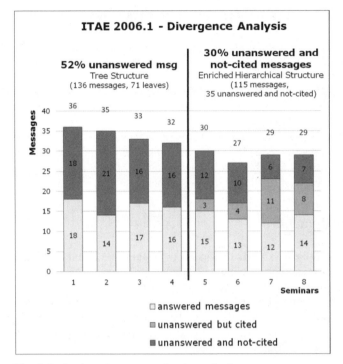

Fig. 4. Reduction of unanswered and not-cited messages percentage

It should be pointed out that perhaps in those seminars the links were established only because that was requested by the mediators (teachers-tutors). There is the intention to investigate in future works if these mechanisms would be used spontaneously, as a means of expression or as a way to converge the discussion, as opposed to being only a demand imposed by the mediators.

It should also be pointed out that in this research one assumes that the participants established the links coherently. However, there may have been a random use of the links without the adequate understanding of this mechanism, aiming only at satisfying the mediators' requests. Further research must be carried out to overcome the limitations here identified.

6   Conclusion

In this article a research on the development of the enriched hierarchical structure forums was presented. It had been identified that in the forum sessions in an online course many messages had been left unanswered – on the average, more than half of the messages had not been chained into the discussion. This problem was ameliorated with the use of the linking mechanism implemented in the Conference service of the AulaNet environment, as described in this article.

The use of this mechanism was studied during the seminars of two editions of the ITAE course. It was verified that the participants succeeded in using the mechanism implemented and this resulted in a significant reduction in the number of unanswered and not-cited messages.

The new tool, developed in the context of this research, opens possibilities for the elaboration of new educational dynamics. For example, in future studies the forum could be initiated with the use of the hierarchical structure, propitiating divergence, and in the final phase ended with the use of the enriched hierarchical structure. In the final phase it might be appropriate to add new categories of messages such as "Synthesis" and "Conclusion".

One future implementation to be investigated should make it possible to add comments to the links so the reader could understand the reason for establishing such link. Another modification to be investigated should make it possible to specify which excerpt of the body of the linked message is being quoted.

Acknowledgments

The AulaNet project is partially financed by Fund. Pe. Leonel Franca and by the Ministry of Science and Technology through its Program Multi-Agent Systems for Software Engineering Project (ESSMA) grant n° 552068/2002-0. It is also financed by individual grants awarded by the National Research Council to: Carlos José Pereira de Lucena n° 300091/2003-6, Hugo Fuks n° 301917/2005-1. Carlos José Pereira de Lucena and Hugo Fuks also receive grants from FAPERJ project "Cientistas do Nosso Estado."

References

1. Moodle, http://www.moodle.org
2. Blackboard, http://www.blackboard.com/
3. AulaNet, http://groupware.les.inf.puc-rio.br
   http:// www.eduweb.com.br
4. Gerosa, M.A., Pimentel, M., Fuks, H., Lucena, C.J.P.: No Need to Read Messages Right Now: Helping Mediators to Steer Educational Forums Using Statistical and Visual Information. In: Proceedings of th 2005 Computer Supported Collaborative Learning, pp. 160–169. International Society of the Learning Sciences, Taiwan (2005)
5. Gerosa, M.A., Pimentel, M.G., Fukś, H., Lucena, C.J.P.: Analyzing Discourse Structure to Coordinate Educational Forums. In: Lester, J.C., Vicari, R.M., Paraguaçu, F. (eds.) ITS 2004. LNCS, vol. 3220, pp. 262–272. Springer, Heidelberg (2004)
6. Stahl, G. WebGuide: Guiding Collaborative Learning on the Web with Perspectives. In: J. Interactive Media in Education, (1), 1-53 (2001)
7. Conklin, J., Begeman, M.L.: gIBIS: A tool for all reasons. J. American Society for Information Science 40(3), 200–213 (1989)
8. Conklin, J., Begeman, M.L.: gIBIS: A hypertext tool for team design deliberation. In: ACM Conference on Hypertext, pp. 247–251. ACM, New York (1987)
9. Conklin, J.: Dialog Mapping: Reflections on an Industrial Strength Case Study. In: Kirschner, P.A., Shum, S.J.B., Carr, C.S. (eds.) Visualizing Argumentation: software tools for collaborative and educational sense-making, pp. 116–136. Springer, London (2003)

10. Kunz, W., Rittel, H.: Issues as elements of information systems. Technical Report S-78-2 (1970)
11. QuestMap, http://www.gdss.com
12. Kanselaar, G., Erkens, G., Andriessen, J., Prangsma, M., Veerman, A., Jaspers, J.: Designing Argumentation Tools for Collaborative Learning. In: Kirschner, P.A., Shum, S.J.B., Carr, C.S. (eds.) Visualizing Argumentation: software tools for collaborative and educational sense-making, pp. 51–74. Springer, London (2003)
13. Fuks, H.: Groupware Technologies for Education in AulaNet, vol. 8(3&4), pp. 170–177. Wiley Interscience, Hoboken (2000)
14. Fuks, H., Gerosa, M.A., Lucena, C.J.P.: The Development and Application of Distance Learning on the Internet. Open Learning - The Journal of Open and Distance Learning 17(1), 23–38 (2002)
15. Fuks, H., Assis, R.L.: Facilitating Perception on Virtual Learningware-based Environments. The Journal of Systems and Information Technology 5(1), 93–113 (2001)
16. Gerosa, M.A., Fuks, H., Lucena, C.J.P.: Analysis and Design of Awareness Elements in Collaborative Digital Environments: A Case Study in the AulaNet Learning Environment. Journal of Interactive Learning Research 14(3), 315–332 (2003)
17. Diestel, R.: Graph Theory, 3rd edn. Graduate Texts in Mathematics, vol. 173. Springer, Heidelberg (2005)
18. Saramago, F.: Suporte Computacional para Discussão Estruturada em Rede: um Estudo de Caso com a Ferramenta de Fórum do Ambiente AulaNet. Master's Thesis, Department of Informatics at Catholic University of Rio de Janeiro (2007)

Understanding Methodological Differences to Study Convergence in Group Support System Sessions

Alanah Davis[1], Victoria Badura[1], and Gert-Jan de Vreede[1,2]

[1] Institute for Collaboration Science
College of Information Science and Technology
University of Nebraska at Omaha
6001 Dodge Street, Omaha, Nebraska 68182
{alanahdavis,vbadura,gdevreede}@mail.unomaha.edu
[2] Delft University of Technology
Faculty of Technology, Policy, and Management
The Netherlands

Abstract. While previous research has studied the process of group divergence extensively, little studies have been published regarding the process of group convergence. Therefore, this research answers a call for more research on group convergence to establish a better understanding of this critical group process. The goal of this study is to answer a methodological question: Does the idea set that a group processes during a convergence activity supported with a Group Support System have to be pre-defined or should it be generated by the group itself? The results of our lab experiment suggest that the results of the convergence process are not significantly affected by whether or not participants use pre-defined ideas or generate their own. This finding has implications for experimental design in collaboration research and participant selection and session design within organizations.

Keywords: Convergence, divergence, brainstorming, thinkLets, collaboration engineering, group support systems, collaboration, facilitation.

1 Introduction

When groups come together to brainstorm solutions for problems they begin with the process of divergence and coming up with as many solutions as possible. Once the initial brainstorming has taken place, groups need to reduce and clarify the ideas until they reach a decision about the next steps for the group to take. This process of reduction and clarification is referred to as convergence. Convergence has been defined as moving from "having many concepts to a focus on and understanding of a few deemed worthy of further attention" [1]. As mentioned, this convergence commonly follows divergence as the ideas generated from the divergence step are then condensed into a subset of categories or concepts for future group activities [2, 3].

Previous research has argued that there is a lack of research and guidance on how convergence activities can be best structured in groups [4, 5]. Therefore, this research presents a foundation experiment for better understanding group convergence in

R.O. Briggs et al. (Eds.): CRIWG 2008, LNCS 5411, pp. 204–216, 2008.
© Springer-Verlag Berlin Heidelberg 2008

Group Support System (GSS) sessions. The goal of this study is to determine whether the idea set that the convergence activity works with has to be pre-defined or should be generated by the group itself. The purpose of this study is to understand the methodological difficulty of the two alternatives. Overall, we want to see which starting point is the best, or if there is a difference at all. The answer to this question can have important implications: If we find that groups that converge with a pre-defined set of brainstorming ideas achieve comparable results to groups that converge on their self-generated ideas, then we can perform future convergence studies using pre-defined idea sets. This will not only enhance our experimental control, but it will also allow us to contain the length of experimental tasks that groups have to be engaged in. However, if we find that groups that converge a pre-defined set of brainstorming ideas as a starting point perform worse than groups that converge on their self-generated idea sets, then we must conclude that experimental convergence studies must include both brainstorming and convergence activities in the same task.

Thus, the outcomes from this research contribute to the theoretical understanding of convergence techniques and their design, as well as assist facilitators who use GSS in order to improve their processes and outcomes. Specifically, our study contributes to research in that we provide guidance for experimental studies in convergence or other patterns of collaboration that use brainstorming data as a starting point. Additionally, we propose a set of measures that can be used to analyze convergence tasks. In terms of a practical contribution, we present guidance for facilitators in situations where not all group members can meet synchronously, thus meaning that some group members may have to work with other people's contributions.

The remainder of this paper is structured as follows: The next section introduces convergence as well as presents the convergence performance measures used for this research. The subsequent sections present the research method as well as the analysis and results. The final section presents a conclusion with implications and areas for future research.

2   Background

2.1   Convergence and Convergence thinkLets

Convergence was originally considered to be one of the five basic patterns of collaboration for modeling collaboration processes [6]. However, later research in the field of Collaboration Engineering (CE) and the patterns of collaboration updated the patterns to include: generate, reduce, clarify, organize, evaluate, and build consensus [7]. In this new taxonomy, the patterns of reduce and clarify replace convergence and represent the process of a group moving from many ideas to a smaller list of ideas. Here, we refer to the concepts of 'reduction' and 'clarification' with the overarching concept of 'convergence.'

The general goal of convergence is to reduce a group's cognitive load in order to better focus on the key concepts which need to be addressed [5]. In many cases, a secondary goal of convergence is to establish a shared meaning of each of the concepts for the group members [5]. Research has found that the combination of these two goals make convergence a challenging and time consuming process for most groups [3]. Therefore, in order to address the challenges with convergence, collaboration engineers have developed a number of thinkLets which can be used in order to facilitate group convergence [5]. A thinkLet describes all the information required for a facilitator to create a predictable, repeatable, pattern of collaboration (i.e., such as convergence) among people working together toward a joint goal [6]. In other words, a thinkLet is a repeatable facilitation technique that leads to a particular pattern of collaboration or combination of patterns. Traditionally, thinkLets have been referred to as existing best practices for facilitation, specified in terms of the tool, the tool configuration, and the facilitation script [8]. A better understanding of process of convergence will enable collaboration engineers to evaluate the various convergence thinkLets and potentially design better thinkLets.

In this research, we attempt to improve our understanding of convergence by specifically evaluating a new convergence thinkLet called FocusBuilder [5]. Davis, de Vreede, and Briggs [5] describe the thinkLet of FocusBuilder as follows:

> "All brainstorm ideas are divided into as many subsets as there are participants. Each participant receives a subset of brainstorm ideas and is tasked to extract the critical ideas. Extracted ideas have to be formulated in a clear and concise manner. Participants are then paired and asked to share and combine their extracted ideas into a new list of concise, non-redundant ideas. If necessary, the formulation of ideas is improved (i.e., the pairs focus on meaning, not merit). Next, pairs of participants work together to combine their two lists into a new list of concise, non-redundant ideas. Again, the formulation of ideas is improved if necessary. The pairing of lists continues until there are two subgroups that present their results to each other. If necessary, formulations are further improved. Finally, the two lists are combined into a single list of non-redundant ideas."

This thinkLet is used in addressing our overall goal of this study. As mentioned earlier, with this research we want to determine whether the idea set that the convergence activity works with has to be pre-defined or should be generated by the group itself. Specifically, we want to understand the methodological implications and considerations of the two alternatives. In order to identify the differences we develop and present a number of convergence performance measures in the following section.

2.2 Performance Measures

Previous research from Davis, de Vreede, and Briggs [5] derived a set of ten performance criteria for convergence thinkLets. This set of performance criteria included *results oriented* criteria including 1) speed, 2) level of comprehensiveness, 3) level of shared understanding, 4) level of reduction, and 5) level of refinement of outcomes [5]. Each of these performance criteria is important for evaluating convergence activities. First, speed or efficiency is important because if more productive brainstorming techniques are devised, their effectiveness should not be offset by overly time-consuming convergence techniques. In some cases, the results of a convergence thinkLet can be considered successful if they are reached quickly. Second, the level of comprehensiveness is relevant as one of the primary goals of convergence is to pare down the output of brainstorming or other activities to a smaller more manageable group. However, this goal may be compromised if during convergence important ideas are lost. Successful comprehensive convergence activities eliminate redundant or less important ideas while keeping and refining the critical ones. Another goal of convergence is the level of shared understanding. The shared context helps the group focus on their context of interest and allows for higher group performance. Shared understanding provides the group members with the ability to discuss, evaluate, and re-propose solutions before advancing, which is important for establishing shared meaning. In some cases it is necessary for a group to reach a minimum level of shared understanding from a convergence thinkLet in order for the process to be considered a success. Fourth, the level of reduction is important, since a primary goal of convergence is to eliminate duplicate ideas it is important to determine which processes (i.e., thinkLets) do this and to what degree. The level of refinement of outcomes is the last results oriented performance criteria. In many group tasks, the results of a convergence activity must represent the intermediate or final deliverables that are reported back to the task owner. This implies that the group has to produce polished, refined outcomes. Therefore, in some cases, the results of a convergence thinkLet can be considered successful if they are sufficiently refined, i.e., in a final version and not draft form.

Additionally, the following *process or experience oriented* performance criteria were established, including 1) acceptance by participants, 2) ease of use for facilitator, 3) ease of use for participants, 4) satisfaction with thinkLet by facilitator, and 5) satisfaction with thinkLet by participants [5]. For our study we developed measures for only two of this second set of criteria. We did not include the process or experience oriented criteria of acceptance by participants, ease of use for facilitator, or satisfaction with the thinkLet by facilitator because as the researchers we acted as the facilitator and therefore could not measure these criteria. However, we did consider ease of use for participants to be important because if participants do not find a thinkLet easy to use they might not be able to complete their task. Secondly, we did consider satisfaction with thinkLet by participants. Researchers have argued that if participants are dissatisfied with a GSS experience, they are less likely to participate in future such efforts [see e.g., 9]. Since GSS workshops can be perceived as a sequence of thinkLets [10], participants' satisfaction with the thinkLets employed in a process is critical. Therefore, for a convergence thinkLet to be considered successful, the participants should be satisfied with the process of the thinkLet and its outcomes.

We began the development of our performance measures by defining each of the criteria (see Table 1). We then developed four survey questions for each of the criteria, which were could be evaluated on a scale of one through seven (1=Strongly Disagree and 7=Strongly Agree). Before beginning our experiment we ran three tests on the final instrument[1], relying on other collaboration engineers that we had access to.

---

[1] Please contact the authors to request a full copy of the performance measure instrument.

Table 1. Definitions of Performance Criteria

Criterion	Definition
Speed/Efficiency	The extent to which the outcomes of the convergence process were put together in a timely and efficient manner.
Level of Comprehensiveness	The extent to which the outcomes of the convergence process include or do not include all concepts that need further consideration in order for the group to successfully complete a task.
Level of Shared Understanding	The extent to which the outcomes of the convergence process include or do not include all concepts that need further consideration in order for the group to successfully complete a task.
Level of Reduction	The extent to which the outcome of the convergence activity are smaller than the input for the convergence activity (i.e. divergence thinkLets).
Level of Refinement of Outcomes	The extent to which the outcomes of the convergence process resemble the final version of what will be reported from the workshop.
Ease of Use for Participants	The extent to which participants found the ThinkLet easy to use.
Satisfaction with thinkLet by Participants	Satisfaction with Process: An affective arousal with a positive valence on the part of an individual toward the convergence process. Satisfaction with Outcomes: An affective arousal with a positive valence on the part of an individual toward the outcomes of the convergence process. [11]

3   Research Method

Our research was set up as a controlled experiment where one half of the participants relied on pre-defined ideas as the basis for the convergence activity and the other half of the participants generated the ideas on their own. However, our experiment is exploratory in that we do not begin with any predefined hypotheses regarding how the differences will impact the performance measures. Our overall purpose is to use this foundation experiment to develop a better understanding of group convergence in GSS sessions. Laboratory experimentation is appropriate for our research as it is the most popular method for evaluating GSS research [12]. Additionally, previous research has suggested that a carefully crafted experiment, such as this one, could be very insightful for assessing the performance of convergence thinkLets, in this case FocusBuilder [5].

3.1   Experiment Task

The task used for this experiment centered on the lack of parking spaces on campus and possible solutions. To begin, participants were informed that the work done in the sessions would be brought to the attention of the vice chancellor's office of academic and student affairs. Participants were then instructed to produce a finalized list of possible solutions for what to do about the lack of parking spaces which they would be happy submitting to the vice chancellor's office. This type of parking problem has been used successfully in various other research experiments due to the high relevance to subjects and use in other studies [e.g., 13].

3.2 Study Participants

The participants for this study included 64 undergraduate students from five different classes. They ranged in age from 18 to 49, with 47 of the participants male and 17 female. Participants worked in four person groups, with a total of 16 groups overall. Gallupe et al. [14] suggests four member groups are suitable for this type of research. The participants were motivated to participate because they were interested in the collaboration technology and the task, described above, addressed a campus issue in which they had a vested interest: the participants' university was undergoing extensive construction projects that limited the availability of parking.

3.3 Experiment Procedure and Process

Participants used the tool *GroupSystems™* and were led through the task with the primary researcher as the facilitator and technographer and the other researchers serving as observers.

The independent variable of the study is the process design with eight groups generating the brainstorming ideas on their own and then converging on those ideas. Then the second set of eight groups relied on those pre-defined ideas from the other participants for the basis of their convergence activity. Therefore, eight groups diverge/generate and then converge and eight groups just converge starting from the brainstorming results from the first eight groups. The groups that began with brainstorming, spent 15 minutes prior to the convergence activity (i.e., FocusBuilder) diverging (i.e., brainstorming) on various solutions to the parking problem. All groups then spent 45 minutes converging.

The facilitator walked the groups through each of the steps of the FocusBuilder thinkLet. To begin all brainstormed ideas were divided into as many subsets as there were participants and they were informed to extract the critical ideas, remove redundancies, and formulate the ideas in a clear and concise manner. Participants were paired to do this step again and then the two groups that were formed were paired to a final group and a final round of the same activity.

3.4 Variables of Study

The primary variable of study was the independent variable, or the two different process designs. Additionally, the seven performance criteria were variables of study, including 1) speed/efficiency, 2) level of comprehensiveness, 3) level of shared understanding, 4) level of reduction, 5) level of refinement of outcomes, 6) ease of use for participants, and 7) satisfaction with thinkLet by participants.

Data was collected from multiple sources for understanding, comparison, and contrast. The first of these sources was the actual session data from *GroupSystems™*. The results of each group session were stored electronically, with the output consisting of all the contributions that the participants made during the convergence activity. The second data source was the questionnaire/performance measure instrument administered directly after each session. Finally, direct researcher observation included notes of critical incidents and participant questions that each of the researchers tracked throughout each session.

4 Analysis and Results

4.1 Data Analysis

Session data was coded to determine the number ideas in each *GroupSystems*[TM] entry; 1) whether each identified idea was on-task or off-task, 2) whether an idea was unique to the set of ideas generated within the group, and 3) whether the idea was critical to the solution set. These variables were then used to calculate two measures of convergence: level of reduction and level of comprehensiveness. Before the data coding began, each of the ideas in the data set had to be identified. For this step, Bouchard and Hare's [15] definition of an idea for the coding scheme was adopted. With this definition, one entry can contain one or more ideas. An entry is limited only by how much a particular participant chooses to type. During the data coding, an entry was counted as having more than one idea if more than one specific benefit or difficulty was identified for the parking problem. For example the idea *"Have students who live on campus park on South Campus and take the shuttle to class"* was counted as one idea even though it mentioned two things. On the other hand, *"Charge based on size of vehicle. Add more parking spaces per lot."* was coded as two ideas because it contained two specific benefits; first fewer cars on campus due to higher fees and second more parking spaces. Either idea could be implemented separately.

For the first step in the data coding, items were coded as On-Task or Off-Task. In order to encourage original ideas to address the parking problem participants had been told to be creative. Therefore ideas were coded as On-Task if they attempted to address the parking problem regardless of feasibility. Tasks that did not address parking in any way were coded as Off-Task.

When coding whether an idea was unique to the set of ideas generated in a group, ideas were coded as Unique and Non-Unique. Ideas were determined to be Unique if they had not appeared in the idea set from that group. The first instance of an idea was coded as Unique. Further examples were coded as Non-Unique unless they suggested a method for implementing a particular solution that has not previously been mentioned. For example *"Buy Elmwood Park and turn it into parking"* was coded as Unique in a group that had not previously mentioned *"buying more land to build parking structures,"* but a later idea of *"buy the church across the street"* was not considered unique.

Finally, an idea was considered Critical if it would be necessary for the group to reach the final solution. However in this instance there is no known solution. Therefore a set of themes was drawn from the session data. An idea was considered Critical if it led to one of these themes. Table 2 lists the themes considered to represent a complete set. The measures of convergence, level of reduction, was based on the number of beginning and ending ideas, and the level of comprehensiveness, was based on the number of ideas in the convergence output that are on-task, unique and critical.

Table 2. Themes included in a Complete Parking Problem Solution Set

Major Themes in Parking Problem Solution Set
Build more parking lots
Build more parking garages
Improve shuttle service
Increase parking fees
Scheduling concerns Class times etc
Evaluation of Faculty to student ratio of parking spaces
No cars allowed
Incentives and Support for carpooling
Practical methods to reduce the number of students on campus
Practical alternatives to driving

Data from one group (49 entries) was coded by one rater who was a primary researcher. A second coder, independent of the study, coded the same group of data. Initial measures of Cohen's Kappa revealed that inter-rater reliability was unacceptable, below .50 in one instance [16]. Agreement was reached through four rounds of re-coding and training on the initial group. Both raters then coded a second set of entries. Due to the small number of observations in each group (between 11 – 49) and the ease of distinguishing for example whether or not an idea was On-Task or Off-Task, full agreement was achieved for *Number of Ideas*, *On-Task*, and *Unique* ideas. Cohen's Kappa was not available for data when the variable becomes a constant, such as all ideas being unique. Cohens Kappa of .824 was achieved for the variable *Critical*. Given the subjective nature of the variable and the conservative nature of Cohens Kappa this level of agreement was accepted. The primary coder coded the remaining 14 groups using the revised rule set.

The second data source concerned the survey instrument that was given at the end of each session. The instrument measured participants' perceptions about the workshop. Participants rated statements about the workshop on a 7 point Likert –type scale, ranging from Strongly Disagree to Strongly Agree. Specifically data was gathered regarding the seven performance criteria defined previously in Table 1.

4.2 Results

Table 3 presents a summary of the idea counts for each group based on the above discussion. These counts were used to calculate the two convergence measures, the *Reduction Ratio* and the *Comprehensiveness Ratio*; which are both needed to understand the convergence process.

The *Reduction Ratio* measure the amount of reduction provided by the activity. The *Reduction Ratio* provides information about the efficiency of a particular convergence activity. The ratio specifically addresses whether or not a group has eliminated all of the ideas it should (e.g., unclear, redundant etc.) from the output. It does not however provide information about the quality of output received from the convergence activity. *Reduction Ratio* is defined as follows:

$$Reduction\ Ratio = \frac{Total\ Brainstorming\ Output - Total\ Convergence\ Output}{Total\ Brainstorming\ Output} \qquad (1)$$

Table 3. Idea Counts for Brainstorming and Convergence Activities

Group	# of Idea in Brainstorm Output	# of Unique & Critical Ideas in Brainstorm Output	# of Ideas in FocusBuilder Output	# of Unique & Critical Ideas in FocusBuilder
1	53	24	5	5
2	11	8	7	5
3	41	19	4	3
4	29	11	3	1
5	19	13	5	5
6	20	11	5	5
7	29	15	8	5
8	34	22	5	5
9	53	24	6	4
10	11	8	6	5
11	41	19	3	3
12	29	11	4	2
13	19	13	6	6
14	20	11	6	6
15	29	15	7	5
16	34	22	2	2

The *Comprehensiveness Ratio* gives an indication of the quality of the output of a convergence activity. The closer the ratio is to 1, the more comprehensive the output. As defined earlier comprehensiveness refers to whether the idea set contains all ideas necessary to reach a solution. Specifically, the ratio provides information about whether or not a group has eliminated ideas that it should not. It is defined as follows:

$$Comprehensiveness\ Ratio = \frac{Total\ Critical\ Ideas\ in\ the\ Convergence\ Output}{Total\ Critical\ Ideas\ in\ the\ Brainstorming\ Input.} \quad (2)$$

Table 4 presents the calculated *Reduction Ratio* and *Comprehensiveness Ratio* for each group. Groups 1-8 diverged (i.e. brainstormed) first and then converged. Groups 9-16 used the brainstorming input from Groups 1-8, but only performed the convergence activity. Since Groups 9-16 used the same initial set of ideas as Groups 1-8 (and were in fact matched up) the observations are not independent. Accordingly, a Paired Samples t-test was used to compare the means of the two groups.

4.3 Summary of Findings

We found that the means of the groups that go through a brainstorming activity will not differ significantly from the means of groups that start with a predefined set of ideas and then converge. Related observations occur when there is some connection between treatment groups [18]. One example of this is pre-test and post-test designs. The treatment groups in this study are related by data sets. Groups that did not diverge started with an idea set developed by a corresponding group that had diverged. In this

Table 4. Convergence Measures for all Groups

	Reduction Ratio	**Comprehensiveness Ratio**
1	0.906	0.792
2	0.364	0.375
3	0.902	0.842
4	0.897	0.909
5	0.737	0.615
6	0.750	0.545
7	0.724	0.667
8	0.853	0.773
9	0.887	0.833
10	0.455	0.375
11	0.927	0.842
12	0.862	0.818
13	0.684	0.538
14	0.700	0.455
15	0.759	0.667
16	0.941	0.909

way each set of brainstorming ideas was used twice; once by the group that created it and once by a group that used it as a starting point for the convergence activity. Related observations are examined using a Paired Sample t-test. The Paired Samples t-test for *Reduction Ratio* and *Comprehensiveness Ratio* provides p-values of .634 and .954 respectively. Since neither p-value is less than $\alpha = .05$ this indicates that means for *Reduction Ratio* and *Comprehensiveness Ratio* between the two treatments are not significantly different. Table 5 provides the Mean differences and p-values for each of the performance variables.

Table 5. Mean Differences and p-values for Performance Variables

	Mean Differences	**p-Values**
Reduction Ratio	-.010	.634
Convergence Ratio	-.002	.954

Analysis of the survey data provides additional support for the idea that it does not matter whether or not a group develops the idea set that they use for convergence activities. Using the same Paired Samples t-test none of the means differ significantly. Table 6 provides the Mean differences and p-values for each of the perception variables. None of the eight paired comparisons has a p-value less than $\alpha = .05$. Therefore we can conclude that the means of all of the measured convergence performance measures do not differ.

Table 6. Mean Differences and p-values for Perception Variables

Criterion	Mean Differences	p-Values
Speed / Efficiency	-.388	.354
Level of Comprehensiveness	-.117	.790
Level of Shared Understanding	-.526	.148
Level of Reduction	-.036	.849
Level of Refinement of Outcomes	-.195	.707
Ease of use for Participants	-.471	.310
Satisfaction with Process	-.409	.147
Satisfaction with Outcomes	-.384	.060

4.4 Limitations

This study suffers from a few limitations. First, the lack of a known solution set means that the critical themes were arbitrarily determined based on the coders' analysis of the ideas generated. Second, the participants in these workshops were undergraduate students that may not always have been positively motivated to participate. In fact, social loafing and distractions as witnessed by non-participating observers was a concern regarding a few participants during one session. Thirdly, participants were allowed to self select into groups. This meant that most participants chose to work with friends or familiar team members, frequently forming groups along cultural lines. This could inhibit group performance. For example, previous research from Valacich et a. [13] have shown that cultural differences among teams can impact individual motivation and participation.

5   Conclusions, Implications, and Future Research

Previous research has studied the process of group divergence, while little research has been done on the process of group convergence [4, 5]. This research takes a cue from previous research which suggests that more research should take place in order to better understand group convergence [5]. The goal of this study was to determine whether the idea set used in the convergence activity can be pre-defined or if it should be generated by the group itself. The results of the lab experiment suggest that it appears that it does not matter whether a group generates the ideas themselves or uses a pre-defined idea set.

Our study has implications for research in that we provide guidance for experimental studies in convergence or other patterns of collaboration that use brainstorming data as a starting point. What we found is that it does not matter if a group diverges before they converge. In other words, it is not necessary for a group to brainstorm the ideas that they will be refining and clarifying during the convergence activity. Additionally, we propose a set of measures that can be used to analyze performance in convergence tasks. In terms of implications for practice, the results offer guidance for facilitators in situations where not all group members can meet synchronously, thus meaning that some group members may have to work with other people's contributions. For example, if people in an organization can only participate in the later stage of collaboration (i.e., only convergence and not brainstorming) is this even useful? Or should they just sit out? Since the results indicate that from the perspective of the performance criteria used in this study it is not necessary for group members to generate the ideas that they will reduce and clarify, managers may decide to include participants later in the process. It is possible for example, that high-level managers might bring unique information or expertise useful to the convergence process. Since high-level managers are often in great demand it may only be possible to involve them in one part of the process. Our research suggests that limited involvement can still be an effective way to contribute to the overall goal of the group work.

This research presents our foundation experiment to enable a better understanding of group convergence in GSS sessions. Future research can use the performance measures developed in this study to complete various other experiments as outlined by Davis, de Vreede, and Briggs [5]. These experiments should be undertaken in order to better understand which convergence thinkLets are useful and if or what new types of convergence thinkLets should be designed. Using the performance measures from this study future research should attempt to identify relevant causal constructs that impact convergence phenomena that are derived from the performance criteria. This would allow for further theorizing on the convergence pattern and help design more effective group processes. Furthermore, future research might test a similar experiment with different task types, for example participants could work on a task that has many unknowns and is not as familiar as the parking problem.

References

1. de Vreede, G.-J., Briggs, R.O.: Collaboration engineering: Designing repeatable processes for high-value collaborative tasks. In: 38th Annual Hawaii International Conference on Systems Science, Los Alamitos (2005)
2. Davis, A., Murphy, J.: An approach to improving creativity and satisfaction in group convergence using a group support system. In: 3rd Midwest Association for Information Systems Conference (MWAIS-03), Eau Claire, Wisconsin (2008)
3. Chen, H., Hsu, P., Orwig, R., Hoopes, L., Nunamaker Jr., J.F.: Automatic concept classification of text from electronic meetings. Communications of the ACM 37(10), 56–72 (1994)
4. Briggs, R.O., Nunamaker Jr., J.F., Sprague Jr., R.H.: 1001 Unanswered research questions in GSS. Journal of Management Information Systems 14(3), 3–21 (1997)
5. Davis, A., de Vreede, G.-J., Briggs, R.O.: Designing thinkLets for convergence. In: 13th Annual Americas Conference on Information Systems (AMCIS-13), Keystone, Colorado (2007)
6. Briggs, R.O., de Vreede, G.-J., Nunamaker Jr., J.F.: Collaboration engineering with thinkLets to pursue sustained success with group support systems. Journal of Management Information Systems 19(4), 31–64 (2003)
7. Briggs, R.O., Kolfschoten, G.L., de Vreede, G.-J., Dean, D.L.: Defining key concepts for collaboration engineering. In: 12th Americas Conference on Information Systems (AMCIS-12), Acapulco, Mexico (2006)

8. de Vreede, G.-J., Kolfschoten, G.L., Briggs, R.O.: ThinkLets: A collaboration engineering pattern language. International Journal Computer Applications in Technology 25(2/3), 140–154 (2006)
9. Reinig, B.A.: Toward an understanding of satisfaction with the process and outcomes of teamwork. Journal of Management Information Systems 19(4), 65–84 (2003)
10. Kolfschoten, G.L., Briggs, R.O., de Vreede, G.-J., Jacobs, P.H.M., Appelman, J.H.: Conceptual foundation of the thinkLet concept for collaboration engineering. International Journal of Human Computer Studies 64(7), 611–621 (2006)
11. Briggs, R.O., Reinig, B.A., de Vreede, G.-J.: Meeting satisfaction for technology-supported groups: An empirical validation of a goal-attainment model. Small Group Research 37(6), 585–611 (2006)
12. Fjermestad, J., Hiltz, S.R.: An assessment of group support systems experimental research: Methodology and results. Journal of Management Information Systems 15(3), 7–149 (1998/1999)
13. Valacich, J.S., Jung, J.H., Looney, C.A.: The Effects of Individual Cognitive Ability and Idea Stimulation on Idea-Generation. Group Dynamics: Theory, Research, and Practice 10(1), 1–15 (2006)
14. Gallupe, R.B., Dennis, A.R., Cooper, W.H., Valacich, J.S., Bastianutti, L.M., Nunamaker, J.F.: Electronic brainstorming and group size. Academy of Management Journal 35(2), 350–369 (1992)
15. Bouchard Jr., T.J., Hare, M.: Size Performance, and Potential in Brainstorming Groups. Journal of Applied Psychology 54(1), 51–55 (1970)
16. Brennan, R.L., Prediger, D.J.: Coefficient Kappa: Some Uses, Misuses, and Alternatives. Educational and Psychological Measurement 41, 687 (1981)

Repeatable Collaboration Processes for Mature Organizational Policy Making

Josephine Nabukenya, Patrick van Bommel, and H.A. Erik Proper

Institute for Computing and Information Sciences, Radboud University Nijmegen
Heyendaalseweg 135, 6525 AJ, Nijmegen, The Netherlands
{J.Nabukenya,P.vanBommel,E.Proper}@cs.ru.nl

Abstract. Organizational policy making processes are complex processes in which many people are involved. Very often the results of these processes are not what the different stakeholders intended. Since policies play a major role in key decision making concerning the future of organizations, our research aims at improving the policies on the basis of collaboration.

In order to achieve this goal, we apply the practice of *collaboration engineering* to the field of organizational policy making. We use the *thinklet* as a basic building block for facilitating intervention to create a *repeatable* pattern of collaboration among people working together towards achieving a goal. Our case studies show that policy making processes do need collaboration support indeed and that the resulting policies can be expected to improve.

1   Introduction

In order to regulate organizational processes, organizations use policies as an instrument to guide and bound these processes. A policy [1] is a guide that establishes parameters for making decisions; it provides guidelines to channel a manager's thinking in a specific direction.

Policies are created in a policy making process, which involves an iterative and collaborative process requiring an interaction amongst three broad streams of activities: problem definition, solution proposals and a consensus based on selection of the line of action to take. The core participants of a policy making process must be involved in complex and key decision making processes within the organization themselves, if they are to be effective in representing organizational interests. Explicit policies are a key indicator for successful organizational decision-making.

The complexity of policy making processes in organizations may be described as having to cope with large problems. Examples include: information technology (IT) procurement, Information Systems security, software testing, etc. These problems may be affected by *(i)* unclear and contradictory targets set for the policy goals; *(ii)* policy actors being involved in one or more aspects of the process, with potentially different values/interests, perceptions of the situation, and policy preferences. This is in line with [2] who also describe complex problems to involve many actors due to the need to mobilize many resources; disagreement about the nature of the problem and the desired solutions due to the many actors involved; and complex decision making because mostly different networks and institutional structures are involved. Policy makers and

R.O. Briggs et al. (Eds.): CRIWG 2008, LNCS 5411, pp. 217–232, 2008.

others involved in the policy making process need information to understand the dynamics of a particular problem and develop options for action [3]. A policy is not made in a vacuum. It is affected by social and economic conditions, prevailing political values and the public mood at any given time, as well as the local cultural norms, among other variables.

A policy making process is a collaborative design process whose attention is devoted to the structure of the policy, to the context and constraints (concerns) of the policy and its creation process, and the actual decisions and events that occur [4]. We aim to examine, and address, those concerns that have a collaborative nature. Such concerns include the involvement of a variety of actors resulting in a situation where multiple backgrounds, incompatible interests, and diverging areas of interest all have to be brought together to produce an acceptable policy result. Due to the collaborative nature of a policy making process, its quality is greatly determined by a well-managed collaborative process. We look towards the field of collaboration engineering to be able to deal with such concerns. Collaboration engineering is concerned with the design of recurring collaborative processes using collaboration techniques and technology [5].

The collaboration technologies that are used to support group work in collaborative problem-solving processes are based on and contain fundamental assumptions (for example, meeting processes should be: open; rational; fair) with regard to how people work together [6]. More examples and details of the assumptions can be seen in [6]. To determine successful application of collaboration technologies, the correctness of these assumptions is a vital aspect. Group Support Systems (GSS) is an example of collaboration technologies that have offered added value in terms of anonymity, and parallel communication, among others, to people working together towards achieving a goal [7]. Inter-organizational policy making networks are an environment where GSS have been applied. It was found out that GSS are most effective in creativity tasks than for preference tasks and mixed motive tasks in such an environment [6]. Our study deals with an exploration of usage of collaborative processes for the realization of good policies in organizational policy making. We use thinkLets to design the collaborative policy making process. To safe guard the GSS principles (assumptions) in the thinkLets we use in this study, we adopt the work of Vreede and Bruijn [6]. For instance, we use GSS principles such as anonymity and parallel work in creativity tasks, while for preference and consensus tasks we apply group-oral discussions.

The main purpose of our paper is to offer a repeatable collaboration process for the realization of good policies in a collaborative policy making effort; and to investigate how this process can be improved by the support of collaboration engineering. The standard repeatable collaborative policy making process presented in this paper is originally designed using a modular approach based on given motivations (see section 5). Nonetheless, we use one standard process due to the constraints in size of the sample population, and the levels of stakeholders involved in implementation of the repeatable collaborative policy making process (see section 4).

The remainder of this paper is structured as follows. Section 2 briefly explains the concepts of collaboration engineering (CE), policy, policy making processes and the collaborative concerns that may arrive from these processes. We then continue in section 3 with an exploration of the potential role of collaboration engineering in addressing these concerns. In section 4 we elaborate on the research method used in our pursuit of developing and implementing the repeatable collaboration process, as well as a brief outline of the four case studies we have performed. Based on these case studies, section 5 discusses the design of the repeatable collaborative policy making process based on the original modular process design. Finally, section 6 provides the conclusion as well as a discussion on further research.

2 CE and Organizational Policy Making Processes

Collaboration Engineering (CE) is an approach to designing collaborative work practices for high-value recurring tasks, and deploying those designs for practitioners to execute for themselves without ongoing support from professional facilitators [8]. Collaboration engineering researchers identified six general patterns of collaboration to enable a group to complete a particular group activity [8, 9]: *i)* Generate – Move from having fewer to having more concepts in the pool of concepts shared by the group. *ii)* Reduce – Move from having many concepts to a focus on fewer concepts that the group deems worthy of further attention. *iii)* Clarify – Move from having less to having more shared understanding of concepts and of the words and phrases used to express them. *iv)* Organize – to move from less to more understanding of the relationships among the concepts the group is considering. *v)* Evaluate – Move from less to more understanding of the relative value of the concepts under consideration. *vi)* Build Consensus – Move from having fewer to having more group members who are willing to commit to a proposal.

The patterns of collaboration do not explicitly detail how a group could conduct a recurring collaboration process, especially with teams who do not have professional facilitators at their disposal. This can be achieved by the key CE concept: the thinkLet. A thinklet is defined by Briggs et al., [9] as "the smallest unit of intellectual capital required to create a single repeatable, predictable pattern of collaboration among people working toward a goal". ThinkLets can be used as conceptual building blocks in the design of collaboration processes, such as improving productivity of and quality of work life for groups by enabling rapid development of collaboration processes [10, 8]. Examples of thinkLets are provided in Table 1. More examples can for example be found in [8].

Table 1. Examples of thinkLets with their respective Collaboration Pattern

ThinkLet Name	Pattern of Collaboration	Purpose
DirectedBrainstorm	Generate	To generate, in parallel, a broad, diverse set of highly creative ideas in response to prompts from a moderator and the ideas contributed by team mates.
BucketSummary	Reduce and clarify	To remove redundancy and ambiguity from broad generated items.
BucketWalk	Evaluate	To review the contents of each bucket (category) to make sure that all items are appropriately placed and understood.
MoodRing	Build Consensus	To continuously track the level of consensus within the group with regard to the issue currently under discussion.

2.1   Organizational Policy Making Processes

With an increase in internal and external business needs, organizations have continuously established organizational policies. Because of their nature, it is important for organizations to create policies for a number of reasons such as: they establish responsibilities and accountability; they help ensure compliance and reduce institutional risk; they may be needed to establish and/or defend a legal basis for action; and they provide clarification and guidance to the organizational community [11]. The concept of *policy* therefore, is defined by Robbins et al., [1] as" a guide that establishes parameters for making decisions", that is, it provides guidelines to channel a manager's thinking in a specific direction. Friedrich [12] regards a policy as" a proposed course of action of a person, group, or government within a given environment providing obstacles and opportunities which the policy was proposed to utilize and overcome in an effort to reach a goal or realize an objective or a purpose." Also, Anderson [13] defines policy as" a purposive course of action followed by an actor or set of actors in dealing with a problem or matter of concern". For our purpose and to integrate the various definitions, we define the concept of a policy *"as a purposive course of action followed by a set of actor(s) to guide and determine present and future decisions, with an aim of realizing goals"* [14].

Organizational policy-stakeholders follow a policy making process to develop and implement a policy. According to Sabatier [4], the process of policy-making includes the manner in which problems get conceptualized and brought to the governing body for solution, these formulate alternatives and select policy solutions; and those solutions get implemented, evaluated, and revised. In other words, the policy-making process connotes temporarily, an unfolding of actions, events, and decisions that may culminate in an authoritative decision, which, at least temporarily, binds all within the jurisdiction of the governing body. In examining the unfolding, attention is devoted to structure, to the context and constraints of the process, and to actual decisions and events that occur. In relation to Sabatier's definition, Mitrof [15] describes policy making as a process of forming, weighing, and evaluating numerous premises in a complex, continually changing and unfolding argument. The premises in these arguments are in effect the assumptions that are made with regard to the stakeholders that are judged to be relevant to the policy issue under consideration. Dunn [16] defines the policy making process as the administrative, organizational and political activities and attitudes that shape the transformation of policy inputs into outputs and impacts. He stresses, that, there is no one single process by which policy is made. Variations in the subject of policy will produce variations in the manner of policy-making. Based on these definitions, we can therefore say that the policy making process can be messy.

2.2   Collaborative Concerns in Organizational Policy Making Processes

Organizational policy processes take a searching, iterative problem solving course. Because of their nature, policy processes have been characterized by complexity. We identify two kinds of complexity in policy making processes: multi-participant complexity, and technical complexity [4, 17]. Both types of complexity have distinguished characteristics/concerns; however, our study focuses only on those concerns/characteristics that have a collaborative nature; and we claim can be met by collaboration engineering techniques. Such collaborative concerns [2, 18, 19, 4, 17, 20]include degree of variance in interests and tasks required, conflicting objectives and criteria, lack of consensus, lack of understanding of the policy problem, lack of a clear methodology/approach, and time pressure.

Having collaborative concerns implies the need to have a standard collaboration process, that is, a well-defined process specification with several choices depending on the context/situation in which a policy needs to be specified, that is referred to when making policies. To achieve this, we turn to collaboration engineering (CE). In the section that follows, we describe how CE can meet these collaborative concerns.

3   Meeting Policy Making Processes Collaborative Needs with CE

The aim of this section is to, given the collaborative concerns from the previous section, refine these to collaborative needs (process requirements) for a collaboration engineer with respect to the organizational policy making process and its context. In other words, we discuss how collaboration engineering can provide for collaborative needs for organizational policy making processes.

– *Policy requirements expectation accommodation* – this need is derived from a number of concerns: the degree of variance in interests and tasks required; conflicting objectives and criteria; and lack of consensus. Policy making stakeholders therefore need a collaborative process that permits them to contribute and the contributions taken into account in policy requirements negotiation. In other words, there is need for a collaborative process that permits stakeholders to arrive at satisfactory (reach for consensus) policy requirements' outcomes without conflicting and compromising overall policy objectives. In the CE approach, execution of collaborative processes permits representation of all the stakeholders in collaborative problemsolving activities by usage of thinkLets [21]. Most thinkLets have built-in rules to ensure equal participation of stakeholders, like in GSS; thereby bettering the chance of their interests being accommodated in the solution.
– *Understanding of the policy process* – this need arises from lack of understanding of the policy process concern. Thus, there is need for a collaborative process that is not complex and is easily understood by the policy making practitioners. In CE, collaboration engineers use building blocks known as thinkLets when designing repeatable collaboration processes. A thinkLet is a facilitation intervention that would improve productivity of and quality of work life for policy practitioners by enabling rapid development of the policy making collaboration processes [10]. In other words, usage of thinkLets would permit policy practitioners to execute the collaboration policy process with ease, hence, making it easily understandable for them.
– *Policy process efficiency* – this need is derived from the time pressure concern. Thus, there is need for a collaborative process in which policy making stakeholders can take less time, effort, and physical resources for attainment of the policy than without the use of a collaborative approach. With collaboration, groups tend to minimize/save on the amount of resources required to attain a goal [5]. To this end, the

CE approach offers a model and guidelines (see [22]) to achieve a balance between efficiency and effectiveness of the process design. In other words, the collaboration process design must make optimal use of the available resources. For example, the time, costs, and effort, policy stakeholders can actually use to achieve the planned policy outcomes in a collaboration session.

– *Structured policy problem solving approach* – this need arises from lack of a clear methodology/approach concern. Thus, there is need for a standard recurring collaborative process that is to be referred to each time policy stakeholders need to tackle complex policy problems. CE is an approach to designing recurring collaboration processes. That is, CE focuses on recurring processes rather than ad hoc processes; where a repeated process if improved, an organization will derive benefit from the improvement again and again; while with ad hoc processes, the value of each process improvement will accrue only once [8]. More so, with the improvement to repeatable processes, the same collaborative policy process could be applied successfully in each policy developing workshop with different groups (policy stakeholders) and focusing on different collaborative policy developing tasks. Also, with the improvement to repeatable processes, practitioners of these processes can learn to conduct them successfully without learning facilitation skills [5].

– *Policy elements identification (with their definitions)* – this need arises from lack of consensus concern. Thus, policy making stakeholders need a collaborative process that enables them to identify and have a common understanding of the policy elements (and their definitions). In CE, the patterns of collaboration 'clarify' and 'consensus building' offer thinkLets support [9] that can enable stakeholders have a common/shared understanding, commitment and consensus of policy elements identified. This means, during collaborative policy process execution, policy stakeholders have the opportunity to perform the tasks collaboratively by support of thinkLets.

In summary, based on the collaborative needs/process requirements formulated above, organizational policy making stakeholders and practitioners need to have repeatable collaborative processes that can enable them solve their policy problems. In the section that follows, we discuss the research method used in our pursuit of developing and implementing the repeatable collaboration process.

4 Research Questions and Approach

In this section, we present the research question and how we addressed it. In coming up with a repeatable collaboration process to meet collaborative needs for organizational policy making processes, the following research question had to be addressed: How can usage of a repeatable collaboration process meet collaborative needs for organizational policy making processes? To achieve this, we followed Zuber-Skerritt's Action research methodology [23]. We used this method in comparison to others, because it appeared to be most appropriate in our context. That is, it allowed us to gain a richer understanding of the workings of our collaboration process in action. Action research also permitted the researchers to intervene in the problem setting, and perform collaboratively [24]. In addition, this method is the most suitable in addressing the "how to" research questions [25], as our research aimed at addressing how to meet collaborative needs for organizational policy making processes using a repeatable collaboration process. Furthermore, the method allowed us to evaluate and improve our problem-solving technique during a series of interventions.

The action research method proposed by Zuber-Skerritt [23] involves four activities/phases that can be carried out over several iterations (in our case four). The first activity 'Planning' is concerned with the exploration of the research site and the preparation of the intervention. The second phases 'Act' involves the actual intervention made by the researcher. In the third phase 'Observe', collection of data during and after the actual intervention to enable evaluation is done. Finally, the fourth activity 'Reflect' involves analysis of collected data and infers conclusions regarding the intervention that may feed into the 'Plan' activity of a new iteration. Following the model described above, the 4 activities were executed as follows: In the 'Planning' activity, we conducted interviews with four organizations that have policy making functions and also performed a literature review to understand organizational policy making. The data collected formed the initial requirements for the repeatable collaboration process.

The 'Act' activity involved actual execution of the repeatable collaboration process in the field both in industrial settings and an inexperienced environment. We applied the generic repeatable collaboration process with three policy types in four case organizations. Below is a description of the cases:

Case 1 - Information Technology (IT) policy document with a team of 5 IT workers of the Ministry of Finance, Planning and Economic Development (MOFPED), Uganda.

Case 2 (Inexperienced environment) - Student portal information system architecture principles with 14 students enrolled in a graduate level Modelleren van Organisaties (Modelling of Organizations) course, Radboud University Nijmegen (RUN), the Netherlands.

Case 3 - IT Security policy document with a team of 6 stakeholders of National Social Security Fund (NSSF), Uganda, involved in formulating IT policies for the organization.

Case 4 - Student portal information system architecture principles with a team of 7 stakeholders of the department of Control, Information, and Finances (CIF) involved in formulating IT business rules, regulations and architecture principles for information systems for RUN, the Netherlands.

To evaluate the performance and perception of the generic repeatable collaborative policy making process by the participants, we collected and analyzed explorative data during the 'Observe' activity. 3 kinds of instruments, that is, observations, interviews and questionnaires comprising of qualitative and quantitative questions, respectively were used for data collection. The tools enabled us to collect and analyze data regards policy requirements expectation accommodation; understanding of the policy process; effectiveness, and efficiency of the policy process and its outcomes; policy elements identification; and policy stakeholders' satisfaction with the process and its outcomes. Evaluation of the generic repeatable collaborative policy making process design was implemented using two procedures. The first three collaborative sessions (cases 1, 2, and 3) were conducted manually, while the fourth session (case 4), we used group support technology (MeetingWorksV7.0) to implement the process, respectively. Results from the cases are presented in section 5.4.

Finally, in the 'Reflect' activity, we tested the process using four cases to allow us to reflect on the process design and improve it continuously. The final design (Figure 1) of the generic repeatable collaborative policy making process was the result of four iterations. The iterations performed earlier were considered less desirable because of perceived inefficiency in the discussion and uneven amount of time required to complete the process for identifying common and priority policy elements with their definitions. For example, in the early iterations, participants executed the policy objectives and policy elements formulation tasks in parallel. This made the process very slow. In other words, participants generated policy elements that were more/less related to the meeting goal, but many of these did not address stated policy objectives/concerns formulated in the previous task. However, sequential execution of the two tasks was deemed necessary for the process as the former task was the basis for the latter (the policy elements being formulated had to address policy objective(s) stated). This also affected the discussion/cleaning-up time and completeness of the process in terms of trying to match the out-of-scope formulated policy elements to stated policy objectives. Also in these iterations, we left policy objectives and policy elements formulation tasks very broad to reduce on the lengthy process execution time. This was, however, forsaken, because not all policy objectives and elements recorded were of priority, consistent and common in order to meet the desired end states.

5   Generic Repeatable Collaboration Process

To design the repeatable collaborative policy making process, we followed the process requirements based on CE techniques as described in Section 2. Even though this approach comprises several design steps, the ones relevant to our research study included decomposing the process into collaborative activities, the classification of these activities into patterns of collaboration, selection of appropriate thinkLets to guide facilitation of the group during the execution of each activity as well as making the design process more predictable and repeatable. The generic repeatable collaborative policy making process presented in figure 1 is originally designed using a modular approach based on given motivations. Nonetheless, we use one standard process due to the constraints in size of the sample population, and the levels of stakeholders that were involved in implementation of the process. Below, we first explain the motivations of using the modular approach. Then we present the design evaluation criteria we followed, and then a description of the generic repeatable collaborative policy making process.

5.1   Modular Process Design

The modular collaboration process contains three modules: *module 1* – deals with formulation and agreeing on policy goals; *module 2* – deals with formulation and agreeing on policy objectives based on the policy goals stated; and *module 3* – deals with formulation and agreeing on policy elements (with their definitions) that address the stated policy objectives. We use the modular approach, because, depending on the kind of policy, stakeholders wish to achieve: the policy scope (its extent/coverage), ambitions (what the stakeholders want to achieve), instruments and their combinations (what resources are required in what phase to achieve a given ambition) vary. In other words, policy making involves different levels of stakeholders who perform different tasks in different phases of the policy process; that is, not all kinds of stakeholders are involved in all the phases of policy making; for instance top level stakeholders are responsible for identifying and formulating policy goals, as well as define its scope. Also given the levels of tasks/phases involved in policy making, different phases may require different instruments or a combination of them, such as sharing of knowledge and information, and expertise on the part of the stakeholders involved. For instance, some policy process phases may require only expert-driven stakeholders, while others may require a combination of both expert-driven and non-expert stakeholders to be involved. Also some process phases may require more time to achieve a given ambition in comparison to others. For instance, formulation of policy elements (with their definitions) and respective implications may require more time as compared to formulation of policy goals. Thus far, in using the modular collaboration process, the policy making process characteristics (collaborative concerns) mentioned earlier on are taken care of; better still, making the collaboration process more flexible.

5.2 Design Criteria

The design of the repeatable collaborative policy making process was derived from a few iterations based on selected design criteria. The criteria selection was made according to the goal of the evaluation itself. Evaluation of the collaboration process aimed at addressing how to meet collaborative needs for organizational policy making processes using a repeatable collaboration process. The following six criteria were considered: *(i)* effectiveness - the repeatable collaboration process should enable policy making stakeholders to achieve their goal, *(ii)* policy process efficiency - the collaboration process should take stakeholders less time for attainment of the policy than without the use of a collaborative approach, *(iii)* degree of applicability (structured policy problem solving approach) - the extent to which the repeatable collaboration process can be applied to formulation of varying policy types, *(iv)* policy elements identification (with their definitions) - the collaboration process should enable stakeholders to have a common/shared understanding, commitment and consensus of the policy elements (and their definitions) identified, *(v)* policy requirements expectation accommodation - the collaboration process should permit stakeholders to contribute and the contributions taken into account in policy requirements negotiation. In other words, the collaboration process should permit stakeholders to arrive at satisfactory policy requirements' outcomes without conflicting and compromising overall policy objectives, and *(vi)* understanding and ease of use of the policy process - the collaboration process should not be complex and should be easily understood by the policymaking stakeholders. That is, the process should be easy for the practitioners to learn and execute routinely.

5.3　Generic Process Design

The collaboration process design shown in Figure 1 was not from scratch. The design
was based on the policy process requirements derived from the explorative field study
with four case organizations that have policy making functions, and also in concurrence
with the policy process discussed by Ford and Spellacy [11]. A typical policy making
process includes six stages [11]. However, our process design only involves the devel-
opment/formation phase of the organizational policy making process; therefore it caters
for a pre-used policy. We use a generic repeatable process and not the modular design
because of the kinds and levels of stakeholders that were involved in the implementation
sessions. Also, we were constrained by the numbers of stakeholders in terms of participation. In addition, the policy types, that the participants were to formulate, did not
necessitate going through the first phase (pre-development) as these were preliminarily
developed by top level stakeholders in respective case organizations. In other words,
not all the kinds and levels of stakeholders that were involved in the collaboration process sessions participated in the preliminary tasks. The participants therefore only had
to discuss, agree and use these elements as prior knowledge to formulating policy objectives and policy elements. It is on this basis, that we merged module-one of the modular
design to "pre-development" phase as refereed to in the generic repeatable process.

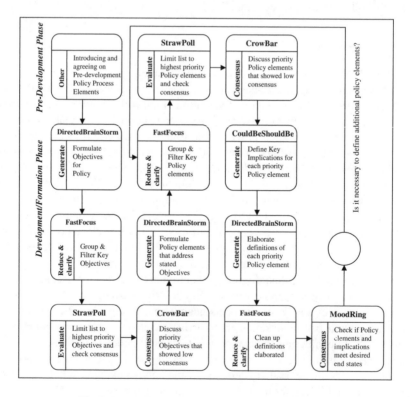

Fig. 1. Repeatable collaborative policy-making process

The repeatable collaborative policy making process underwent four iterations prior to deriving the final process design. The four iterations of the earlier versions of the process were applied in the four cases described in section 4. The final process design shown in Figure 1 presents the steps required to develop/form a policy document, and the patterns of collaboration with related thinkLets used to guide the group to execute each step.

The development/formation phase of the collaboration process in Figure 1 has two main parts: *part 1* – pre-development /meeting phase, and *part 2* – the development phase. We refer to part 1 as pre-development phase based on the fact that, the top level policy stakeholders though in consultation with other middle to low level stakeholders preliminarily develop these elements. This means that, prior to the actual development of the policy, top level policy making stakeholders have various policy meetings to gather information on the kind and the need for the policy. This phase involves familiarizing and discussing the following pre-development elements: the problem to be solved; the ambitions (goals) of the policy, the policy scope, the relevant information to be used to develop the policy; a legal framework to support the policy to be developed; the ownership of the policy; leadership positioning i.e. who is to spearhead the process; who are the stakeholders (internal and external); technical resources to facilitate the process. The second part, the development phase, involves different policy stakeholders (irrespective of levels) to identify and agree on policy objectives; then the identification of and agreement on common policy elements with their definitions and respective implications /terms that should suit the desired end state (policy objectives). These activities (process) should finally generate a policy document which clearly articulates solutions.

In the brainstorm activity that follows, guided by the DirectedBrainstorm thinkLet, participants are required to formulate policy objectives. The result from this activity is a brainstormed list of policy objectives.

Using the FastFocus thinkLet, the activity that follows requires participants to organize the resulting list by extracting only the key policy objectives. They do this by grouping and filtering ideas, as well as eliminating any redundancies. The result from this activity is a cleaned list of key policy objectives. The participants then use these results to evaluate/limit the cleaned list to the highest priority objectives. They do this by rating the key objectives using a given criteria. The evaluation activity is guided by the Straw-Poll thinkLet followed by a CrowBar thinkLet to discuss ideas that may have low consensus. The outcome of this activity is a list of priority key policy objectives.

In the activity that follows, guided by the DirectedBrainstorm thinkLet participants are asked to formulate common policy elements that address the key priority policy objectives. The result of this activity is a brainstormed list of policy elements. Using the FastFocus thinkLet, the participants organize (clean-up) the resulting brainstormed list by grouping and filtering only the key common policy elements. The result of this activity is a cleaned list of key policy elements. Based on the results from this activity, and using the StrawPoll thinkLet followed by a Crowbbar thinkLet, participants are then required to evaluate/limit the list to the highest key priority policy elements. The outcome of this activity is a list of priority key policy elements that address the stated policy objectives.

The activity that follows involves defining key terms/implications for each of the key priority policy elements. Using the CouldBeShouldBe thinkLet, participants brainstorm implications that they 'could' consider as appropriate for each priority policy element. Using the brainstormed list of implications, participants then choose implications they 'should' take as key to each priority policy element. The activity that follows requires participants to elaborate/define each of the priority policy elements. This is guided by the DirectedBrainstorm thinkLet, followed by a FastFocus thinkLet.

Finally, the activities above result into a Policy document. Using the MoodRing thinkLet, participants are required to check completeness of the policy document by reaching consensus. They do this by voting on a YES/NO basis, where a YES is voted if the priority policy elements (with their definitions) and respective implications meet the desired end states (i.e. address the stated policy objectives) and a NO if they do not. A verbal discussion is held to address issues identified as incomplete, until some sort of consensus on completeness is reached.

5.4   Results

Satisfaction is defined as an affective response with respect to the attainment of goals (process outcomes; and the process by which the outcomes were attained). To measure this construct, we used the 7-point Likert scale general meeting survey questionnaire where participants can strongly disagree to strongly agree. The instrument validation and theoretical underpinnings can be seen in [26]. Results in Table 2 are from the questionnaire we used, and they indicate that the participants were reasonably satisfied with the repeatable collaboration process outcomes, and the process by which the policies were formed.

Table 2. Satisfaction with process and outcome

	1	2	3	4
Satisfaction with process				
Score	4.800	3.838	4.500	4.800
Standard deviation	1.376	0.995	1.366	1.053
Satisfaction with outcome				
Score	5.160	4.363	5.367	5.486
Standard deviation	1.310	1.094	0.908	0.598

In other words, the participants felt that the results from the workshops were useful to them as they gave better understanding of what issues they found vital to the policy. They also observed this process as an interactive and better method/approach of formulating policies. For example, most positive comments received from the workshops included "the results are useful for me, because they give me a better understanding of the things users of the policy find important", "the process can be very useful for my work; trying to formulate issues about a variety of subjects and with different groups of people", "I liked the process because it forces you in a direction in which you are obliged to perform some actions in a specified order". However, not all participants in the first three cases were happy with the way of executing the process. Most of their negative comments had to do with "the lack of a tool causing problems such as time delay, and noise". From the researchers' perspective, the process was satisfactory because most participants indicated that they were mainly interested in the outcome/results of the process than the way it was executed.

We define *policy process efficiency* as the degree to which there is savings of the amount of resources (for example time, costs, and effort) required for attainment of the goal. In other words, the collaboration process should take participants less time and effort for attainment of the policy than without the use of a collaborative approach. To measure this construct, we considered the execution duration (timing) of each stage of the process; and also how well the participants understood the process tasks (used less effort) for successful execution in order to realize/come up with a policy.

Though the majority of the participants felt that the process execution was efficient in terms of cognitive load/less effort and time, a few were not happy with the time length particularly with some activities such as in the grouping and filtering of key policy issues. For example one participant said "I believe to fully realize satisfactory results from specific activities of the process, it requires a more in-depth session". Such remarks were corrected in subsequent workshops and also taken along in the final process design. In addition to execution time, participants being able to execute the collaboration process with less effort, (for instance there were less to none questions of how to do things) made the researchers conclude that the participants clearly understood the collaboration process (understanding of the policy process).

Effectiveness is defined as the extent to which there is effort for policy stakeholders to achieve their goal. We measured this construct by how well the participants managed to come up with a policy at the end of the policy process execution. From our observations, it was noted that the participants effectively managed to formulate respective policy types. This was demonstrated during the consensus stage of the process, and also based on results from satisfaction with the process outcomes (see Table 2). In the consensus stage, participants were required to check if the policy document met the desired objectives for which it was intended for. They did this by voting on a YES/NO basis, where a YES was voted if the policy elements (with their definitions) and respective implications/terms met the desired end states and a NO if they did not. Based on the feedback from the voting sheets (see Table 3), it was observed that the participants achieved satisfactory results, that is, they managed to form a policy based on the desired end states. For those that voted a NO, a verbal discussion was held to re-address their issues until some sort of consensus was achieved.

Table 3. Voting consensus results

	Yes	No
Case 1	4 (80%)	1 (20%)
Case 2	12 (75%)	4 (25%)
Case 3	5 (83%)	1 (17%)
Case 4	5 (71%)	2 (29%)

Policy requirements expectation accommodation is defined as the ability of the process to accommodate awareness of each stake holder's desired policy preferences. In other words, the process should permit stakeholders to arrive at satisfactory policy requirements' outcomes without conflicting and compromising overall policy objectives. To measure this construct, we used consensus levels (Table 3) and satisfaction results

(Table 2) in addition to feedback from data session logs transcribed by domain experts. From our observations, it was noted that participants were able to contribute and the contributions taken into account in policy requirements negotiation. The consensus activity enabled participants to discuss and arrive at satisfactory policy requirements' outcomes in relation to overall policy objectives. The same results were also used to measure *policy elements identification* (with their definitions). We define this construct as the extent to which the collaboration process should enable stakeholders to have a common/shared understanding, commitment and consensus of the policy elements (and their definitions) they have identified. Based on these results, it was observed that the participants perceived it as having a common/shared understanding of the policy elements identification.

We define *degree of applicability* (structured policy problem solving approach) as the extent to which the repeatable collaboration process can be applied to formulation of varying policy types. To measure the degree of applicability, we implemented the collaboration process to four cases with different policy types. These included formation of an Information Technology (IT) policy, Architectural Principles for a student Information System Portal, and a Security policy for an IT-Driven organization. It was observed that the collaboration process was flexible in terms of its applicability in formation of three different types of policies.

Over all, the process proved to be reasonably successful across all the four cases. This is reflected in the 'observe' activity results. For instance, using the results in table2, it can be seen that satisfaction levels, both with process and outcomes are higher for participants (in cases 1, 3 and 4) that are more experienced in formulation of policies and have interest in the process path, i.e. working from top to bottom and giving thorough attention to precise definitions and formulations. Participants in case 2 specifically the students were inexperienced and to them they felt that the process was more useful to the policy experts that were in their workshop. More so, the participants in case 4 specially commended the efficiency of the process because of the process outcome, and their ability to generate many ideas during the creativity tasks in few minutes due to the support of the Meetingworks collaborative software. This is consistent with some observations in GSS studies for policy making [6, 17]. More so, particular thinkLets such as the DirectedBrainstorm thinkLet and CouldBe-ShouldBe thinkLet enabled ease of execution of the creativity tasks. None the less, there was a minimal difference between the over all policy outputs that cases 1 and 3 made in comparison to case 4. The data logs (though not attached, but can be shown on request) and participants' feedback from the questionnaires and interviews strengthens our observations.

6   Conclusions and Further Research

In this paper, we have discussed a generic repeatable collaboration process for organizational policy making. This process design was refined in four iterations using feedback from observations, questionnaires, and interviews in an action research paradigm. The generic repeatable collaborative policy process enables stakeholders to: identify and agree on policy objectives, and common policy elements with their definitions and respective implications; and also generate a policy document that articulates solutions.

Our results based on the four cases we conducted suggest that the idea of developing policies using a repeatable collaborative policy making process is feasible. In other words, the affirmative feedback received from our participants in terms of satisfaction (with process outcome), process effectiveness, efficiency and applicability suggest that the CE approach has indeed the potential to support organizations in developing quality policies. We consider these findings remarkable due to the fact that this study was resource constrained, and as such we could not adequately test the process to a successful conclusion. The first limitation was the number and category of subjects that were used in each of the four pilot studies as seen in section 4. As a result, there was a significant variation in the experience of the groups. The second limitation was the time availability. Much as many ideas were generated, discussed, and evaluated in the time stipulated to complete the process (that is, 2 hours); this time was still not enough to actually develop an inclusive policy document. The third limitation was the procedure we used to conduct the process in the field. Specifically the manual way of doing things slowed down the process execution; in addition to the inadequacy of the group support software that we used as it has its own limitations. Thus, based on the analysis from the results of this study, they can be used as avenues for future research. First, we need to test and validate the modular design in the field. In addition, in terms of time usage, we need to determine which thinkLets and in which order would be the most effective and efficient. Secondly, we are also working towards a more theoretical underpinning of our results. In other words, we aim to more explicitly rationalize design decisions taken in collaborative policy making processes. We aim to do so by explicitly relating the goals of the collaborative policy making process (its why), the requirements of the process following from these goals (its what), to the situation in which it needs to be executed (its within).

References

1. Robbins, S., Bergman, R., Stagg, I.: Management. Prentice Hall Australia Pty Ltd., Prentice-Hall (1997)
2. Koppenjan, J., Klijn, E.: Managing uncertainties in Networks. A network approach to problem solving and decision making. Routledge, London (2004)
3. Buuren, M.v., Edelenbos, J., Klijn, E.: Managing knowledge in policy networks. In: International Conference on Democratic Network and Governance, Copenhagen, Denmark (2004)
4. Sabatier, P.A., et al.: Theories of the Policy Process. West view Press, Boulder, Co. (1999)
5. Briggs, R., de Vreede, G., Nunamaker Jr., J.F.: Collaboration Engineering with Thinklets to Pursue Sustained Success with Group Support Systems. Journal of MIS 19, 31–63 (2003)
6. de Vreede, G.J., Bruijn, H.d.: Exploring the Boundaries of Successful GSS Application: Supporting Inter-Organizational Policy Networks. Database Journal 30, 111–131 (1999)
7. Nunamaker, J., Dennis, A., Valacich, J., Vogel, D., George, J.: Electronic Meeting Systems to Support Group Work. Communications of the ACM 34, 40–61 (1991)
8. de Vreede, G., Briggs, R.: Collaboration Engineering: Designing Repeatable Processes for High-Value Collaborative Tasks. In: Dickson, G., DeSanctis, G. (eds.) Proceedings of the 38th HICSS. IEEE Computer Society Press, Los Alamitos (2005)
9. Briggs, R.O., Kolfschoten, G.L., de Vreede, G.J., Dean, D.L.: Defining Key Concepts for Collaboration Engineering. In: Proceedings of 12th AMCIS, Mexico (2006)

10. de Vreede, G.J., Kolfschoten, G.L., Briggs, R.O.: Thinklets: a collaboration engineering pattern language. International Journal of Computer Applications in Technology 25, 140–154 (2006)
11. Ford, M.T., Spellacy, P.: Policy development: In theory and practice. In: Baltimore, M.D. (ed.) National Assoc. of College and Univ. Business Officers, Annual Meeting (July 2005)
12. Friedrich, C.: Man and His Government. Wiley, New York (1963)
13. Anderson, J.E.: Public Policy-making: An Introduction, 5th edn. Houghton Mifflin, Boston (2003)
14. Nabukenya, J.: Collaboration Engineering for Policy Making: A Theory of Good Policy in a Collaborative Action. In: Bounif, H. (ed.) Proceedings of the 12th Doctoral Consortium, held in conjunction with CAiSE 2005 (2005)
15. Ian Mitrof, I.: Stakeholders of the Organizational mind: Toward a new view of organizational policy making. Jossey-Bass Inc., Publishers, San Francisco (1983)
16. Dunn, W.N.: Public Policy Analysis: An Introduction. Prentice-Hall, Englewood Cliffs (1981)
17. van den Herik, C.W.: Group Support for Policy Making. PhD thesis, Delft University of Technology, Delft, Netherlands (1998)
18. van den Riet, O.: Policy Analysis in Multi-Actor Policy Settings: Navigating between negotiated non-sense and superflous knowledge. PhD thesis, TU Delft, Netherlands (2003)
19. Roelofs, A.: Structuring Policy Issues: Testing a mapping technique with Gaming/Simulation. PhD thesis, Katholieke Universiteit Brabant, Tilburg, Netherlands (2000)
20. Eden, C., Jones, S., Sims, D.: Messing about in Problems: an informal structured approach to their identification and management. Pergamon Press, Oxford (1983)
21. Kolfschoten, G.L., Briggs, R.O., Appelman, J.H., de Vreede, G.-J.: ThinkLets as Building Blocks for Collaboration Processes: A Further Conceptualization. In: de Vreede, G.-J., Guerrero, L.A., Marín Raventós, G. (eds.) CRIWG 2004. LNCS, vol. 3198, pp. 137–152. Springer, Heidelberg (2004)
22. Kolfschoten, G.L., de Vreede, G.-J.: The Collaboration Engineering Approach for Designing Collaboration Processes. In: Haake, J.M., Ochoa, S.F., Cechich, A. (eds.) CRIWG 2007. LNCS, vol. 4715, pp. 95–110. Springer, Heidelberg (2007)
23. Zuber-Skerritt, O.: Action research for change and development. Gower Publishing, Aldershot (1991)
24. Hult, M., Lennung, S.-A.: Towards a definition of action research: A note and bibliography. Journal of Management Studies 17, 241–250 (1980)
25. Baskerville, R.L.: Investigating information systems with action research. Communications of the Association for Information Systems 2 (1999)
26. Briggs, R.O., Reinig, B., de Vreede, G.J.: Meeting Satisfaction for Technology Supported Groups: An Empirical Validation of a Goal-Attainment Model, Small Group Research (in press, 2006)

Comparing Usage Performance on Mobile Applications

Luís Carriço, Luís Duarte, António Broega, and Diogo Reis

LaSIGE & Department of Informatics, University of Lisbon,
Edifício C6, Campo Grande, 1749-016 Lisboa, Portugal
lmc@di.fc.ul.pt,
{lduarte,abroega,dreis}@lasige.di.fc.ul.pt

Abstract. This paper presents an analysis tool for comparative and collaborative evaluation of mobile artefact usage. Three scenarios were envisioned for the comparative dimension covering both multiple user performance analysis and single-user evolution analysis through three different settings: result browsing, interaction replay and online monitoring. The collaborative dimension is detailed according to two settings: existence of a public display and the use of shared spaces to exchange information between analysts. A couple of analysis sessions were performed by end-users under group psychotherapy and educational domains to assess how the tool fits in such scenarios.

Keywords: Groupware, User Performance & Monitoring, Mobile Devices.

1 Introduction

The use of mobile devices has demonstrated to improve traditional ways of performing tasks in distinct areas, such as healthcare, education or therapy [17][18][24]. As they penetrate these and other domains, it is often the case that the results of mobile application usage need to be assessed and frequently compared. Discussing and comparing students' homework, patients' tasks or therapists/doctors notes, assessing the evolution of an individual's performance or monitoring how people work in real time tasks are but a few examples where this usage analysis may come handy. Moreover, as these devices become smaller, lighter, autonomous and pervasive, sustaining new interaction forms, new form-factors and higher computing power, researchers must also focus their attention on understanding the gains, obstacles and opportunities of using such technology. Again, the comparative analysis of mobile application usage may shed some light on strait usability issues, usage context and context change impact, environmental influences, etc.

Qualitative analysis of application usage is by itself a demanding, time-consuming process. When comparing multiple tasks, from the same (over time) or multiple users, then the complexity tends to augment. If we add to that the real-time dimension, for instance comparing multiple users' activities while they perform them, then the requirements often exceed the capacity of a single human analyst. Collaboration is needed.

Current groupware approaches typically ignore comparative analysis, particularly when real-time considerations are required. On the other hand, support to comparative analysis of mobile applications usage is naive in general, and often disregards

R.O. Briggs et al. (Eds.): CRIWG 2008, LNCS 5411, pp. 233–247, 2008.
© Springer-Verlag Berlin Heidelberg 2008

collaboration. The direct application of these three dimensions, comparative analysis, collaboration and real-time, to complex scenarios, such as those revealed in group psychotherapy or other forms of group evaluation, is even rarer.

As such, we propose CATMA, an analysis tool that covers both the comparative and collaborative analysis for mobile application usage. The tool has initially emerged from the JoinTS project [5]. Its aim was to offer computational support for psychological group therapy. JoinTS itself derived from a previous project, SCOPE [23] that provided computational support for cognitive-behavioural therapy (CBT). SCOPE provided computational support to individual therapy sessions, allowing both patients and therapists to use mobile devices to create, fill-in and analyze digital artefacts (e.g. questionnaires, forms). JoinTS added a collaborative dimension to the process of exposing ideas during therapy sessions. The tool presented in this paper takes a step further and provides therapists with the means to fully analyze the patients' actions and results of the artefact filling-in process. This analysis covers: individual and multiple user analysis (including artefact filling-in result comparison and interaction replay); collaborative analysis (with the aid of a second analyst) through simultaneous observation of results, live monitoring of the patients' actions and through the sharing of annotations. In CBT patients fill their homework, using PDAs, everywhere (including during sessions), and bring results to sessions with a therapist. On the group version, sessions have several patients and sometimes more than one therapist discussing, monitoring and analysing patient's performance. A thorough requirements analysis, including therapists was made, for both single CBT and group CBT. We then observed that similar requirements were patent in other domains, from education to other forms of healthcare, or simply/specially in users and usability evaluation.

The paper follows with a related work discussion. Then we briefly describe SCOPE's previously developed tool-set, followed by a deeper presentation of CATMA along the comparative and collaborative analysis dimensions. The case studies are presented and the paper ends with conclusions and future work.

2 Related Work

Analysing usage activities is one of the main tasks on usability evaluation. Several tools have been developed to support it [2][6][11][16][26]. In general they offer the ability to annotate, synchronously, usage video recordings or interaction logs. In mobile application analysis and in critical scenarios, however, video recording tends to be unfeasible, for practical or ethical reasons. Interaction log analysis, on the other hand, is particularly suited for mobile and field-studies. Tools such as Playback [16] adopt this approach. Crowe [6] goes a bit further and provides real-time monitoring of applications' usage, combined with other data gathering techniques and maintaining the annotation capability. Overall, however, all these tools lack a multi-user analysis environment or any feature that supports collaborative analysis.

Regarding groupware applications, Greenberg presents SharedNotes [9], a system in which users are able to create annotations for their digital artefacts. These may be later published in a shared space while in a meeting, focusing more on the transitions between private and public notes. Notable [3] is another annotation system, focusing more on document (and respective annotations) search and on the separation of the document visualization and annotation taking platforms. These two works, despite providing valuable design cues, do not cover a comparative dimension of the used artefacts. NotePals [7] is another annotation sharing system which allows users to aggregate notes to artefacts and allow other users to access them. Unlike the other examples, there is a clear attempt at providing a certain degree of comparative analysis in addition to its collaborative facet. However, the static nature of the used artefacts (e.g. the lack of a rule engine associated with artefact's triggered events; the lack of editing support; the impossibility of attaching an annotation to a specific trench in artefact usage) does not promote the employment of the system for usage evaluation ends, thus not accomplishing the goals we propose. The Pebbles project [15] focuses more on the collaborative use of mobile devices. Users operate their PDA's connected to a PC to remotely send input data, thus enabling direct manipulation of the same display by multiple users. While this solution is a good example of a collaborative application it doesn't integrate any kind of comparative features, hence not covering our goals.

Pinelle presents and discusses a set of design practices for groupware tools in [19]. In addition, a prototype for homecare is presented which allows clinicians and patients to share documents allowing direct access to these using a timeline. This work shows the closest features to our approach. However, no emphasis is given to a comparative dimension. The lack of annotation support is another feature in which it differs from our approach, hence not fully covering the comparative and collaboration dimensions. Other uses of technology in the healthcare field are usually related with the use of mobile devices to share artefacts between clinicians, namely patient records or prescriptions. [4][14] are examples of such applications. These were demonstrated in hospital settings and use mobile devices to share the documents (typically electronic patient records) in spontaneous meetings. Although the approaches show interesting features, they require complex infrastructure support, are confined to the settings they were built on and do not provide feedback on the digital artefact manipulation process, merely allowing live monitoring in the best case. Xu [27] presents a biofeedback system which allows for multimodal data archiving, real-time annotation and information visualization. Despite having a rich set of mechanisms to perform collaborative analysis, no mention is made to the existence of a comparative dimension in the application, thus not achieving all goals we propose. Most of the existing references focus more generic healthcare scenarios, as presented earlier and also diagnostic procedures [4]. Unfortunately, these are usually too generic to be successfully applied to more specific contexts, such as psychotherapy in our case. Other researches are related to how the information about patients is visualized [12] and how therapists may interact with it using multiple devices [1], a feature we propose in our work. These approaches have limited results, as usually the information is displayed without using any filters, becoming too complicated for both patients and clinicians to understand it. Interaction with multiple devices is also usually restricted to controlling desktop computer applications with mobile devices, thus not promoting real collaboration or interaction between different applications using different types of devices.

3 Prototyping Framework

CATMA builds on a mobile prototyping framework, named MobPro [20][21]. Mob-Pro comprises a set of libraries and tools that allow: (task 1) the rapid creation and adjustment of mobile digital artefacts; (task 2) its navigation and manipulation on mobile devices; (task 3) and its subsequent usage analysis. CATMA focuses on the last two tasks, as described bellow. For explanation purposes, we will refer as "users" those individuals that manipulate/fill the artefact (involved in task 2) and "analysts" those that utilize the analysis components (involved in task 3). The artefact creation component (task 1) includes two fundamental tools, both requiring no programming skills: a desktop tool for artefact design and a mobile one for in-situ artefact adjustment [20]. MobPro was developed in C#, for Microsoft platforms. An earlier simplified version is available for J2ME, running on PalmOS.

3.1 Artefacts

Artefacts are more or less sophisticated abstract entities used in our system to simulate, among other, paper forms, prototypes, applications, etc. As such, artefacts contain pages, each one with a set of elements (e.g., text boxes, pictures, radio buttons, etc.). They also contain rules that determine the artefact behaviour. Specific events, such as selecting a certain answer, a time-out or a next page request, trigger rules that have associated actions (e.g., pop a help message, skip a set of pages or disable an interaction element).

Artefacts are instantiated by the tools of our framework and have two basic external representations: within a central SQL database (used by the desktop creation tools) and as XML-files (used everywhere else).

3.2 Manipulation Tool

The manipulation tool enables users to interact with the artefacts. It instantiates them and provides an interface for page navigation (see top navigation bar, on Fig. 1). The bar allows users to sequentially access pages or jump directly to a specific one (central box and button). Upon user action, the actual loaded page depends on both the issued navigation command and the possibly triggered associated artefact's rules.

The middle part of the figure corresponds to the artefact instantiation. In the example, it comprises two textual labels, a text box and a track bar. On the latter duo, the user is able to enter data that is kept in a results XML file. Optionally, the tool also registers the history of all user interactions, time-stamped, in a log file, also in XML. The bottom menu bar enables access to application's functions such as opening and closing artefacts, viewing summaries, locking and saving results, etc.

3.3 Single-User Analysis Tool

The single-user analysis tool is available for both desktop and mobile platforms. Two operation modes are provided: result's viewer and log player. The result's viewer mode is a simplified version of the Manipulation tool (see above) and enables the analyst to browse through the results entered by the user.

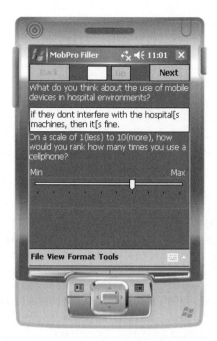

Fig. 1. Manipulation Tool   **Fig. 2.** Single-User Analysis Tool

The log player uses the information recorded in the log files and reproduces the user's interaction according to the timestamps associated with each interaction. Fig. 2 shows the tool in the log player mode. The menu bar, besides loading results, allows the analyst to set the reproduction speed. A time-based navigation bar substitutes the structural one. The analyst is able to play, pause and stop, and to advance and recede to the time when the user changed the page. Note that this navigation is history-based and not stack-based. The status bar shows a timeline and the total time that the user spent when manipulating the artefact.

4   CATMA

CATMA (Comparative & collaborative Analysis Tool for Mobile Artefacts) aims at supporting comparative and collaborative analysis of mobile applications/artefacts in various configuration settings. The mobile applications/artefacts targeted by CATMA are those managed by MobPro tools. CATMA introduces two central units to the framework: virtual spaces where multiple artefacts can be instantiated, manipulated and discussed; and a communication subsystem (Fig. 3) that supports complex group management and messaging delivery, through diverse network configurations.

As for the MobPro presentation, we will refer to "users" as those individuals browsing and filling-in the artefacts, whereas "analysts" are individuals that use CATMA to analyse users' activities.

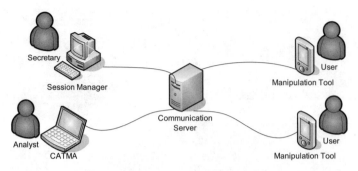

Fig. 3. CATMA's Communication Subsystem

CATMA's communication subsystem ensures not only the correct flow of messages in the system but also the group management during analysis sessions. The Communication Server acts as a message dispatcher, receiving and forwarding messages to the right recipients using a publish / subscribe mechanism The server typically runs on an isolated PC and all session's participants are required to connect to it prior to joining an existing session. The Session Manager allows an analyst or an auxiliary secretary to create, initiate and end analysis sessions. It is also responsible for controlling user entrance in sessions. The manager may access either an SQL database or a set of XML files to check on the expected users for a specific session. All messages exchanged between the Communication Server, the Session Manager and the participants' tools (CATMA and the Artefact Manipulation Tool) use a simple XML schema.

4.1 Comparative Analysis

Within this dimension, we present the mechanisms CATMA provides to visualize and comment upon multiple users' interaction with mobile artefacts. Three fundamental scenarios are considered: results' browsing, interaction replay and online monitoring.

4.1.1 Results Browsing

In this scenario, the analyst is able to browse synchronously through a set of results (final filling-in status). Its normal usage includes comparing results entered on the same type of artefact by: (i) different users; (ii) the same user, on different occasions. Both pertain to comparative analysis scenarios, though the former aims the performance comparison among individuals, whereas the latter assesses user evolution. Other usages are possible, e.g., analysing results from different artefact types, while expectedly less common. Fig. 4 displays CATMA during a results' browsing task. On the right, the figure shows a querying user interface that permits the analyst to select the results from a repository (SQL database or XML file-set). Artefact type, timeframe and user, are some of the filters available for result selection. Once selected, the artefacts/results may be instantiated in a virtual space, shown at the middle of the figure. Two artefacts are presented, along with an isolated annotation. The left artefact and its container are shown in more detail in Fig. 5.

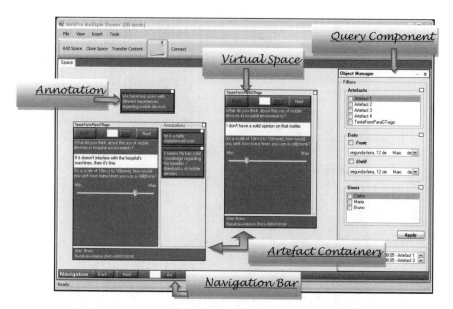

Fig. 4. Results Browsing

The container includes the artefact instantiation (similar to the results' viewer mode of MobPro tools), an annotations section (Fig. 5 on the right) and a status information bar. On top of the artefact is a navigation bar, providing independent browsing for each artefact on the virtual space, i.e., if the analyst presses "next" on a container navigation bar, then only the related artefact advances to the next page. The annotations' section in the figure shows two opened annotations associated to the current page of the artefact. Browsing through pages will change the annotations accordingly.

The overall annotation process usually starts with a virtual space annotation (e.g. the isolated one, seen in Fig. 4). If the annotation represents a note about the whole analysis process or if it pertains to the entire artefact, then it stays at the virtual space level. If the annotation refers to a specific artefact page, it can be dragged into the corresponding annotation section.

CATMA synchronized-navigation bar (at the bottom of the tool, on Fig. 4) adapts its user interface to the virtual space content. In the results' browsing case (shown in Fig. 6), it presents mechanism to quickly navigate simultaneously on all the artefacts in the virtual space. The analyst may use "back"/"next" to recede/ advance the current page on all artefacts; synchronize artefacts on a specified page using the "go" button (set all artefacts to page X); or synchronize them with a selected artefact. Artefacts become desynchronized (showing different current pages) when a specific container navigation bar is used.

4.1.2 Interaction Replay

In this situation, the analyst may observe multiple log-files simultaneously. Again, its normal objective involves comparison of interaction histories over similar artefacts, from different users, or from the same users on different occasions.

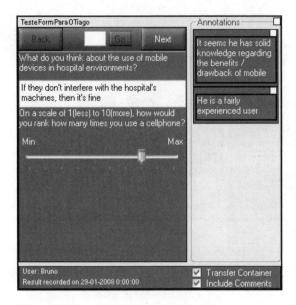

Fig. 5. Artefact Container

Fig. 6. Results Browsing Navigation Bar

The tools' interface is similar to the previous scenario (see Fig. 4). The exceptions are the navigation related bars and the behaviour. The bars associated with each artefact, within the containers (as in Fig. 5), are substituted by a time-based navigation bar and a timeline similar to those of the log player mode of MobPro's Single-User Analysis Tool (see Fig. 2).

Fig. 7. Interaction Replay Navigation Bar

Fig. 7 shows the CATMA synchronized-navigation bar for the interaction replay scenario. Actions on this bar allow the analyst to start/pause all the logs at their current time (weather it is or it is not the same); advance/recede to the next/previous page transition; or stop and rewind the whole set. Annotations are also possible. However, they may further refer to specific time-reproduction periods. In that case, they will reappear during the reproduction of that period, during a play operation.

4.1.3 Online Monitoring

In this mode, the analyst is able to monitor users' interactions with the artefacts while they manipulate them on their own mobile devices. Connection between CATMA, running on the analyst's desktop PC, and the MobPro manipulation tools, running on nearby or distant users' devices, is done through the communication subsystem. Common uses for online monitoring range from follow-up to coordination of group activities. On the former, for instance, CATMA can be used to simultaneously appreciate differences between using artefacts in stationary positions or on the move, in crowded settings.

CATMA monitoring scenario differences from previous ones in two basic dimensions: First, it no longer works as a stand-alone application. Accordingly, a simple user/device selection component substitutes the querying interface (previously shown on the right of Fig. 4). This component allows the analyst to select from the connected users (within a session), who will be monitored in the virtual space.

Fig. 8. Online Monitoring Navigation Bar

Secondly, the navigation control on the monitored artefacts has no direct match on the MobPro tools. The analyst may enable/disable navigation on users' devices through a navigation bar as that shown in Fig. 8. Moreover, the analyst can lock interaction with the current page using the "lock…" button. The bars on the artefacts' containers are similar.

4.2 Collaborative Analysis

Within this dimension, we present the mechanisms that CATMA provides for an analyst to work with other analysts or/and with users in order to visualize and comment on users' interaction with mobile devices. We envisioned two basic settings: public space dissemination and shared space collaboration. Note that the latter is an extension of the former. Moreover, they are orthogonal to the comparative analysis dimension. In fact, each collaborative setting supports the three comparative analysis scenarios, which can also exist without collaborative setting.

4.2.1 Public Display Dissemination (PDD)

In this simple setting no real support for collaboration is provided at the tool level, apart from the existence of a public space that is perceived by all participants in the analysis session. Usually, in each session there is a single analyst who takes on a facilitation and coordination role and a group of users. Note that the users may (online monitoring scenario) or may not (other scenarios) be connected to CATMA, hence their intervention in the analysis process is not done with the tool.

CATMA support to PDD setting is two folded: the availability of multiple virtual spaces and the corresponding data-transfer mechanisms; the support for a distributed public space. Multiple virtual spaces are available in tabular format, as depicted in Fig. 9. Only one virtual public space can be created in each instance of CATMA, even though several private ones can be used to stage data. Transfer between spaces is done by selecting artefact containers (with all its content – both artefacts and annotations), artefacts (without associated annotations) or/and annotations in the origin space and pushing the transfer content button (or dragging into another "tab").

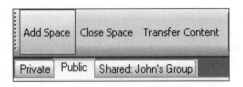

Fig. 9. Virtual Space Management

The distributed public space is a set of simplified instances of CATMA, henceforward referred as public-CATMA, synchronized with the analyst's version. On the latter the analyst prepares data on the private virtual spaces, transfers it to the public one, where he/she moves and changes artefacts and public annotations as needed. All changes on the public virtual space of the analyst's CATMA are propagated to the interconnected public-CATMA instances. A public-CATMA has a single virtual space (public) and by default allows no interaction. The underlying communication subsystem is responsible for that propagation.

Two basic configurations are envisioned: (1) a large public display showing a public-CATMA instance is present in the meeting room; (2) each participant (user) has his/her own copy of the a public-CATMA. The latter usually involves a PC per participant, instead or in complement to a PDA (e.g., in the monitoring scenario). A variant with tablets is particularly interesting for outdoor meetings (e.g. for in exposure therapy, or field studies). A third variant may use an interactive large public display [5]. However, in this simplified setting only one virtual public space allows data manipulation.

4.2.2   Shared Space Collaboration (SSC)

This setting envisages the existence of scenarios with more than one analyst. Moreover, the SSC setting foresees situations where several sub-groups coexist. Each sub-group is composed by several analysts that share information (annotations and artefacts) and can act upon it (manipulating and annotating artefacts). Each analyst can participate in several sub-groups within a collaboration session. Users may or may not be involved (with a role similar to the one in the above setting) and the existence of a public space is also optional.

Two new mechanisms are added to the previous setting: the support for confined and distributed shared spaces; a data access mechanism within those spaces. The first is again available trough CATMA's virtual spaces (see Fig. 9, last "tab"). Contrary to the public space, each CATMA instance is able to support more than one shared space. Upon creating a shared virtual space, the analyst is allowed to invite others, from those connected to the current session, to form a sub-group. Upon approval, a virtual shared space is created on the CATMA instance that accepts the invitation, thus entering that sub-group. The distributed shared space is composed by all the virtual spaces that adhered to the sub-group. The communication subsystem ensures the propagation of changes on either virtual space through the whole distributed space. Transfers between spaces work similarly to the previous setting.

At this stage, the data access mechanism is comprised by a simple lock, being essentially applied on annotations. If an annotation is open for edition, no other instance is able to edit it. Annotations perform an important role in this setting as they also provide the communication support for multiple users.

5 Case Studies

Two case studies were performed to evaluate CATMA. The first one pertains to the CATMAs' genesis, which is the group therapy scenario, whereas the second refers to education and particularly to the usability evaluation of MobPro's manipulation tool and to some extent to tool for document reading.

5.1 Group CBT

Group Cognitive Behavioural Therapy (CBT) provides a particular adequate test-bed for CATMA: first, because it was on the genesis of the tool itself; secondly, because it provides real scenarios with rich requirements. CBT is a form of psychotherapy where patients and therapists engage in a series of meetings to assess the former's current problems [25][28]. This type of therapy relies heavily on the patients' commitment to the process instead of immediately offering the solutions for the focused pathology. Therefore, patients are often required to keep a diary where they write their thoughts or emotions when confronted with situations related to their problems. In addition, therapists often supply questionnaires patients are advised to fill-in autonomously as homework. All these tasks and associated results are analyzed cooperatively in each session by both therapist and patient. CBT has been proven to be a reliable treatment, especially regarding anxiety and depression disorders, albeit patients are required to be participative in the process for it to take a positive effect.

Group CBT takes its foundation on the individual counterpart's characteristics, encompassing theme discussion or questionnaire filling-in processes from the patients' perspective and annotating or monitoring activities from the therapists' point of view. However, the increased number of participants in therapy sessions brings added complexity to the process management. In addition, a second therapist is often required whose roles include preparing data for discussion, alerting for specific issues regarding session or home activities and taking notes. The main therapist usually acts as a facilitator, promoting discussions, proposing new discussion themes and also taking notes about relevant issues. Collaboration between therapists is paramount, although it must be kept concealed from the discussion with the patients. Finally, it is also frequent to perform in-exposure sessions [10] in the outside world in addition to traditional office therapy sessions.

5.1.1 The Experiment

One of the major drawbacks of health-related domains is that experiments should only take place after a thorough evaluation. Moreover, due to ethic reasons, it is impossible to record real therapy sessions. However, with the help of a team of therapists, a small therapy group was arranged to simulate a group session.

We used a group of 5 students, aged 20-24 to act as patients while a couple of therapists played their own role. A set of general and therapy questionnaires, the latter defined by the therapists, were delivered to the students in a PDA. They answered them using the MobPro manipulation tool. The students got acquainted to the tool and the process, including their role in the group session. The therapists were instructed on how to use CATMA on a previously arranged training session.

The experiment was performed in a room, with a large oval table; a small table in the back; a large display with a corresponding projector; a dedicated wi-fi network and a PC. The PC ran the communication subsystem, a server to a database containing the artefacts, the results and logs (from students usage), and an instance of a public-CATMA that was projected in the large display. The therapists were given a laptop each running CATMA. The secondary therapist sat on the back while the principal sat on the oval table near the large display. Students kept the PDAs that they used previously while filling-in the questionnaires.

The experiment ran in two phases: on the first one, the previously filled questionnaires were discussed; on the second one a new questionnaire (artefact) was answered by the students during the session. In both phases therapists created their own private space, a distributed shared space (between them) and a public space that included the public-CATMA. The session followed at the will of the therapists. The therapist interaction with CATMA was observed and filmed.

5.1.2 The Results

Despite a few connection stability issues, the experiment ran with few technical problems. A post-experiment interview was made with the therapists, after visualisation of the recorded video. No major difficulties were reported, apart from the initial setup of the environment. Some of these emerged from the connection problems, others from the setup of the distributed shared space. The need for recreating the virtual spaces on the second phase of the project was not well understood. The technical reason for that is related with the use of online monitoring spaces instead of results' browsing.

On the positive side, therapists pointed out the ability for cooperating and communicating over a digital medium and the direct annotation of artefacts and pages under analysis as a major breakthrough. The preparation and transition of notes and artefact to successive spaces was also welcomed as well as the ability of having public discussion space (not common in paper based therapy). On the second phase (online monitoring), therapists referred as highly helpful the awareness they had on the patients/students hesitations while filing the artefacts as well as the ability to control navigation on patients activities. Overall, therapists felt that the tool is strongly beneficial in real group therapy settings.

5.2 Education and Usability Evaluation

This case study builds on the analysis of a real evaluation process conducted in a faculty course, Hypermedia. Students' assignments were to comment and summarize articles related to the course subject matter. Afterwards students were asked to answer to a test in digital format using MobPro's manipulation tool, first without consulting their notes and then accessing them. Finally students answered two usability questionnaires, again using MobPro: one about the tool they used to read/listen and comment the articles (called Rich Book Player - RBP [8]) and the other to assess the usability of MobPro itself. Two articles, "The Dexter Hypertext Reference Model" and "AHAM: A Dexter-based Reference Model for Adaptive Hypermedia", were delivered to the 33 students. The first article was read, commented and the questionnaires were filled in a controlled room, whereas the seconds' work was performed at students' home. At the time, the analysis focussed on the evaluation of student's performance in the subject matter and in the usage of the RBP. In both cases the work was done by hand, without computational support, except for the quantitative analysis and statistics. The need of a qualitative analysis that enabled the comparison of results and especially of interaction logs was felt.

5.2.1 The Experiment
The case study in focus in this paper uses the results and interaction logs that emerged from filling-in the above mentioned questionnaires. Three sets of results were considered: the ones from the assignment evaluation (test); the other from the evaluation of the reading/listening and commenting tool; and the last one from the evaluation of the MobPro tool. Three teachers were involved in the analysis of the all sets, on three sessions. The sessions were filmed for later study. On the first session, each teacher stood in his own office, thus remotely connected to the communication server.

One of the teachers prepared two private spaces, one with the results and the other with the logs, started a session at a previously arranged time and created two shared spaces. The other two teachers joined the session and accepted the invitation to join the distributed shared spaces. The analysis started with the first teacher publishing the prepared contents into corresponding shared spaces. First the results were analysed. Then logs were selectively published by the first teacher that also controlled the playback. In both cases teachers were encouraged to enter their own comments.

On the second session, on the evaluation results of RBP, the same teachers worked in the same room with a large public display. A public-CATMA, besides the instances running on the laptops of each teacher, was running on a dedicated PC connected to the projector. The session procedures were similar, except for the replacement of the shared spaces by a single public one. A different teacher worked as the session facilitator (preparing and publishing the material).

On the third session, where the MobPro's manipulation tool usage was analysed, the setting was the later one. However, each teacher had direct access to the results and logs, thus preparing and working in their own private space, in the first place. No facilitator was previously appointed. Teachers published their own comments and target results/logs as they felt.

5.2.2 The Results
Some interesting results were found on the first two sessions. Teachers, other than the facilitator, created their own private spaces where they copied the contents shared by the first teacher. This allowed them to create their own private comments before publishing them to the shared/public space. On the first case subjects felt that the invitation and consequent acceptance were unnecessary, since there was no one else in the meeting. A mechanism similar to the public space was suggested. On the other hand, on the second experiment, the need for reformatting the public space from results analysis to logs inspection was felt clumsy. The suggestion here was to have more than one public space available, although that raises an issue for the public-CATMA version.

On the last session, in particular, difficulties were felt in the control of different published copies of artefacts and comments. Sometimes more that one teacher publish to the public space the same result set, and their own comments, and the tool was not able to merge them into the same container with the multiple comments. In this case two containers were crested in the public space with the same result set.

6 Conclusions and Future Work

In this paper we presented CATMA, a tool for comparative and collaborative analysis of mobile applications usage. The work motivated by the support to psychological group therapy, clearly encounters broader application areas, not only in practical domains, but also in the study and evaluation of the mobile artefacts themselves. CATMA's major contribution is the confluence of comparative analysis with collaboration and real-time monitoring. An experiment was performed that qualified its applicability in the envisaged scenarios. In G-CBT the experiment results clearly point to an enhancement of the therapy process and an obvious acceptance on the therapists' side. In the Education scenario, we were able to determine interesting behaviours by the subjects using our software as well as listing.

In the near future, we aim at integrating multimodal features to CATMA, opening new frontiers and broadening even further the covered domains. We are also keen on extending CATMA's own evaluation to all involved actors.

Acknowledgements

This work was supported by FCT, through project JoinTS, through the Multiannual Funding Programme and through Individual Scholarship SFRH / BD / 39496 / 2007.

References

1. Alsos, O.A., et al.: Interaction Techniques for using PCs and Handhelds Together in a Clinical Setting. In: Procs. of the 4th Nordic Conference on Human-Computer Interaction: Changing Roles, Norway, pp. 125–134 (2006) ISBN: 1-59593-325-5
2. Badre, A.N., et al.: A user interface evaluation environment using synchronized video, visualizations and event trace data. Journal of Software Quality 4 (1995)
3. Baldonado, M., et al.: Notable: an Annotation System for Networked Handheld Devices. In: CHI 1999 Extended Abstracts on Human Factors in Computing Systems (1999)
4. Camacho, J., et al.: Supporting the Management of Multiple Activities in Mobile Collaborative Working Environments. In: Dimitriadis, Y.A., Zigurs, I., Gómez-Sánchez, E. (eds.) CRIWG 2006. LNCS, vol. 4154, pp. 381–388. Springer, Heidelberg (2006)
5. Carriço, L.: Managing Group therapy Through Multiple Devices. In: Procs. of the 12th International Conference on Human-Computer Interaction HCI International (2007)
6. Crowe, E.C., et al.: Comparing interfaces based on what users watch and do. In: Procs. of the 2000 Symposium on Eye tracking research and applications, pp. 29–36 (2000)
7. Davis, R., et al.: NotePals: Lightweight Note Sharing by the Group, for the Group. In: Procs. of the SIGCHI, USA, pp. 338–345 (1999)
8. Duarte, C., et al.: Evaluating Usability Improvements by Combining Visual and Audio Modalities in the Interface. In: Procs. of HCI International 2007, China, pp. 428-437 (2007) ISSN: 0302-9743
9. Greenberg, S., et al.: PDAs and Shared Public Displays: Making Personal Information Public and Public information Personal. Journal of Personal and Ubiquitous Computing 3, 54–64 (1999)

10. Heimber, R.G.: Cognitive-Behavioural Group Therapy for Social Phobia: a Treatment Manual (1991)
11. Hilbert, D.M., et al.: Extracting usability information from user interface events. ACM Computing Surveys 32(4), 384–421 (2000)
12. Lanzenberger, M., et al.: Applying Information Visualization Techniques to Capture and Explore the Course of Cognitive Behavioral Therapy. In: Procs. of the 2003 ACM symposium on Applied Computing, USA, pp. 268–274 (2003) ISSN: 1-58113-624-2
13. McLoughlin, E., et al.: MEDIC - MobilE Diagnosis for Improved Care. In: Procs. of the 2006 ACM symposium on Applied Computing, France, pp. 204–208 (2006) ISBN: 1-59593-108-2
14. Mejia, D., et al.: Supporting Informal Co-Located Collaboration in Hospital Work. In: Haake, J.M., Ochoa, S.F., Cechich, A. (eds.) CRIWG 2007. LNCS, vol. 4715, pp. 255–270. Springer, Heidelberg (2007)
15. Myers, B., et al.: Collaboration using Multiple PDAs Connected to a PC. In: Procs. ACM CSCW 1998, USA, pp. 285–294 (1998)
16. Neil, S.A., et al.: Playback: A method for evaluating the usability of software and its documentation. IBM Systems Journal 23(1), 82–96 (1984)
17. Newman, M.: Technology in Psychotherapy: An Introduction. Journal of Clinical Psychology 60(2), 141–145 (2003)
18. Perry, D.: Handheld Computers in Schools. In: British Educational Communications and Technology Agency (2003)
19. Pinelle, D., et al.: Task Analysis for Groupware Usability Evaluation: Modeling Shared-Workspace Tasks with the Mechanics of Collaboration. ACM Transactions on Computer-Human Interaction 10(4) (2003)
20. Sá, M., et al.: A Framework for Mobile Evaluation. In: Procs. of (CHI 2008), SIGCHI Conference on Human Factors in Computing Systems, Italy, pp. 2673–2678 (2008)
21. Sá, M., et al.: A Mixed-Fidelity Prototyping Tool for Mobile Devices. In: Procs. of the International Working Conference on Advanced Visual Interfaces AVI 2008 (2008)
22. Sá, M., et al.: Detecting Learning Difficulties on Ubiquitous Scenarios. In: Procs. of HCI International 2007, China, pp. 235–244 (2007) ISSN: 0302-9743
23. Sá, M., et al.: Ubiquitous Psychotherapy. IEEE Pervasive Computing, Special Issue on Healthcare 6(1), 20–27 (2007)
24. Smith-Stoner, M.: Uses for Personal Digital Assistants. Home Health care Nurse 2(12), 797–800 (2003)
25. Timms, P., et al.: Cognitive Behavioural Therapy. In: Leaflet for the Royal College of Psychiatrists Public Education Editorial Board (March 2007)
26. Weiler, P.: Software for the usability lab: a sampling of current tools. In: Procs. of INTERACT 1993 and CHI 1993, Conference on Human Factors in Computing Systems, pp. 57–60 (1993)
27. Xu, W., et al.: Multimodal Archiving, Real-Time Annotation and Information Visualization in a Biofeedback System for Stroke Patient Rehabilitation. In: Procs. of CARPE 2006, USA, pp. 3–12 (2006)
28. Cognitive Behavioural Therapy. In: Explanation page by the National Association of Cognitive Behavioural Therapists

Coordination Patterns to Support Mobile Collaboration

Andrés Neyem, Sergio F. Ochoa, and José A. Pino

Department of Computer Science, Universidad de Chile,
Blanco Encalada 2120, Santiago, Chile
{aneyem,sochoa,jpino}@dcc.uchile.cl

Abstract. The increasing popularity of portable devices and advances in wireless communication technologies push the development of mobile groupware applications. Mobile applications are challenging for software designers because the use of centralized components is not recommended, the communication service cannot be ensured and the software must run on computer devices with little hardware resources. Frequently, data and services interoperability is also required for collaborators. A design patterns system is presented as a way to deal with these modeling requirements; it is intended to help modeling the coordination services required to support mobile collaboration. These patterns serve as educational and communicative media for developers, students or researchers on how to design services for mobile collaborative applications. They also foster the reuse of proven solutions.

Keywords: Coordination patterns, groupware mobile applications, design guidelines, mobile collaboration.

1 Introduction

Collaborative systems provide support for groups of persons while they communicate and coordinate their activities to reach a common goal [10]. Both communication and coordination are required to support collaboration. Communication refers to the information exchange among cooperating group members, and coordination relates to coordinate group tasks.

Building collaborative systems has always been a complex undertaking because it involves issues that are not relevant while developing single-user systems, such as human-to-human communication and awareness, group dynamics, users' social roles, group memory, and other organizational and social factors. Trying to deal with these issues, the CSCW community has designed solutions to support communication and coordination in stable communication scenarios (wired networks) and the obtained results are highly applicable. However, the current advances in wireless communication and mobile computing have brought new challenges, needs and opportunities for collaboration in several scenarios [27].

Most communication and coordination solutions designed to support collaboration on wired networks are inapplicable to wireless networks. This occurs because wireless networks have dynamic topologies and thus, it is impossible to ensure availability of communication among collaborators or access to centralized data/services [2], [20].

R.O. Briggs et al. (Eds.): CRIWG 2008, LNCS 5411, pp. 248–265, 2008.

In turn, the design of coordination services that support collaboration in mobile scenarios gets complicated. Examples of these coordination services are user/session management, roles support, shared data space management and message delivery. These services now should not use centralized components [20], should allow mobile users to be autonomous and collaborate on-demand [22], [26], should allow interoperability of data and services [20], [22] and probably these services will have to run on mobile computer devices with little hardware resources [2], [20].

This paper proposes a design patterns system [28] intended to help software designers to model the coordination services required to support mobile collaboration. Next section describes the mobile collaboration considering four different work scenarios, and the requirements to design the coordination services for each one. Section 3 presents the related work. Section 4 describes the proposed patterns system to deal with the design challenges of coordination services. Finally, section 5 presents the conclusions and future work.

2   Mobile Collaboration

Mobile collaboration has increasingly become an important issue in CSCW. However, efforts to understand the implications that mobile work and mobile collaboration have on groupware design are still a research subject [1], [2], [11], [13], [24]. Mobile groups are highly varied in the ways they organize work, in the physical dispersion of mobile workers, and in the styles of collaboration that take place among workers [3], [16], [30]. While trying to make sense of this diversity, there exist efforts to describe and classify these variants by focusing on specific types of mobility [15], types of physical distributions that occur in mobile groups [16], and levels of coupling among mobile collaborators [25], [8].

These research contributions show that mobile workers are those who have to work out of office, move around locally or remotely. Their tasks must be performed on the site within a specific timeframe or they have to work at different locations with a carried "portable office" with limited resources. They often work based on an irregular schedule with narrow time windows for action. Mobile workers have to cope with great uncertainty and interference. Therefore, in order to understand the implications that mobile work and mobile collaboration have for the design of coordination services, we found it is necessary to provide a conceptual framework to analyze and explore the vision of time and place in the context of mobility (Fig. 1). This classification is a variant of the traditional time space taxonomy proposed by Ellis at al. [10].

This conceptual framework lets developers analyze the interactions among several mobile workers, considering time and space dependence when carrying out an interaction. Next sections describe each of the quadrants of this framework.

2.1   Different Time / Different Place

The first quadrant of Fig. 1 indicates the interactions between two mobile workers can be done independently of the timeframe they have available to collaborate and the place where they are located. This type of collaboration is possible just if there are

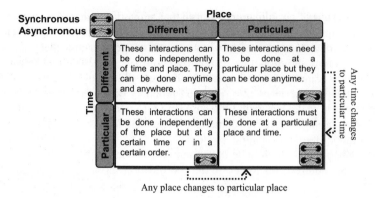

Fig. 1. The conceptual framework to analyze the mobile workers interactions

one or more intermediary components (a server or mobile collaborative units) able to keep and communicate a set of ordered messages between the sender and the receiver.

Let us assume a mobile worker wants to collaborate with a partner working at a distant location and in a different timeframe. Then, the requester can send a coordination message to the destination actor through the (intermediary) mobile workers that are close to him/her. Such message will eventually be propagated when mobile actors physically move; thus, the coordination message eventually could be delivered to the destination actor. The coordination message typically has information allowing these actors to hold a meeting. For example, a message indicating the reason (or need) to collaborate, and a list of places and times frames where the requester will be in the future or a phone number to coordinate a meeting.

2.2 Different Time / Particular Place

This quadrant means the interactions among collaborators can be done anytime, but at a specific place. An example of such type of interactions occurs when inspectors work in construction scenarios. Typically each construction site has a main contractor, which outsources several parts of the construction project, e.g. electrical facilities, gas/water/communication networks, painting and architecture. Some of these sub-contracted companies likely work during different time periods but in the same place. While inspecting, engineers revise various parts of the facilities and record the advances on a mobile groupware application. The main contractor is in charge of keeping track of the construction project updates. In order to get a whole view of the state of the project, all inspectors of sub-contracted companies need to go to a particular place to synchronize their information with the main contractor. Thus, they are able to detect incomplete or contradictory information with other sub-contractors.

2.3 Particular Time / Different Place

This quadrant indicates interactions among actors must be done during a specific time frame, but including different places. The delivery of healthcare services to patients at their homes presents an example of this interaction scenario. Each patient is typically treated by a team, including therapists, nurses, social workers, and home health aides.

Workers physically move and work out at different locations in a particular time. They spend most of the day in the community and they may only spend minimal time in the office. Typically these persons have particular timeframes during the labor day to collaborate among them in order to coordinate care plans for some patients. The collaboration process could be done by phone or through a mobile groupware system.

2.4 Particular Time / Particular Place

The fourth quadrant means some interactions need to be done in certain places and within a certain period. The relief efforts at a disaster site make an example of this interaction scenario. When a disaster occurs in an urban area, there is minimal availability of communications services [22]. Therefore, the participating organizations (police, firefighters, medical personnel and government agencies) have to go to the command post to get basic information about the affected buildings (e.g. maps, probable people locations and vulnerable points), exit routes, resources deployed in the area and tasks assignments. If these first responders update the shared information received from the command post, i.e. because it was wrong, then the updated information should be carried to the command post and communicated to the disaster decision-makers in order to disseminate the new information.

2.5 Requirements to Support Mobile Collaboration

The previous conceptual framework shows there are many work scenarios where mobile collaboration need to be supported. Provided the actors' mobility, the interaction scenarios can change from one quadrant to another one. This section summarizes the groupware requirements that are need to support the interaction scenarios described by the conceptual framework.

Autonomy. Collaborative mobile applications should work as autonomous solutions in terms of communication, data and functionality. This is because wireless networks have a high disconnection rate and their communication threshold is short [20].

Interoperability. Since mobile workers could include unknown persons trying to do casual or opportunistic collaboration, their mobile groupware applications should offer communication, data and services interoperability.

Shared information availability. Shared information supporting collaborative applications in these scenarios need to be highly replicated due to frequent disconnections (even using access points).

Variability of the work context. Since users are on the move to carry out their activities, their work context can frequently change. Some attributes, such as network topology and the Internet/servers access, will change from one location to the next one.

Use of hardware resources. Collaborative mobile applications should operate, in many cases, with constrained hardware resources; e.g., the case in which these solutions need to run on Personal Data Assistants (PDAs). Therefore, the communication and coordination services should be lightweight.

Low coordination cost. Tasks are often strongly partitioned among workers. This partitioning minimizes coordination demands and it allows people to work autonomously and in parallel [24]. Ideally, the coordination process should be unattended [22].

Awareness of users' reachability. Mobile workers need to know when a particular user is reachable, because they do on-demand collaboration. Hence, awareness mechanisms indicating user reachability should be embedded in mobile groupware applications.

Deployment ease. An important factor is the speed of having the device ready to operate. A quick boot-up time will let workers productively use dead times. Applications – in particular, these workspaces – should also self-configure automatically after boot-up [22].

3 Related Work

There are several experiences reporting the use of collaborative mobile applications [20]. Although some of these applications are fully-distributed, they do not describe or evaluate the strategies used to support coordination in mobile collaborative scenarios. Thus, the potential design solutions cannot be evaluated when they are formalized through design patterns or reused in future applications. Schümmer and Lukosch argue that groupware reuse should focus on design reuse rather than code reuse [28]. These researchers also propose a patterns system for groupware for stationary scenarios, therefore they do not consider the users mobility.

Jørstad et al. have proposed and describe a set of generic coordination services for distributed (but stable) work scenarios [14]. These services include locking, presentation control, user presence management and communication control. Other researchers have also proposed similar solutions to support coordination on fixed networks [4], [6], [12]. However, the contextual variables influencing the collaboration scenario (e.g. communication instability and low feasibility to use servers) and the mobile work (e.g. use of context-aware services and support for ad-hoc coordination processes) make such solutions unsuitable to support mobile collaboration. Next section presents the patterns system we propose to structure the coordination services in order to support mobile collaboration.

4 Patterns System

We propose the use of a layered and fully-distributed architecture for mobile groupware applications since collaboration is based on communication and coordination [10]. The advantages of the layered architecture have already been discussed and recognized by the software engineering community [5], [7], [9]. Figure 2 shows the *crosslayer* pattern, which structures the basic functionality of a mobile groupware application.

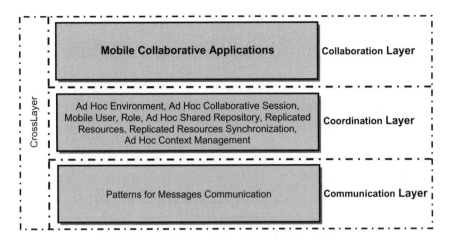

Fig. 2. Layered architecture to support mobile collaboration

The coordination patterns system proposed by this article focuses on the services of the coordination layer. These patterns use (through the API) the services provided by the communication layers. Communication services were also grouped and specified in a patterns system for ad-hoc communication [18]. Next section briefly describes these patterns using the description language proposed by Schümmer and Lukosch [28].

4.1 CrossLayer

Context. Groupware applications separate functionality in three basic concerns: communication, coordination and collaboration. Each layer provides services and records data related to such services. These services are different in term of concerns and granularity. The interaction between services related to different concerns is hierarchical: communication <-> coordination and coordination <-> collaboration. Interoperability among these services is required to support mobile collaboration, because frequently the service provider and the consumer run on two different computing devices.

Problem. If the services provided by the groupware system are not well structured, the system will be limited in terms of scalability, maintainability and adaptability.

Solution. The services related to different concerns can be grouped in different layers of a software architecture. Designers can thus separate concerns and increase the system scalability, maintainability and adaptability. The platform also gets sound to deal with interoperability issues. The architecture should be fully replicated to cope with the mobile users' autonomy. Thus, we can view the collaboration scenario as a dynamic mesh without centralized components.

Groupware services and public data structures belonging to each layer should be accessible through an API in order to keep the services independence and the access control. The interaction protocol between services is part of each layer, and it can be dynamically selected based on contextual information. For example, mobile devices with little hardware resources require lightweight mechanisms for data sharing or peers discovery. If an application running on a laptop must interact with a service running on a PDA, contextual information about the PDA's hardware resources (stored in a particular layer) will be needed to dynamically adapt the interaction protocol between them.

Related Patterns. The most related pattern is *layer* [9] but it is not focused on distributed/groupware systems or services interoperability.

Related Mobile Groupware Requirements. Autonomy, interoperability and deployment ease.

4.2   Coordination Patterns

The coordination patterns concern the provision of services required by mobile workers' applications to coordinate the operations on the shared resources (e.g. files, sessions and services). This coordination is made individually (per mobile unit) and it generates a consistent view of the group activities. Figure 3 shows this patterns system.

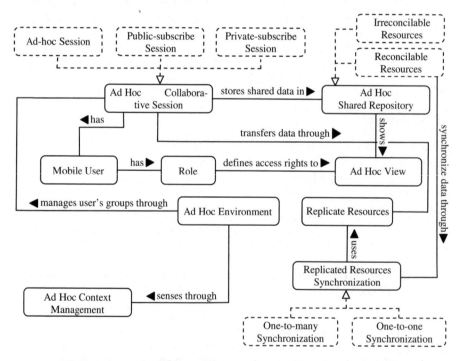

Fig. 3. Coordination Patterns System for Mobile Groupware Applications

4.2.1   Ad Hoc Environment

Context. Mobile collaborative applications can require several work sessions. Each session groups users and shared data and services. Mobile users need to know which sessions are currently available in order to try to access those relevant ones for them, or otherwise to create a new work session to collaborate with team mates.

Problem. It is not possible to use a centralized list of the current available work sessions due to the mobile collaboration process feature mentioned in section 3. It means the list of work sessions with the respective participants must be kept in a distributed way and the system has to ensure its information integrity.

Solution. This pattern proposes an ad-hoc environment to deal with this problem. It contains a fully distributed list of work sessions available for each mobile user. Each mobile unit (and mobile user) has an instance of this list and a set of services to reconcile the local list with the list of team mates.

The local ad hoc environment maintains information with the environment ID (there is one for each user), description, creator and the list of available work sessions available. It also contains the list of sessions where the local user is member. The integrity of the information stored in each ad hoc environment can be kept using a reconciliation service such as the one proposed by Messeguer et al. [18]. This information can be used by the mobile groupware application to implement a mechanism of user connection awareness or support to on-demand collaboration. The advantages of this solution are the following ones:

- One mobile application may support several work sessions composed of several mobile users.
- A session member's work should not interfere with the work of other session's members, even if they are working on shared objects.
- The environment can provide general services to support shared workspaces. Examples of the services are files transfer, message delivery, peer detection and user/session awareness.

Related Mobile Groupware Requirements. Awareness of users' reachability, variability of the context, use of hardware resources and deployment ease.

4.2.2 Ad Hoc Collaborative Session

Context. Mobile users require on-demand collaboration based on some goals, e.g. common work or similar interests. Typically they share data, knowledge or services as part of this collaboration process. The interaction among mobile users should be protected in order to avoid unauthorized access to resources shared among them.

Problem. Similar to the ad hoc environment, mobile collaboration requires managing sessions in a distributed way and keeping the integrity of the information shared by the session members. Access control to shared resources based on each mobile user's role for a session is also required. Given the users' mobility, work sessions should be dynamically splitable or unifiable depending on the availability of a communication link among session members.

Solution. The solution to this problem is to use an ad-hoc collaborative session. The management of these sessions is done in a fully distributed way; therefore each mobile unit has to do it locally and keeping synchronized with the rest of the session members. Similar to traditional groupware sessions [12], ad hoc collaborative sessions have a list of supported roles (rights to access the shared resources), users members (with roles), a shared dataspace, and a session type considering the access control for users (ad hoc, public or private session).

A work session is created when the first user is registered as member of it and it is deleted when the last user is unregistered. A session is potentially alive even if no users are currently connected, but there are registered users. The work session types matching mobile collaboration are the following ones: ad-hoc, public-subscribe and private-subscribe. The ad-hoc session is an open public resource that can be accessed by any user connected to the wireless network. The public-subscribe session involves a simple subscription process. Typically, users request a session subscription and automatically obtain the right to access it. Finally, private-subscribe sessions require a subscription process carried out by invitation. Each invitation has associated a user role. If the mobile worker accepts the invitation, then s/he will play such role in that session. The strategy for session management must allow mobile users participate in more than one session.

Every mobile user must have a local private and a shared repository for each session s/he belongs. It allows her/him to share resources on-demand. When a user logs in a session, s/he becomes visible and s/he can access the shared resources of such session. At that moment, the user's local shared resources become visible to the rest of the session members. When a user leaves a session, the local private and shared resources are kept available for him/herself, by allowing the user work asynchronously.

Typically, not all users have the same rights to access shared resources. The rights are related to the user's role for each session s/he is working on and indicates the user capability to carry out certain operations or processes on the shared resources. Mobile users usually have many work sessions with certain assigned role. Therefore, they need a mobile environment organizing and eventually coordinating multiple working sessions or user groups playing several roles. Sessions, users and roles management should be fully-distributed since the mobile environment should be autonomous.

Related Mobile Groupware Requirements. Autonomy, shared information availability, awareness of users' reachability, low coordination cost and deployment ease.

4.2.3 Ad Hoc Shared Repository

Context. Team members doing mobile collaboration produce information as a result of the individual and collaborative work. These persons are frequently disconnected, perform activities autonomously and work in parallel; therefore they need instances to share and synchronize their information.

Problem. Since mobile workers have to be autonomous, the resources required by them during an activity should be reachable all the time. Nevertheless, the mobile groupware applications are not able to support collaboration accessing centralized dataspaces. Therefore, the shared dataspace should be fully distributed. It means the shared information will be replicated in the mobile units used by the work session members. This replication adds inconsistency to shared resources and the reconciliation process increases its complexity.

Solution. A solution to this problem involves the use of an ad-hoc data shared repository. This component is a fully distributed dataspace embedding two mechanisms to keep consistency of the shared resources among session members.

Typically session members have a local (private) repository to store the private resources and a shared (public) repository to store the resources they want to share with the partners. The shared repository contains two types of information resources: irreconcilable and reconcilable. The irreconcilable resources are those pieces of information that the system has no information about their internal structure. The consistency among these resources is kept just through file transfer. On the other hand, a reconcilable resource is a piece of information with a well-known internal structure; therefore it can be synchronized with other copies of such resource (from other mobile users) in order to obtain a consistent representation of it. The sharing process and the data consistency are done using synchronization processes.

Since the structure of XML documents is flexible and it can be dynamically analyzed, we recommend using such format to implement the shared resources and to reconcile information. Every mobile host has to maintain two types of XML documents: versions and editions. Versions contain changes that have been performed locally (without communicating them to the other hosts). Editions are, in a sense, stable versions; they contain changes that have been agreed with another host, after a reconciliation process. Therefore, an edition can have both versions and editions as directed descendents in the version graph, whereas a version does not have descendants, because first it has to be upgraded to an edition. Consequently, versions always contain the most recent information.

Related Mobile Groupware Requirements. Autonomy, interoperability, shared information availability, low coordination cost, and deployment ease.

4.2.4 Replicated Resources Synchronization

Context. Mobile users work autonomously most of the time and they carry out sporadic on-demand collaboration processes to keep updated the local dataspace. Even if the collaboration process is tightly coupled, the users' mobility may cause disconnections and inconsistencies on the shared information.

Problem. Data consistency in fully distributed scenarios usually involves synchronization processes. These processes define which replicas exchange updated packets and in which direction the exchange will be done. When the synchronization process has to be done using a Mobile Ad hoc Network (with dynamic topology) including heterogeneous devices, the synchronization processes will be affected by several factors. Examples of such factors are: bandwidth between mobile devices, computing power of the involved devices, network topology and latency of changes. In addition, this synchronization process must be done in a short time period, because frequently reconciliations are done as unattended (background) processes triggered while the user is on the move. Since the period of contact among mobile collaborators cannot be ensured, the reconciliation process should be as fast as possible.

Solution. The reconciliation algorithm proposed to perform the synchronization of shared resources is simple; it uses a mechanism to control possible conflicts between different replicas. The main design goal of this algorithm was to minimize data transfers, as a way to reduce the synchronization process duration. The algorithm transmits just the differences between data structures and, at the same time, is able to reconstruct diverging replicas from a common previous edition on the same host in order to reconcile them locally. Then, the result of the reconciliation is propagated to the other hosts, communicating only the changes done on the common latest edition.

Figure 4 shows an example of synchronization of two documents using the reconciliation algorithm. The reconciliation process starts when Host A sends a reconciliation request to Host B. It receives the request and starts a local reconciliation using the information sent by A. We refer to the copy of the document stored on Host A as Doc_A and that maintained on Host B as Doc_B. Let us also suppose that, after the execution of the first part of the protocol, the document Doc_{CE} has been choosen as Latest Common Edition [19]; i.e., Doc_{CE} is base document of the XMLTreeDiff algorithm [17]. Host A computes XMLTreeDiff with Doc_{CE} and Doc_A as arguments (Doc_{CE} is the base document, whereas Doc_A is the modified document). The output of the execution of this method will be the "diff" document Doc_{diff}, which will be sent to Host B. After receiving Doc_{diff}, B executes XMLTreeMerge with Doc_{CE} and Doc_{diff} as arguments in order to reconstruct Doc_A locally. Therefore, Host B now has a local copy of Doc_A and, naturally, Doc_B. Thus, the reconciliation between these two documents is performed on Host B without exchanging information with Host A during the execution of the algorithm.

It is possible to reconcile Doc_A and Doc_B using the XMLTreeReconcile component. The arguments of this component are: the local copy of the document Doc_B, the remote copy Doc_A and the latest common edition Doc_{CE}. The output will be a "reconciled document" called Doc_{CEn}. The final step is the generation of the reconciled

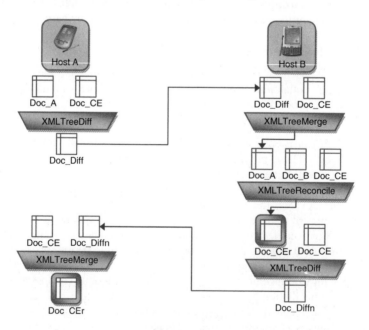

Fig. 4. The reconciliation algorithm

document on Host B. This action is executing the XMLTreeDiff again with Doc_{CE} and Doc_{CEn} as arguments, in order to compute a new "diff" document. Afterwards, the document Doc_{diffn} is sent to Host A, and XMLTreeMerge executes with Doc_{CE} and Doc_{diffn} as arguments. Now, Hosts A and B store the reconciled copy, which will be the new latest common edition.

Related Mobile Groupware Requirements. Shared information availability, low coordination cost, and use of hardware resources.

4.2.5 Replicate Resources

Context. Users produce data as a result of the mobile collaboration process. This data is stored in local files that mobile users share to support collaboration.

Problem. Users' mobility causes high disconnection rate when transfering a file between mobile units. This disconnection rate forces to design a robust mechanism to support this service. Moreover, the service should be fast and simple enough to run on small computing devices.

Solution. The solution to this problem is to provide a FileTransfer component which is in charge to manage all the file transfer processes. This component implements a transparent way to share these files through multicast or unicast transmission among users interacting in a work session. The file transfer is based on the distribution of a set of small information pieces which can be sent in any order from the sender to the receiver. When a user decides to download certain remote file, the component creates a download request. Then, the file transfer manager uses the contextual information (i.e. hardware features of the interacting mobile computing devices, and the distance between them) to determine the appropriate block size in which the file will be broken down before being transmitted. The block size is relevant to consider because it directly influences the performance of the file transfer process.

Furthermore, increasing the file transfer performance typically involves several other mechanisms. These include reusable data channels to eliminate the high startup costs during the secure authentication, but most importantly, multiple data channels for parallel file transfers. In case only part of the file is needed on the remote user, partial file transfers are allowed in either block or striped mode.

Related Mobile Groupware Requirements. Shared information availability, and low coordination cost.

4.2.6 Mobile User

Context. Users participating in a mobile collaborative process need to be uniquely identified regardless of the computing device they are using. Provided this is an on-demand process, mobile users need to know the identities of the potential collaborators who are currently available.

Problem. Users need a transparent mechanism allowing them to be uniquely identified regardless of the mobile device they are using. A similar identification is required for the potential collaborators (other mobile users in the same area). The information about users' and neighbors' IDs should be managed in a fully distributed way, due to the aforementioned restrictions.

Solution. The solution to this problem is to have a data structure, that we have called *mobile user*, containing the local user information required to support the mobile collaboration and to implement user presence. This structure is local to each mobile device and it is shared among users in order to keep a common view from the users participating in a work session.

Fig. 5. Mobile user data structure

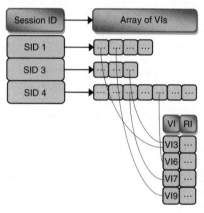

Fig. 6. Matching VIs and RIs

The mobile user data structure contains the mobile unit ID, the virtual (user) ID, the user's role, the visibility attribute and the list of neighbors (Fig. 5). The user's virtual identity (VI) is a unique ID which is linked to the IP address of the user's device. This VI is linked to the real identity (RI) which is the permanent user's ID. User sessions can be implemented as dynamic arrays of virtual identities (Fig. 6). On the other hand, the user visibility attribute allows implementing privacy policies, and awareness of user roles and user availability. The list of neighbors includes the set of potential collaborators available during a particular period.

This list can be updated by two mechanisms: (1) peers discovery and (2) list synchronization. Peers discovery involves sending a message to a peer destination. If the destination is reached, a message is returned to the sender indicating the list of interim visited nodes. Such data is used to update the local list of reachable mobile units and neighbors. Then, a change-propagation mechanism can be triggered to the rest of the session members. In that case, the list update is done using a typical synchronization process.

Related Mobile Groupware Requirements. Awareness of users' reachability, autonomy, and low coordination cost.

4.2.7 Role

Context. Mobile collaborative applications usually require support for mobile users with different rights to access the shared information.

Problem. Users having the same access rights should be treated in the same way by the groupware system. Fully distributed access control management to shared resources is needed because mobile collaboration processes require autonomy.

Solution. The solution involves assigning a *role* to each mobile user in each session that he/she belongs to. The user's role is linked to the user's VI and RI (Fig. 5) and it defines the access rights of that user over the shared resources. Taking into account the reusability of this proposal, we can consider the role as a class maintaining information related to its name, the session to which it belongs and list of access rights to the shared resources (data and services). The role class has to implement methods to store an instance, to erase an instance, to check if a role exists, to check if a mobile user has enough rights to access a shared resource, and to request a list of roles available in certain sessions.

Related Mobile Groupware Requirements. Awareness of users' reachability, shared information availability, and low coordination cost.

4.2.8 Ad Hoc View

Context. The user's role sets the user's access rights on the shared resources (data and services); thus, users with the same role should have access to the same resource list.

Problem. Since the shared resources in an ad hoc session are distributed but no fully replicated, frequently users with the same role have access to different lists of shared resources. Mobile collaboration requires keeping the coherence of the access to shared resources as much as possible, in order to avoid data islands (generating unnecessary parallel work) inside a work session.

Solution. The solution to this problem is to use an *ad hoc view* of the shared resources. This view contains a list of resources with their access grants, which are available for all users having a specific role. There is a view per role. Users with the same role should have access to the same list. These lists are reconcilable as a way to keep the coherence of each view. The only difference that is allowed between the lists of two users having the same role is the resources availability. Although all shared resources are visible, some of them are reachable (if they are locally stored or they are replicable from a neighbor's dataspace) and other ones are unreachable (if neither the current mobile unit nor its neighbors have the resource). In order to increase the availability of the shared resources, a user can ask for a particular view that tries to replicate (in the local shared dataspace) the remote resources that are currently visible but unavailable for him/her.

The ad hoc view can also be considered as a class interacting with the role class presented in the previous section. This class should provide methods to store and delete an instance, to check if a view exists, and to refresh and reconcile a view.

Related Mobile Groupware Requirements. Shared information availability.

4.2.9 Ad Hoc Context Management

Context. By context we mean the variables that can influence the behavior of mobile applications; it includes computing devices internal resources (e.g. memory, CPU speed or screen size) and external resources (e.g. bandwidth, quality of the network connection, and mobile hosts' location and proximity). Both types of variables are relevant to support coordination processes. However, the external variables are more dynamic in mobile scenarios than the internal ones; therefore, it is usually very challenging to sense, store and appropriately use the information they contain. Mobile applications need to be aware of the context in which they are being used to be able to adapt to heterogeneity of hosts and networks as well as variations in the user's environment. Furthermore, context information can be used to optimize application behavior compensating resource scarcity.

Problem. Contextual information is changing all the time while doing mobile collaborative work. Mobile collaborative applications have to sense it, store it and appropriately use it. Since this information is used by the groupware system to dynamically adapt its behavior, such information has to be available all the time and it has to be as complete as possible. Usually there are computing devices participating in the collaboration process which are not able to sense some context variables; however, they are able to use this information if another device provides it to them.

Solution. The solution to this problem involves the creation of an *ad hoc context manager*. This component has to be fully distributed and it must store, update and monitor current status of the context. Mobile groupware applications will adapt their functionality based on that information to cope with the changes in the work scenario (e.g., a mobile worker gets isolated or networking support is not available anymore). For instance:

- If you want to provide a service which is dependent on the place where the user is located, then the context manager needs to implement a model of each place as a full-fledged object and assign a set of command objects with corresponding services to that object.
- If you want to adapt the application behavior according to different time intervals, then the context manager must use condition/action rules to support the behavioral adaptations.
- If you want to extend existing software to add context-aware behaviors, then the context manager must have a functionality that wraps the corresponding class with an object which delegates the request to the component implementing the adaptation (e.g. a rule object or rule manager). This solution uses the Decorator pattern to unobtrusively add new code.

It must be noted the context manager has to be carefully engineered in order to reduce the use of limited resources, such as battery, CPU, memory or network bandwidth. A service-oriented approach can be useful to design and implement this component, because it deals with the heterogeneity of computing devices and resources shortage.

Related Mobile Groupware Requirements. Variability of the context, use of hardware resources, and awareness of users' reachability.

4.3 Summary

It is possible to draw a correspondence matrix considering the proposed patterns and the requirements for mobile collaboration presented in section 2 (Fig. 7). This matrix allows developers to select one or more design patterns in order to deal with a particular requirement. Furthermore, it is important to mention the proposed fully distributed

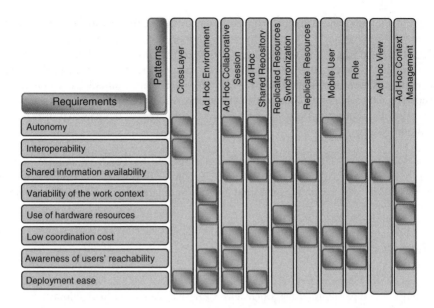

Fig. 7. Correspondence matrix

architecture provides autonomy to the groupware solutions. Also, the separation of design concerns in several layers provides flexibility and scalability.

The proposed patterns system has been implemented on a middleware platform and a variety of applications are currently using these coordination services [21], [22]. These applications include mobile collaborative software to support disaster relief operations, to conduct inspections in construction sites and to manage exams in computer science courses [21], [23].

5   Conclusions and Further Work

Mobile collaboration has brought the opportunity to support work activities in scenarios where workers have to be on the move to carry out a job. Several researchers have envisioned a positive impact on the productivity and the quality of work when users follow a mobile collaboration strategy [29], [27]. However, the features of these collaborative activities bring new challenges to groupware system designers. Requirements, such as user autonomy, low coordination cost and high availability of shared resources, impose several constraints on the communication and coordination services required to support mobile collaboration. For example, no centralized components can be used because the users' mobility can make these resources inaccessible.

This paper presents a patterns system to support the design of coordination services required by mobile collaborative applications. These patterns deal with most of the stated requirements. These patterns serve as educational and communicative media for developers, students or researchers on how to design coordination mechanisms for mobile collaborative applications. They also foster the reuse of proven solutions.

At the moment, these patterns have shown to be useful to design both, mobile groupware applications and a middleware to support collaborative systems [21]. The reuse of these designs and the implementation of the proposed solutions have been quite simple. However, the authors have been involved in each one of these testing experiences. Therefore, future work is required to carry out evaluations with external groupware developers in order to determine the real contribution of this proposal. Moreover, the patterns system should be extended to support devices that use mobile telephone systems.

Acknowledgments. This work was partially supported by Fondecyt (Chile), grants N°: 11060467 and 1080352 and by MECESUP (Chile) Project N°:UCH0109. Andrés Neyem's work was partially supported by the Scholarship for Thesis Completion from Conicyt.

References

1. Alarcon, R., Guerrero, L.A., Ochoa, S.F., Pino, J.A.: Analysis and Design of Mobile Collaborative Applications using Contextual Elements. Computing and Informatics 25(6), 469–496 (2006)
2. Aldunate, R., Larson, G., Nussbaum, M., Ochoa, S., Herrera, O.: Understanding the Role of Mobile Ad hoc Networks in Non-traditional Contexts. In: IFIP Int. Conf. on Mobile and Wireless Comm. Networks, pp. 199–215. Springer, Boston (2006)
3. Andriessen, J.H.E., Vartiainen, M.: Mobile virtual work: A new paradigm? Springer, Heidelberg (2006)
4. Arvola, M.: Interaction Design Patterns for Computers in Sociable Use. International Journal of Computer Applications in Technology 25(2/3), 128–139 (2006)
5. Avgeriou, P., Zdun, U.: Architectural Patterns Revisited - A Pattern Language. In: 10th European Conference on Pattern Languages of Programs, pp. 1–39. UKV Konstanz, Konstanz (2005)
6. Avgeriou, P., Tandler, P.: Architectural Patterns for Collaborative Applications. International Journal of Computer Applications in Technology 25(2/3), 86–101 (2006)
7. Buschmann, F., Meunier, R., Rohnert, H., Sommerlad, P., Stal, M.: Pattern orinented Software Architecture - A System of Patterns. John Wiley & Sons, Chichester (1996)
8. Churchill, E.F., Wakeford, N.: Framing mobile collaboration and mobile technologies. In: Brown, B., Green, N., Harper, R. (eds.) Wireless world: social and interactional implications of wireless technology, pp. 154–179. Springer, New York (2001)
9. Clements, P., Bachmann, F., Bass, L., Garlan, D., Ivers, J., Little, R., Nord, R., Stafford, J.: Documenting Software Architectures: Views and Beyond. Addison-Wesley, Reading (2003)
10. Ellis, C.A., Gibbs, S., Rein, G.L.: Groupware: some issues and experiences. Communications of the ACM 43(1), 38–58 (1991)
11. González, V.M., Tentori, M.E., Morán, E.B., Favela, J., Martínez, A.I.: Understanding mobile work in a distributed information space: implications for the design of ubicomp technology. In: 2nd Latin American Conference on Human-Computer Interaction, pp. 52–63. ACM Press, New York (2005)
12. Guerrero, L.A., Fuller, D.: A Pattern System for the Development of Collaborative Applications. Journal of Information and Software Technology 43(7), 457–467 (2001)

13. Guerrero, L.A., Ochoa, S.F., Pino, J.A., Collazos, C.A.: Selecting Computing Devices to Support Mobile Collaboration. Group Decision and Negotiation 15(3), 243–271 (2006)
14. Jørstad, I., Dustdar, S., Van Thanh, D.: Service Oriented Architecture Framework for collaborative services. In: 14th IEEE International Workshops on Enabling Technologies: Infrastructure for Collaborative Enterprise, pp. 121–125. IEEE Press, New York (2005)
15. Kristoffersen, S., Ljungberg, F.: Mobility: From Stationary to Mobile Work. In: Braa, K., Sorensen, C., Dahlbom, B. (eds.) Planet Internet, pp. 137–156. Studentlitteratur, Sweden (2000)
16. Luff, P., Heath, C.: Mobility in collaboration. In: ACM Conference on Computer-Supported Cooperative Work, pp. 305–314. ACM Press, New York (1998)
17. Mascolo, C., Capra, L., Zachariadis, S., Emmerich, W.: XMIDDLE: A Data-Sharing Middleware for Mobile Computing. Journal on Personal and Wireless Communications 21(1), 77–103 (2002)
18. Messeguer, R., Ochoa, S.F., Pino, J.A., Navarro, L., Neyem, A.: Communication and Coordination Patterns to Support Mobile Collaboration. In: 12th International Conference on Computer Supported Cooperative Work in Design, pp. 565–570. IEEE CS Press, New York (2008)
19. Neyem, A., Ochoa, S., Pino, J.: A Strategy to Share Documents in MANETs using Mobile Devices. In: 8th International Conference on Advanced Communication Technology, pp. 1400–1404. IEEE CS Press, New York (2006)
20. Neyem, A., Ochoa, S.F., Pino, J.A.: Designing Mobile Shared Workspaces for Loosely Coupled Workgroups. In: Haake, J.M., Ochoa, S.F., Cechich, A. (eds.) CRIWG 2007. LNCS, vol. 4715, pp. 173–190. Springer, Heidelberg (2007)
21. Neyem, A., Ochoa, S.F., Pino, J.A.: Integrating Service-Oriented Mobile Units to Support Collaboration in Ad-hoc Scenarios. Journal of Universal Computer Science 14(1), 88–122 (2008)
22. Ochoa, S., Neyem, A., Pino, J., Borges, M.: Supporting Group Decision Making and Coordination in Urban Disasters Relief Efforts. Journal of Decision Systems 16(2), 143–172 (2007)
23. Ochoa, S.F., Neyem, A., Bravo, G., Ormeño, E.: MOCET: A MObile Collaborative Examination Tool. In: Smith, J.M., Salvendy, G. (eds.) HCI 2007. LNCS, vol. 4558, pp. 440–449. Springer, Heidelberg (2007)
24. Perry, M., O'hara, K., Sellen, A., Brown, B., Harper, R.: Dealing with mobility: understanding access anytime, anywhere. ACM Transactions on Computer-Human Interaction 8(4), 323–347 (2001)
25. Pinelle, D., Gutwin, C.: A Groupware Design Framework for Loosely Coupled Workgroups. In: 9th European Conference on Computer-Supported Cooperative Work, pp. 65–82. Springer, Netherlands (2005)
26. Pinelle, D., Gutwin, C.: Loose coupling and healthcare organizations: adoption issues for groupware deployments. Computer Supported Cooperative Work 15(5-6), 537–572 (2006)
27. Schaffers, H., Brodt, T., Pallot, M., Prinz, W.: The Future Workplace - Perspectives on Mobile and Collaborative Working. Telematica Instituut, The Netherlands (2006)
28. Schümmer, T., Lukosch, S.: Patterns for Computer-Mediated Interaction. John Wiley & Sons, West Sussex (2007)
29. Tarasewich, P.: Designing Mobile Commerce Applications. Communications of the ACM 46(12), 57–60 (2003)
30. Wiberg, M., Ljungberg, F.: Exploring the vision of anytime, anywhere in the context of mobile work. In: Malhotra, Y. (ed.) Knowledge Management and Virtual Organizations, pp. 157–169. Idea Group Publishing (2001)

LeadFlow4LD: Learning and Data Flow Composition-Based Solution for Learning Design in CSCL

Luis Palomino-Ramírez, Miguel L. Bote-Lorenzo, Juan I. Asensio-Pérez, and Yannis A. Dimitriadis

School of Telecommunication Engineering, University of Valladolid, Camino del Cementerio s/n, 47011 Valladolid, Spain
lpalomin@ulises.tel.uva.es, {migbot,juaase,yannis}@tel.uva.es

Abstract. IMS-LD is the de facto standard for learning design (LD) specification which typically comprises an activity flow and a data flow. Nevertheless, the specification of the data flow between tools is an open issue in IMS-LD, especially in collaborative learning. In such case, handling shared data derived from individual and collaborative tools is error-prone for learners who suffer an extra cognitive load. Additionally, problems in the collaborative data flow specification affect the reusability of the whole learning design. In this paper, we present LeadFlow4LD, a solution of specification and enactment for LD in CSCL in order to address the aforementioned issues in an interoperable and standard way. Such a solution is based on approaches for the composition of the activity flow specified in IMS-LD and the data flow specified in a standard workflow language, such as BPEL. An architecture and a prototype for validating the propose solution through a case study based on a significant CSCL situation are also presented.

Keywords: Data Flow, Learning Design, Workflow, CSCL, IMS-LD.

1 Introduction

The LD approach [1] has evidenced and promoted a major shift in technology enhanced learning, since it pays special attention to the process of teaching and learning, instead of the previous approach to the delivery of educative contents, as learning objects. Thus, according to LD, a learning design formally defines a sequence of learning activities in which students and teachers play roles, individually or in groups, through the use of tools and services with the aim of accomplishing their learning goals [2]. With the aim of providing interoperability, the IMS-LD language [1, 3] has come up as the de facto standard in LD for a variety of pedagogical approaches including collaborative learning [4], although there are other competing non-standard Educational Modelling Languages [5].

LD shares several common features with workflow system [1], since it deals with the coordination of activities. Nevertheless, a fluent design and enactment of a learning design by the users (learners, tutors) of an e-course requires a major automated

R.O. Briggs et al. (Eds.): CRIWG 2008, LNCS 5411, pp. 266–280, 2008.

support with regard to the sequence of learning activities (the learning flow) and the flow of information (the data flow), as in document-oriented workflow systems [6]. However, the specification of data flow between tools is still an open issue for IMS-LD [7] [8] [9] [10] [11] [12]. The main problem lies in the fact that IMS-LD supports a human-oriented data flow approach [11], which means that the user, not the system, is responsible for managing the data flow between tools.

Although this human-oriented data flow approach may be partially valid for individual learning, serious drawbacks have been identified in CSCL situations [11] [12]. Due to the complex interactions between users and data, locating and handling shared data in different group structures (see e.g. the expert and super groups in a situation based on the jigsaw Collaborative Learning Flow Pattern (CLFP) [3]) is error-prone for users who suffer an additional cognitive load. Moreover, the reusability issue in collaborative data flow specification becomes essential [12].

As a response to these issues, LD and Workflow may be considered and exploited as similar but at the same time as complementary approaches supported by the corresponding technologies [11], due to their focus on human activities and system tasks, respectively. On the one hand, by delegating the responsibility of managing the shared data to a workflow-based solution (design formalization and enactment engine), it would be feasible to eliminate error-prone situations for users as well as to eliminate the excess of their cognitive load. On the other hand, by separating the declarative-level specification from the instance-level one, the reuse of both the learning and data flow specifications could be facilitated. Therefore, this so-called composition-based approach requires the coordinated execution of the learning flow specified in IMS-D and the data flow specified in a standard workflow language [11]. However, a concrete solution based on this approach is still missing [11] [12], due to the existence of several workflow standards and the complexity of the proposed approach. Such a concrete solution, the implementation of a prototype and its validation through significant case studies may enable a deeper understanding of the approach and foster a shared standards-based solution by the wider community.

In this paper, we present LeadFlow4LD (LEarning And Data FLOW composition-based solution FOR Learning Design) as a solution for addressing all the aforementioned issues related to the data flow problem of IMS-LD in CSCL situations. LeadFlow4LD comprises approaches for the learning flow specification in IMS-LD and the data flow specification in the selected workflow language. Furthermore, it defines an approach for the coordination of both streams, as well as a logical architecture for the enactment of such learning designs. Moreover, in order to validate the proposed solution, a prototype of the enactment architecture has been implemented, while distinct configurations of a significant case study have been carried out.

The paper is structured as follows. In section 2 the data flow problem of IMS-LD in collaborative learning through a case study is presented. In section 3 we describe LeadFlow4LD as a technology independent approach. Besides, the implementation of a prototype for the enactment of learning designs specified according to Lead-Flow4LD is also described. In section 4 the evaluation of the proposed solution through different configurations of the case study is presented, while in section 5 the related work to the IMS-LD data flow problem is reviewed. Finally, in section 6 we present our conclusions and future work.

2 The Data Flow Problem of IMS-LD in Collaborative Learning

As already known, tools and services used in individual or collaborative activities may need data as input or output. Such data artifacts can be, for example, the conclusions of a discussion, the answers of a test, the specification data for a simulator, etc. Since these tools may span through various activities or be handled by different actors in the same or different activities, a data flow is generated that is related to the learning flow. However, such data flow cannot be specified through the mechanisms provided by IMS-LD [8]. Instead, since IMS-LD follows a human-oriented approach, the participants of the learning process should be responsible for handling this data flow [11].

In this section we are going to illustrate the data flow problem in collaborative learning settings, which has been documented in literature [12], through a case study that is both significant and authentic. Then, the presentation of the proposed solution can be validated on this case study in a later section, and subsequently support and shed light to the posterior discussion on advantages and limitations.

The case study, called *CNS2* (Collaborative Network Simulator 2) and illustrated as an activity diagram in Fig. 1a, corresponds to a collaborative learning scenario in which learners perform different activities based mainly on the well-known network simulator *ns-2* [13], in order to evaluate and analyze different network protocols.

In the first activity, a simulation *Tcl/Tk* script (D1) is generated individually by each learner using a generic editor. Then, the generated *Tcl/Tk* script becomes an input to the *ns-2* tool during the simulation (second) activity. As a result, two types of

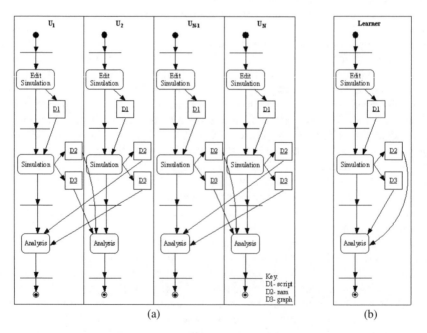

(a)   (b)

Fig. 1. Activity diagram for the case study CNS2 (a) at instance-level (b) at declarative-level (not sufficient)

output files are generated, named the *graph* (D2) and *nam* (D3) files. The first file contains different measures taken during the simulation that can be plotted as *x-y* graphs, while the second one contains the behavior of different elements defined in the scenario that can be visualized using a network animator tool [14]. Finally, the simulation analysis activity is carried out, in a collaborative way, since each learner analyzes the simulation results of one of his peers. For example, the user *u1* should analyze the *graph* and *nam* data files that belong to the user *u2*, and so on.

In IMS-LD, such collaborative data flow specification is carried out through a verbal (textual) instruction to each user included in the activity description regarding the path where the right artifacts are located. This is clearly an error-prone approach for users [12], since the instructions could be misunderstood, badly applied, ignored or forgotten by the users [15]. The error-prone character of this approach reveals an important issue, since the learning objectives may not be accomplished as expected. Furthermore, this approach produces an additional cognitive load to users, who should be concerned about understanding well the instructions, locating and retrieving the right artifact, instead of an automatic selection, retrieval and delivery by the system [12].

In other terms, in order to distinguish each user data, instance-level data flow specification (see Fig. 1a) within the learning design is necessary [12]. Then, data flow design cannot be specified at declarative-level (see Fig. 1b) and therefore it cannot be used several times for different instances of users and CSCL situations [16]. Therefore, the instance-level collaborative data flow specification affects the reusability of the whole learning design, which is also a relevant issue [12].

In order to address the issues illustrated in the above case study, while keeping interoperability with IMS-LD, LeadFlow4LD is presented in the next section, which is a composition-based solution in which the learning flow is specified in IMS-LD while the data flow is specified in a standard workflow language.

3   The Proposed Solution: LeadFlow4LD

3.1   Overview

An overview of the proposed solution is illustrated in Fig. 2. As we can see, Lead-Flow4LD consists of two main approaches. The first approach is related to the learning design definition which conforms to LeadFlow4LD, while the second one is related to the enactment of such learning designs. On the other hand the learning design definition consists of a declarative-level definition and an instance-level one for both learning and data flow. This separation allows the learning design definition be reused for different contexts and situations [16], so as the learning and data flow definitions can be reused for different users, groups and shared data logic. In addition, a coordination definition is also proposed in order to enact coordinately both streams.

Although this paper focuses on the approach of learning design definition, enactment and validation, it is also considered for completeness. Moreover, we pay attention to a technology independent approach which describes how the distinct abstract components interact with the learning design (see Fig. 3). Even though the validation of the approach through a prototype is illustrated through concrete technologies, the detailed specifications for each technology are out of scope of this paper.

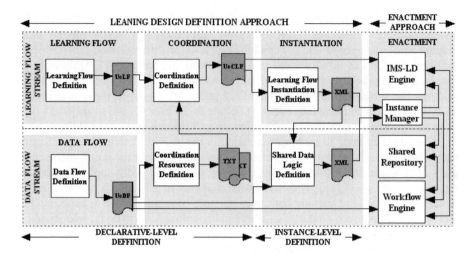

Fig. 2. Overview of LeadFlow4LD

3.2 Learning Design Definition Approach

Overview. A learning design which conforms to LeaDFlow4LD consists of different definitions as we will describe throughout the next subsections. For a better understanding of the approach, an overview of the interactions among the involved elements is illustrated in Fig. 3, which are also referenced throughout these subsections.

Defining a list of tools. The learning flow defines the sequence of learning activities, meanwhile the data flow defines the sequence of tool invocations and the data flow among them (see Fig. 3). Nevertheless, in order to keep interoperability with IMS-LD, LeadFlow4LD does not demand that all tools must be called upon in the data flow. Instead, tools can also be defined as resources in the learning flow, or even not be defined at all, which means that the instructional designer does not provide tools support to the activity, but the learners employs their own tools. In such cases, output data from these tools may require to be imported into the data flow, or data defined in the data flow may require be exported outside it. Therefore, the learning and data flow definitions are not disjoint at all, but it must be previously known whether the tools will be defined in the learning flow or called upon in the data flow in order to define separately both streams. With this purpose, the definition of a list of tools as a previous step for a learning design definition is necessary.

Learning flow definition. Since IMS-LD fails in data flow specification in collaborative learning [1], LeadFlow4LD proposes that it should only be used for the learning flow definition, but not for the data flow one, which will be specified separately in a standard workflow language. Nevertheless, the aim of the learning flow definition approach is not to describe how this sequence of learning activities in collaborative learning should be specified with IMS-LD, since it has already been worked in literature [2], but to describe the implications in the learning flow definition due to the

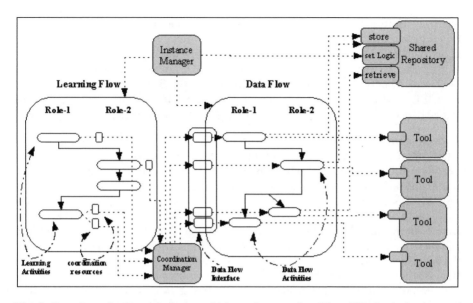

Fig. 3. Overview of a learning design which conforms to LeadFlow4LD. In the background white the learning design-related components are represented, while in background gray the enactment-related ones can be found.

separation of both streams. With this purpose, LeadFlow4LD demands that IMS-LD properties are not used for artifacts and users specification, since the learning flow definition should focus on the flow of activities at declarative-level but not at the instance-level. Moreover, it should also focus on defining activities-related resources including tool definitions, as well as activity descriptions. Finally, as shown in Fig. 2, the resulting learning design definition must be packaged as unit of learning (UoL) according to the standard IMS-CP [3]. In our case, it is denominated unit of learning flow (UoLF) in order to emphasize the fact that it is just focused on the flow of learning activities.

Data flow definition. The aim of the data flow definition is to sequence the invocation of tools and to manage automatically the data flow among them. Although the data flow in collaborative learning requires an instance-level specification [1], since users need to share their data with other users belonging to the same or different roles (groups), the data flow must be defined first at a declarative-level in the same way as the learning flow is defined. Therefore, the approach to define the data flow considers aspects related to the sequence of invocations of tools (known here as sequence of data flow activities); the data flow interface; and the roles in which such data flow activities are associated. First, the sequence of data flow activities is related to the sequence of learning activities in which the tools have been called upon (see Fig. 3). Nevertheless, both of them do not have to match necessarily the same sequence, since some learning activities may not be associated with tools or more than one, in which case these tools should be sequenced concurrently in the data flow. Secondly, the data flow actually performs as an indirection layer, since while users perform learning activities, they are supported by tools which could be invoked through the data flow interface. Thirdly, LeadFlow4LD demands that the data flow activities must be associated to the same roles than to the learning activities in which the tools are called upon. As it is shown in Fig. 2, the resulting workflow definition of the data flow must be packaged as what we call unit of data flow (UoDF) in order to emphasize the data flow issue. Finally, in order to define the instance-level data flow specification, Lead-Flow4LD proposes a mechanism to define in instantiation time what we call the shared data logic definition, which will be treated later in this paper.

Coordination definition. Once that learning and data flow have been defined separately, the coordinated execution of both streams is necessary. For this purpose, LeadFlow4LD proposes a synchronous master-slave coordination model: while the learning flow is the master stream, the data flow is the slave. That is, while the data flow invokes one or more tools, the learning flow remains blocked, so users cannot move on to the next learning activity, until the tasks carried out by the invoked tools have been finished. With this aim, the so-called coordination resources are defined, and these are added to the learning activities as IMS-LD resources in order to invoke the data flow interface, either to invoke tools, import or export data from/to the data flow (see Fig. 3). Finally, the resulting coordination definition must be packaged as a UoL according again to the standard IMS-CP. In our case, it is denominated unit of coordinated learning flow (UoCLF) in order to emphasize the fact that it is actually a UoLF which has been completed with the so-called coordination resources.

Defining instantiation. Once the learning design which conforms to LeadFlow4LD has been defined at a declarative-level, the instantiation defines the context in which it will be carried out. This context spans both the learning flow and the data flow. On one hand, the instantiation of the learning flow defines groups, users, assignment of users to groups and users to roles [4]. On the other hand, LeadFlow4LD proposes that the instantiation of the data flow defines the shared-data logic among users belonging to the same or distinct roles (groups). Through this approach to define the instantiation, the learning design definition can be reused for different users, groups, group's size and what is our contribution: for different shared data logic. Finally, both instantiations are defined using XML schemas so as to be interpreted by an instance manager, which in turn will interact with the proper enactment engine (see Fig. 2). Next, a description of our proposal for the shared data logic definition is presented.

Shared data logic definition. In order to separate the declarative-level data flow definition from its instantiation, it is necessary to have a shared data logic definition into an external component. For this purpose, a shared repository could be used to store users' data in order to be retrieved later for the same or other users according to the shared data logic defined at instantiation time.

Suppose that S users are part of the same collaborative group (role): $\{u1,u2,...us\}$. In order to define the shared data logic between them, LeadFlow4LD defines the tuple:

$$L (D, S, P)$$

Where:
L- The shared data logic
D- The shared (declarative) data
S- The group's size
P- The peer, who data is accessible for the user ui, $0<=P<S$.

For example, the shared data logic: $L(D,S=2,P=1)$ defines a typical peer sharing logic. That is, working in pairs, each user gets access to the data belonging to his unique peer. Now consider the shared data logic defined by $L(D, S=3,P=1)$, which means that in a group of three people, each user has access to the data belongs to his first peer (in a triplet a user has two peers). Finally, consider the case when the shared data logic is defined by a null peer: $L(D,S,P=0)$, which means that there are not shared data at all, because each user only has access to his own data, which is the case of individual learning.

Nevertheless, data within the shared repository are stored and therefore retrieved regarding each user, not to the number of the peer. So, in order to retrieve data regarding to some user ui according to his peer P defined in the shared data logic, an evaluation of the matrix function $getUser(ui,S,P)$ shown in Fig. 4 is necessary. Note that particular cases for $S=2$ and $S=3$ are shown in Fig. 4b and 4c respectively.

(a) General case

$U_i \backslash P$	0	1	2	...	S-1
U_1	U_1	U_2	U_3	...	U_S
U_2	U_2	U_3		...	U_1
U_3	U_3		...		U_2
...		...	U_S	...	...
U_{S-1}	U_{S-1}	U_S	U_1	...	U_{S-2}
U_S	U_S	U_1	U_2	...	U_{S-1}

(b) S=2

$U_i \backslash P$	0	1
U_1	U_1	U_2
U_2	U_2	U_1

(c) S=3

$U_i \backslash P$	0	1	2
U_1	U_1	U_2	U_3
U_2	U_2	U_3	U_1
U_3	U_3	U_1	U_2

Fig. 4. Matrix function $getUser(ui,S,P)$ used to get the user who shares his data with the user ui according to the defined shared data logic $L(D,S,P)$

For example, the shared data logic definition given by $L(D1,S=3, P=2)$, means that the user $u1$ must get access to the retrieved data from the shared repository according to the tuple $(u3, D1)$, since $u3$ is the result of evaluating the function $getUser(u1,S=3,P=2)$ shown in Fig. 4c. Although LeadFlow4LD requires a shared data logic definition among users from the same or different roles, the current approach only covers the first, so an extension of the current approach is necessary.

3.3 Architecture for the LeadFlow4LD Run-Time Environment

The enactment of a learning design which conforms to LeadFlow4LD requires the integration of different components and specifications (see Fig. 3). Taking advantage from service-orientation, we used services as the basic components of service oriented architectures (SOA) [5]. In this context, we have proposed a three-tier logical architecture for the enactment of such learning designs as illustrated in Fig. 5. So, once a UoCLF and its corresponding UoDF have been defined and stored in the proper repository, the learning flow and data flow instance manager and their clients are responsible for deploying, instantiating, monitoring and terminating the enactment of learning design instances (both learning flow and data flow instances). Beside this, the data flow instance manager client is also responsible for setting the shared data logic through the shared repository service. Then, the learning flow engine is responsible for playing the active learning flow, while its client, the so-called enactment client indicates the proper activity that should be carried out by each user, as well as the corresponding resources. Finally, when a user invokes a service (either an importing or exporting data service), the client manager provides the client with the proper service to invoke operations regarding to the data flow service. In the next section a prototype based on this architecture is implemented.

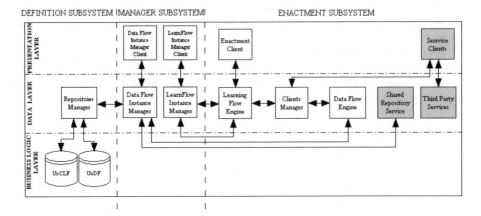

Fig. 5. Logical architecture for the LeadFlow4LD run-time environment. The external elements are represented in gray background.

3.4 Prototype Implementation

One of the main decisions related to the implementation of the prototype refers to the selection of BPEL [6] as the workflow language for the data flow specification. Although this is not the only choice, BPEL seems to be an adequate one for several reasons; It is a commonly accepted standard widely used by the industry and especially for the composition, orchestration and coordination of web services, a popular technology based on the service orientation, thus fostering interoperability and integration [7]. Furthermore, specifying the sequence of service invocations and the data flow among them can be easily implemented with BPEL through appropriate mechanisms.

Current prototype implements the following elements that correspond to the logical architecture shown in Fig. 5: the enactment client, the learning flow engine, the data flow engine and the client manager. On one side, *Coppercore* [8] has been used to implement the learning flow engine, whereas *WebPlayer*, distributed together with *Coppercore*, has been used to implement the enactment client. On the other side, *ActiveBPEL* [9] has been used to implement the data flow engine, whereas the *JNLP (Java Network Launch Protocol)* application manager of *Java Web Start* has been used to implement the client manager. *JNLP* [10] is a standard specification that allows Java applications to be automatically launched from an application server in order to be executed locally in the client machine. Therefore, during the coordination definition, the so-called coordination resources are specified conform to the *JNLP* specification. The files that correspond to the coordination resources contain the address of the application server where the resources required for the execution of the proper application are located. Then, these files are added as resources to the activities into the learning flow and as result a *UoCLF* is defined. Finally, the shared repository service was implemented as a web service using *MySQL* data base management system to storage the state information.

4 Validation of the Proposed Solution through the Case Study

In this section we present preliminary results regarding the validation of the proposed approach that are illustrated through the case study *CNS2* presented in section 2. Table 1 shows the specific information provided by the designer with regard to the specific tools used in this CSCL scenario. Thus, it can be seen that e.g. the *Tcl/Tk* editor tool has been specified as an IMS-LD resource to be employed in the simulation edition activity. On the contrary, the *ns-2* tool will be invoked as a third-party web service, while the network animator and *x-y* plotter have not been prescribed by the designer, thus it allows the users to choose the most appropriate ones.

Table 1. Tools, data and activities for the case study *CNS2*

Tool	Specified as	Input data	Output data	Activity
Tcl-Tk Editor	Resource in *IMS-LD*	-	Sim. script	Edit simulation
ns-2	Third-party service	Sim. script	*nam, graph*	Simulation
Network Animator	-	*nam*	-	Sim. analysis
x-y Plotter	-	*graph*	-	Sim. analysis

An overview of the learning design which conforms to LeadFlow4LD for the case study *CNS2* is shown in Fig. 6. The prototype has been tested in several configurations, related to number and size of groups, as well as data sharing logic. Therefore, we could determine the flexibility of the proposed approach and the current prototype, especially with regard to the complex characteristics of collaborative learning.

The snapshot shown in Fig. 7 presents an activity description from the user perspective. Note that, even in this simple peer sharing case, the user's cognitive load is reduced notably, since he is not concerned in handling adequately data flow according to instructions. Instead, a click in the corresponding input of the system environment window is sufficient.

On the other hand, in Fig. 8, we can see what error-prone situations are avoided, due to the automatic delivery of the artifacts, when the user *u1* invokes the service for downloading (exporting) the *nam* file regarding to his partner defined in the shared data logic (*u2*).

Reusability of the whole learning design at the definition level has been shown through the evaluation of several runs in different situations, corresponding to different users, number of groups, group's size or even the shared data logic. In all runs, except one, no changes were necessary with regard to the learning design definition (both learning and data flow), which shows that LeadFlow4LD fosters reusability. However, in the case of learners who have to access the data of two peers at a time, instead of one, a change in the learning and data flow definitions were necessary, since current shared data logic definition approach is limited to users having access to the data of one peer at a time. Even in this case, such change could be easily made in the learning and data flow definitions but an extension of the shared data logic mechanism is necessary.

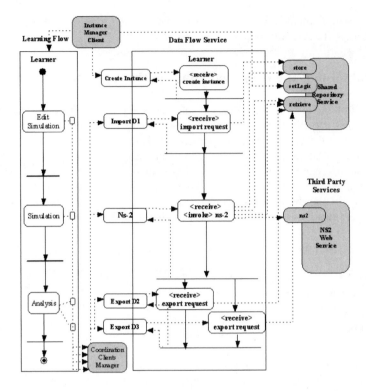

Fig. 6. Overview of the learning design which conforms to *LeadFlow4LD* of the case study *CNS2*. (White background and gray background denote design components, and enactment components respectively.)

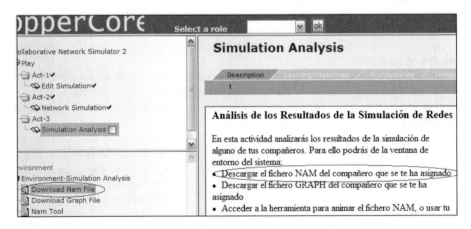

Fig. 7. Snapshot of the *WebPlayer* showing the activity description (center frame) and the system environment window (bottom left frame)

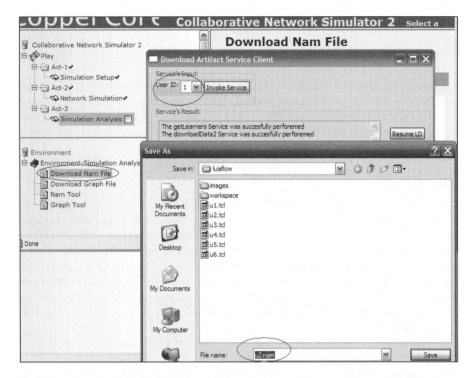

Fig. 8. Snapshot of *WebPlayer* for showing that the *nam* artifact regarding to the assigned peer is retrieved automatically by the workflow engine

5 Related Work

The data flow problem of IMS-LD has been dealt in many ways for different authors in literature. Peter and Vantroys [11] state that IMS-LD lacks of data flow management, but do not go further in defining the associated defects. Wilson criticizes in [12] IMS-LD since it does not consider whether the results of a service are going to be exported to other services; however, he does not provide any solution to this issue. Dalziel states in [13] that IMS-LD requires mechanisms in order to pass information between tools with a possible information processing between them, but the proper mechanism for data flow between tools is not proposed. Furthermore, Miao et al. mention in [14] that IMS-LD has no means to specify the relation between data and tools, and therefore they propose a new scripting language. Nevertheless, this is not an interoperable solution as LeadFlow4LD is. Moreover, they do not consider the global data flow problem in collaborative learning [1] since they mainly address a data flow automation issue. Vantroys and Peters, propose in [15] a mapping approach between IMS-LD and the standard workflow XPDL, but according to [16] both of them are complementary approaches, and therefore a mapping mechanism between them is not enough in order to address the data flow problem. According to [16], a solution space is described in order to resolve the data flow problem of IMS-LD, that considers the substitution, mapping and composition approaches. However, in [16], no concrete solutions and results are provided with regard to the composition-based approach based on learning design and workflow standards. On the other hand, the initial proposal for the coordination mechanism is based on a third independent stream, i.e. a Petri Net stream. However, the two-streams master-slave coordination mechanism proposed in LeadFlow4LD is simpler to implement than the 3-streams coordination mechanism proposed in [16]. Finally, in [1] a case study in order to evaluate the data flow problem of IMS-LD in collaborative learning is presented but a concrete solution to the problem is missing.

6 Conclusions and Future Work

IMS-LD, as the de facto standard for learning design supports a user-oriented mechanism for data management and especially handling input and output data related to tools. Nevertheless, this approach has special serious drawbacks in CSCL situations. Due to the need of sharing data among users, learners suffer a high cognitive load as well as error-prone situations when they locate and handle data from other users that may potentially affect the accomplishment of the learning objectives Furthermore, another relevant issue refers to reusability of the whole learning design since the shared data logic cannot be specified at declarative level, but instance-level data flow have to be specified.

In this paper we propose an interoperable solution to this problem, called Lead-Flow4LD, which consists of approaches for the specification of both learning and data flows, using respectively IMS-LD and a standard data flow workflow language. Furthermore, an approach for coordination of both streams has been also presented. Moreover, we have presented an architecture related to the run-time environment, necessary to enact learning designs which conforms to LeadFlow4LD. We have also implemented a prototype based on this architecture and several runs of the same case study have been carried out. Results indicate that LeadFlow4LD solves all the issues mentioned in this paper about the data flow problem of IMS-LD in CSCL situations. On one side, by delegating the responsibility for shared data management to the workflow engine, a learner is not responsible for locating and handling the shared data, therefore reducing error-prone situations and the associated cognitive load. On the other side, through the separation of the declarative-level learning design definition from its instance-level definition the learning design can be reused for distinct contexts including users, groups, groups'size or even the shared data logic. For this purpose, we have proposed a mechanism to define the shared data logic in an independent way. This approach fosters reusability of the data flow and consequently of the whole learning design. Finally, it is interesting to see LeaDFlow4LD as the in between approach of a future integration of learning design and workflow streams.

Future work covers several issues. Firstly, an extension of the shared data logic mechanism is necessary so as to include shared data among users belonging to different roles as well as a user having to access the data of more than one peer at a time, and therefore it should be explored and incorporated in LeadFlow4LD. Evaluation needs to go further, than the significant *CNS2* case study presented in this paper. More complex case studies have to be carried out and evaluated from a technological and educational perspective in authentic environments, in order to identify advantages and drawbacks of the proposed approach. Of course a complete adoption of LeadFlow4LD requires the development of authoring tools that support its use by educational practitioners in the same direction as Collage [17]. Also, integration of LeadFlow4LD with tailorable service-oriented educative systems, such as Gridcole [18], is currently under development, as well as the generation of specific documents for the guidelines of the distinct approaches of LeadFlow4LD regarding to the selected technologies.

Acknowledgments. This work has been partially funded by the EU Kaleidoscope NoE FP6-2002-IST-507838, Spanish Ministry of Education and Science project TSI2005-08225-C07-04, Autonomous Government of Castilla y León, Spain (projects VA009A05, UV46/04 and UV31/04), Tecnológico de Monterrey Campus Guadalajara, and Fundación Carolina. The authors would also like to thank Henar Muñoz Frutos as well as the rest of EMIC/GSIC research group at the University of Valladolid for their support and ideas to this work.

References

1. Palomino-Ramírez, L., Bote-Lorenzo, M.L., Asensio-Pérez, J.I., Dimitriadis, Y., de la Fuente-Valentín, L.: The Data Flow Problem in Learning Design: A Case Study. In: 6th International Conference on Networked Learning (NLC 2008), Halkidiki, Greece (2008)
2. Hernández-Leo, D., Asensio-Pérez, J.I., Dimitriadis, Y.: Computational Representation of Collaborative Learning Flow Patterns using IMS Learning Design. Journal of Educational Technology & Society 8, 75–89 (2005)
3. IMS, C.P.: IMS Content Packaging Specification (2003)
4. Hernández-Gonzalo, J.A., Villasclaras-Fernández, E.D., Hernández-Leo, D., Asensio-Pérez, J.I., Dimitriadis, Y.A.: InstanceCollage: a Graphical Tool for the Particularization of Role/group Structures in Pattern-based IMS-LD Collaborative Scripts. In: Proceedings of the 8th International Conference on Advanced Learning Technologies (ICALT 2008), Santander, Spain, pp. 1–5 (2008)
5. Papazoglou, M.P., Georgakopoulos, D.: Service-Oriented Computing. Communications of the ACM 46, 25–28 (2003)
6. Andrews, T., Curbera, F., Dholakia, H., Goland, Y., Klein, J., Leymann, F., Liu, K., Roller, D., Smith, D., Thatte, S.: BPEL4WS Specification: Business Process Execution Language for Web Services Version 1.1 (2003)
7. Juric, M.B., Mathew, B., Sarang, P.: Business Process Execution Language for Web Services. Pakt (2004)
8. Vogten, H., Martens, H.: CopperCore 2.2. 4. Heerlen: Open University of The Netherlands (retrieved on August 9, 2005), http://www.coppercore.org
9. ActiveBpel, L.L.C.: ActiveBPEL, the Open Source BPEL Engine, http://www.activevos.com/community-open-source.php
10. Schmidt, R.: Java Network Launching Protocol (JNLP) Specification v1.0.1 (2001)
11. Peter, Y., Vantroys, T.: Platform Support for Pedagogical Scenarios. Journal of Educational Technology & Society 8, 122–137 (2005)
12. Wilson, S.: Workflow and web services. CETIS White paper (2005) (Last retrieved January 22, 2007), http://www.e-framework.org/resources/SOAandWorkflow2.pdf

13. Dalziel, J.R.: Lessons from LAMS for IMS Learning Design. In: Proceedings of the 6th International Conference on Advanced Learning Technologies (ICALT 2006), Kerkrade, The Netherlands, pp. 1101–1102 (2006)
14. Miao, Y., Hoeksema, K., Hoppe, H.U., Harrer, A.: CSCL scripts: modelling features and potential use. In: Proceedings of the International Conference on Computer Support for Collaborative Learning (CSCL 2005): the next 10 years!, Taipei, Taiwan, pp. 423–432 (2005)
15. Vantroys, T., Peter, Y.: COW, a Flexible Platform for the Enactment of Learning Scenarios. In: Favela, J., Decouchant, D. (eds.) CRIWG 2003. LNCS, vol. 2806, pp. 168–182. Springer, Heidelberg (2003)
16. Palomino-Ramírez, L., Martínez-Monés, A., Bote-Lorenzo, M.L., Asensio-Pérez, J.I., Dimitriadis, Y.A.: Data Flow between Tools: Towards a Composition-Based Solution for Learning Design. In: Proceedings of the 7th International Conference on Advanced Learning Technologies (ICALT 2007), Niigata, Japan, pp. 354–358 (2007)
17. Hernández-Leo, D., Villasclaras-Fernández, E.D., Asensio-Pérez, J.I., Dimitriadis, Y., Jorrín-Abellán, I.M., Ruiz-Requies, I., Rubia-Avi, B.: COLLAGE: A collaborative Learning Design editor based on patterns. Journal of Educational Technology & Society 9, 58–71 (2006)
18. Bote-Lorenzo, M.L., Gómez-Sánchez, E., Vega-Gorgojo, G., Dimitriadis, Y., Asensio-Pérez, J.I., Jorrín-Abellán, I.M.: Gridcole: a tailorable grid service based system that supports scripted collaborative learning. Computers & Education 51(1), 155–172 (2008)

A Version Control System as a Tool and Methodology to Foster Children's Collaboration in Spatial Configuration Decision Tasks

Filipe Santos[1], Benjamim Fonseca[2], Leonel Morgado[3], and Paulo Martins[3]

[1] ESE, Instituto Politécnico de Leiria, Campus 1, Apartado 4045,
2411-901 Leiria – Portugal
fsantos@esel.ipleiria.pt
[2] UTAD/CITAB, Apartado 1013, 5001-801, Vila Real, Portugal
[3] UTAD/GECAD, Apartado 1013, 5001-801, Vila Real, Portugal
{benjaf,leonelm,pmartins}@utad.pt

Abstract. Nowadays a growing number of research and projects emphasize a culture of childhood that sees the child as an active participant in social life and, in particular, in the planning and decision processes of the spaces where they live most of their time. Under these projects children are asked to have a direct role on the configuration of several spaces, such as public and school playgrounds. This paper refers to a work in progress where a version control system for a multiuser virtual world is being developed to address some of the challenges that this kind of task and participants bring. Particularly this tool is being designed and prototyped to foster children's involvement in primary school activities where a class participation and consensus are requested on the several school spatial configurations.

Keywords: Virtual Worlds, Children, Groupware Development Frameworks, Workflow Management Systems.

1 Introduction

Urban Geography knowledge on man-environment relationships has long helped the planning of spaces to meet certain criteria, such as safety, beauty and economic opportunities, among others. Nevertheless, this activity has always been made by adults for adults as children have been seen as passive social actors, as the necessary competencies were seen as still not fully developed. But recently a growing number of research and projects emphasize a culture of childhood that sees the child as a competent actor in social life, as they are recognized to have the capability of constructing meanings of the world that surrounds them and make decisions [1]. Under this perspective some urban policies are already focusing on "work toward a new governmental philosophy of the city, engaging the children as parameters and as guarantors of the needs of all citizens" [2]. Children are participating in the planning and decision processes of spaces such as public playgrounds [3][4] or helping to define paths for walking to school autonomously [5][6].

R.O. Briggs et al. (Eds.): CRIWG 2008, LNCS 5411, pp. 281–288, 2008.

As nowadays schools are one of the places where children spend most of their time [4][7] children are asked to have a more participative role on the decision processes about the configuration of its several spaces, such as the classroom and playground. As school is also the privileged place where society expects them to learn the several competencies that they will later need for their adult life this space is also the privileged one to learn such collaborative competencies and to participate in decision making processes. School also offers a rich variety of spatial configuration decisions as many aspects of the daily school activities require spatial configurations (tents in an encampment, scenery in a theater play, etc.). Teachers may therefore use such activities in their curricula as a way to foster these competencies.

One important aspect about framing these activities in school is that they can't collide with the teacher's pedagogical model as teachers insert these activities in their work methodology. But currently we are seeing teachers embracing more "active pedagogies" that lay in the notion of the competent child and where children are asked to have an active role. In Portugal one pedagogic model that is spreading and that put great emphasis in collaborative work is Portuguese Modern School Movement in which classroom activities require the student's active participation either in negotiation and decision-making processes or others [8][9].

Under this paradigm and perspective teachers are defining and adopting several strategies to make this collaborative process happens. This brings several difficulties as children in primary school context still need to develop the cognitive and social skills to such a collaborative process.

2   Spatial Configuration Decision Processes: Virtual Worlds as a New Medium, Tool and Methodology

Multi-user Three-Dimensional Virtual Worlds (referred here simply as "virtual worlds") have been used with success in many educational scenarios and purposes [10] [11] [12]. They offer an excellent simulation of three-dimensional (3D) spaces and as they allow the simultaneous presence of multiple users they may be used by teachers and children to mediate 3D spatial configuration collaboration processes with new strategies. As some of these worlds offer scripting mechanisms that allow the creation of personalized multi-user 3D applications and tools to better meet a task, methodologies can be defined to take advantage of this medium.

2.1   Empirical Observations in a Portuguese Primary School

We have been conducting empirical studies at Amélia Vieira Luís primary school, in Lisbon, with a teacher that uses the Portuguese Modern School Movement model of pedagogy with his 2nd year children. He promotes discussion with children on several issues that affects them as a class but as children are still learning the dynamics of such a process and competencies to use in this, it is not an auto-sustained process and the teacher still plays an important role. Therefore implementing such a process in a virtual world requires a software tool that helps the teacher under this new medium.

When the class is discussing a spatial configuration the teacher uses several pedagogical materials to support such an activity. The maquette is the most used one as it offers a small scale 3D version of the space that is being discussed. But one limitation of this material is that "as each child collaborates, by proposing his own configuration, changes are done in the maquette that may radically change the configuration proposed by the child that intervened before him – therefore destroying the vision of this child. In this sense, each child may have difficulties to believe that it gave a positive contribution to the collaborative process (if any at all) by seeing a final configuration which is possibly very different from his own. " [13]

Based on several observations of the participation processes that take place in this class we propose and describe a tool based on a version control system (VCS), that will fit to a particular methodology to address the problems of such activity.

2.2 Version Control Systems as Data Structures

In our first proposal we thought of a visual 3D interface that could help children understanding abstract notions such as collaboration and a decision making process by metaphorically giving them a concrete representation. The collaboration process, being a dynamic abstract entity that evolves over time, should have a visible interface so that it has a concrete existence.

Inspired on a successful version control system that helps communities collaborate and reaching a consensus, the one at the core of Wikipedia, we have thought of how one could be adapted for 3D media content. We proposed therefore an "historic tool" that would save the several versions of a space during a collaboration process in a linear tree data structure which could be represented visually (fig.1):

Fig. 1. A linear tree representing several versions of the configuration of a space

One promising methodology for using such a mechanism is starting by an initial configuration (the "r" – root) and give children a sequential order for suggesting small changes to that configuration. These successive changes would constitute new versions which would be saved under each child's name (fig.1 assumes that 4 children, A, B, C and D are intervening). One advantage of this strategy is that it may help each child "realize that it gave a positive contribution to the process as it was his/her configuration that served as the basis of the next child's configuration. In this sense, the teacher may help the children see that the final configuration may be seen as the product of all interventions combined." [13].

This "Version tree" was well accepted by the teacher as it may be used to represent the three time notions that children know well: past, present and future. Such a visual element can be used pedagogically as it visually show to children that a collaboration process is a dynamic entity that evolves over time.

2.3 Version Control Systems as a Tool to Promote Confrontation

As children are in a cognitive development stage where they lack some of the necessary competences to the negotiation processes such a methodology and strategy must be simple, and more demanding user requisites which were later observed can be difficult to solve with a simple linear structure. So the next stage was to define how to visually represent complex information of such a complex process with such a simple structure.

One complexity of the process is egocentrism: young children have difficulties in giving up their own perspective as they do not totally understand the relations that involve "giving and receiving" [14]. Therefore a mechanism where different versions could be easily confronted to promote discussion was necessary (fig. 2).

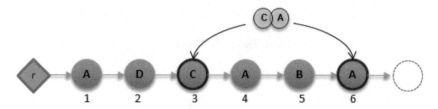

Fig. 2. Confrontation of two versions

We found the "time travel" metaphor appealing as it didn't break the linear version tree which we think it is easily understood by children. The teacher can use such tool to take children in a "time travel" visit to a previous 3D configuration and where discussion – by confrontation with the current version - can be encouraged.

2.4   Version Control Systems as a Tool to Manage Errors

Along our observations we have inquired the teacher for user requisites. One requisite the teacher asked - as it was already used by the teacher in his classroom collaborative processes with high success – was the most challenging for our team, as it disrupted the notion of a linear version tree: the consensus points (fig. 3).

Under this philosophy the teacher and children identify at a certain point of the discussion that the collaboration process has reached a "dead end" (a situation where a consensus is not possible) and jointly decide to use the time travel possibility to travel back to a point where the discussion was doing well to "fork the future" from that point. This means that new editions (spatial configurations) could be made from that point, creating a new line of states in the future.

This feature brought complexity to the process but we have designed it not to add more entropy to the process increasing exponentially the number of configurations: on the contrary, it is a way that the teacher uses precisely when entropy is becoming visible. Under this perspective the version tree continues to be linear as the future isn't forked, a path is erased and another one is created and continued in a new line of thought. As already mentioned we tried to represent a complex tree data structure in a simple representation in the virtual world. "Forking" may have an excessive cognitive load for small children. Our task was, therefore and concluding, to define a 3D version control interface that would be based in linear trees in a way that children see clearly just the representation of the 3 time points they know intuitively – past, present and future.

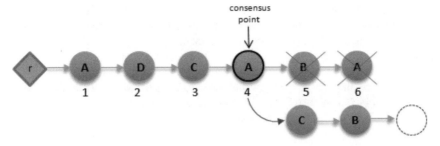

Fig. 3. Definition of a consensus point in a "past" version, where a new "future" is defined

2.5   Framework Proposal

Fig 4 shows the framework developed for a tool and methodology to help the teacher manage the decision making process. It is based on a Control Version System which is called "the historic" following the metaphor of a time travel line. The teacher has access to a "history management tool" as he/she wants to take children back in time to version confrontation. A "confrontation module" is responsible for showing children the two selected versions for comparison. In practice this module may be implemented on the prototype in several pedagogical ways, either by showing two versions side-by-side (splitting the user screen 50%-50%) or showing just one world with the objects of a version clearly marked and the objects of the other version showing some degree of transparency (as if they were "ghost objects" from the past). More studies need to be conducted to see what will be the confrontation module strategy more suitable for small children.

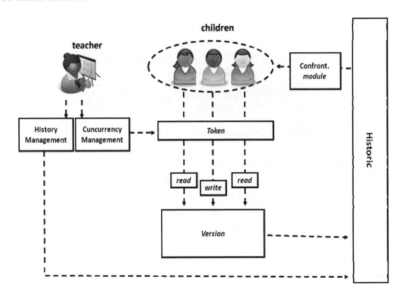

Fig. 4. Final framework for a tool and methodology based on a Control Version System

Also this strategy asks for some mechanisms that prevent concurrent editing, as each child wants to feel that he/she is the solo author of the configuration/version he's/she's proposing. A "concurrency management tool" is given to the teacher that, in practice, corresponds to giving the active child a token (it may be a pleasurable object as a magic wand) that gives the child editing privileges. We see in fig. 4 that the child in the middle is the one that has "write" privileges in the world while the others are mere spectators ("read" privileges). As the child finishes his/her proposal, it is saved as a "version" and added in the historic tool.

3   Prototype Implementation

Currently there are several platforms that allow or facilitate the creation of 3D Virtual Worlds. For the implementation of the prototype we have chosen the OpenCroquet [15] software development environment, as it offers a set of characteristics that were appealing to our project. Built in its core for synchronous peer-to-peer applications, it offers a large set of classes designed especially for virtual world creation which reduced the time needed to deliver a functional prototype. The supported decentralized peer-to-peer based network architecture is appealing as most primary schools in Portugal don't have a dedicated server to their projects. Finally, as it is open source, it can be distributed freely to all schools that manifest interest in it. Fig. 5 shows a screenshot of the developed prototype where children and teacher interact as avatars with the 3D models of objects in the playground:

Fig. 5. Children and teacher interact as avatars in the prototype

As explained before the teacher's avatar has special privileges, as he has the responsibility to manage all the activity. No chatting/communication mechanisms are required to mediate the communication as this environment is to be used in presence scenarios, inside the class: teacher and children are in the same physical space and communicate face to face – it is just the editing process that is technologically mediated.

3.1   Implementation of the Version Control System

The version control system is visually interfaced in the virtual world as a linear tree, representing past, present and future states (fig 6):

Fig. 6. A version tree in an initial stage of a discussion process

As we can see, the version tree interfaces the virtual world as a sequence of iconic spherical buttons that by lining up give the impression of a time line. Each iconic button will hold the "avatar face" of the child that edit one configuration. This visual representation of the tree is needed as it is both the tool the teacher uses to "travel back in time" and it is the concrete visualization of abstract concepts – collaboration, discussion and flow processes – which help young children learn some of the intrinsic complexities of a collaboration process.

4   Conclusions and Future Work

In this article we have presented a version control system for virtual worlds which are specially designed to mediate and support collaboration processes that involve spatial configurations in a classroom activity. We have used a visual linear tree that simplifies the collaboration process (normally represented by complex tree data structures) by turning it visually accessible by a simple linear tree, more easily understandable by children in primary schools. We intended to use linear trees as metaphors for processes such as collaboration that evolve dynamically over time and are therefore easily represented by linear past, present and future states. We also expect that such a visual element, used in the activity, may help children comprehend some of the dynamics and skills needed for an abstract concept – collaboration – by turning it into a concrete object which can be seen and interacted with visually.

The version tree also helps the "pedagogy of confrontation" as it allows the comparison of different proposed spaces through the time travel metaphor. We believe therefore that this data structure is in one of the simplest forms possible to use as a version control mechanism with pedagogical purposes.

We are currently modelling several 3D aspects of the school infrastructure into the virtual world and debugging/testing the prototype in lab conditions to finally begin the testing phase with children. By the end of the school year we expect to have the first results.

References

1. Lourenço, O.: Psicologia de desenvolvimento cognitivo, 2nd edn. Almedina, Coímbra (2002)
2. The City of Children,
   http://www.lacittadeibambini.org/inglese /progetto/
   motivazioni.htm

3. Ferré, M.B., Guitart, A.O., Ferret, M.P.: Children and Playgrounds in Mediterranean Cities. Children's Geographies 4(2), 173–183 (2006)
4. Unicef Innocenti Research Centre: Cities with Children – Child Friendly Cities in Italy. Unicef, Siena (2005)
5. The City of Children,
   http://www.lacittadeibambini.org/inglese/pubblicazioni/
   articoli/a_scuola_eng.doc
6. Rissotto, A., Tonucci, F.: Freedom of Movement and Environmental Knowledge in Elementary School Children. Journal of Environmental Psychology 22, 65–77 (2002)
7. Sarmento, T., Marques, J.: A Participação das Crianças nas Práticas de Relação das Famílias com as Escolas. Interacções 2, 59–86 (2006)
8. Niza, S.: O Modelo Curricular de Educação Pré-Escolar da Escola Moderna Portuguesa. In: Formosinho, J.O. (ed.) Modelos Curriculares para a Educação de Infância, 2nd edn., Porto Editora, Porto, pp. 137–159 (1998)
9. Grave-Resendes, L., Soares, J.: Diferenciação Pedagógica. Universidade Aberta, Lisboa (2002)
10. Dickey, M.D.: Three-Dimensional Virtual Worlds and Distance Learning: Two Case Studies of Active Worlds as a Medium for Distance Education. British Journal of Educational Technology 36(3), 439–451 (2005)
11. Johnson, L.F., Levine, A.H.: Virtual Worlds: Inherently Immersive, Highly Social Learning Spaces. Theory Into Practice 47(2), 161–170 (2008)
12. Dickey, M.D.: Teaching in 3D: Pedagogical Affordances and Constraints of 3D Virtual Worlds for Synchronous Distance Learning. Distance Education 24(1), 105–121 (2003)
13. Santos, F., Fonseca, B., Morgado, L., Martins, P.: Children as Active Partners: Strategies for Collaboration in Spatial Tasks through Virtual Worlds. In: Sixth International Conference on Creating, Connecting and Collaborating through Computing, pp. 73–76 (2008)
14. Selman, R., Selman, A.: Children's ideas about friendship: a new theory. Psychology Today 71–114 (October, 1979)
15. OpenCroquet, http://www.opencroquet.org

Facilitating Audio-Based Collaborative Storytelling for Informal Knowledge Management

Stephan Lukosch[1], Michael Klebl[2], and Tanja Buttler[3]

[1] Delft University of Technology, Faculty of Technology Policy and Management, Systems Engineering Department, Jaffalaan 5, 2628BX Delft, The Netherlands
s.g.lukosch@tudelft.nl
[2] FernUniversität in Hagen
Department for Cultural and Social Sciences
58084 Hagen, Germany
michael.klebl@fernuni-hagen.de
[3] FernUniversität in Hagen
Department for Mathematics and Computer Science
58084 Hagen, Germany
tanja.buttler@fernuni-hagen.de

Abstract. The increased demand for audio books and the rise of podcasting indicate a comeback of listening. On this basis, audio-based collaborative storytelling functionalises the act of telling stories in groups. In this paper we identify the requirements for informal knowledge management by means of audio-based collaborative storytelling. After reviewing the state of the art, we present our solution which addresses these requirements and supports a process for audio-based collaborative storytelling. Our solution consists of a storytelling client application which supports nomadic work as well as a web portal which aims at building a storytelling community. We present the storytelling client application and the web portal along our process for of collecting, structuring, linking and using audio clips, and finish with a report on first experiences and an outlook on future advancements.

1 Introduction

Telling stories is not only a given phenomenon of human practice; it is purposefully used as a method or a procedure in different areas of application under the designation *storytelling*. Collaborative storytelling aims at the development of a common understanding within a group by coordinated narrating activities (when each person contributes his or her own knowledge and his or her own interpretation of a common experience), in order to make implicit knowledge explicit. In doing so, the social relations between the persons taking part in the process are specified, e.g. in the form of roles as initiators, producers, co-producers and recipients. This results in symmetrical or asymmetrical relations. In situations of collaborative learning including related teaching activities both relations can be

R.O. Briggs et al. (Eds.): CRIWG 2008, LNCS 5411, pp. 289–304, 2008.
© Springer-Verlag Berlin Heidelberg 2008

found: cooperation between students on the same level as well as informational activities directed from teachers (experts or lectures) to students (thought of as laymen or novices) [1]. If we add as a third perspective that also those, which have fewer possibilities for communication are facilitated to express their knowledge and their requirements, three approaches to collaborative storytelling can be differentiated:

- *peer-to-peer*: Approaches focussing on equal discourse are predominantly found where collaborative storytelling is used pedagogically. This is, where student invent and tell stories in order to acquire or construct knowledge in the learning process. These forms of a discourse also show in areas of the knowledge management, e.g. in narrative reflection on completed projects [2].
- *top-down*: Approaches to storytelling addressing knowledge transfer or dissemination of information are found in educational contexts as well, where stories are told for learners, not by learners. In addition, the utilisation of the storytelling in knowledge management related to corporate goals presupposes an asymmetrical relationship between the tellers of stories and their recipients, where the storytellers have means to define knowledge [3] [4].
- *bottom-up*: Approaches to storytelling targeting participation or qualification for participation are promoted by grassroots media [5], e.g. by the concept of community radio (i.e. open channels on local radio) [6,7]. However, even forms of communication between experts and laymen do not serve exclusively the dissemination of knowledge from experts to laymen. Alleged laymen, e.g. future users of a technical product, are experts for the area of application. As such their narrations contribute to a deepened understanding of knowledge in the targeted domain. For example, the use of *Storycards* in *eXtreme Programming (XP)* benefits from this fact [8].

A special quality in collaborative storytelling arises from the restriction on verbal telling of stories, i.e. from the restriction on auditive production and perception. Firstly, spoken language is the basis for telling stories, even if in the western industrial societies literary, i.e. written, forms of stories are far common. However, quite recently a comeback of listening is to be determined, recognizable from the risen demand for audio books [9,10] and from the rise of podcasting [11,12]. Secondly, spoken language is an essential and quite natural part of human communication. The act of telling stories in groups ties to the everyday experience of discussing collectively remembered episodes. Here, narrative structures do not develop exclusively as a creation of an autonomously narrating person, but are formed by demands, additions, references, interpretation offers and much more from listening persons, who become co-tellers [13]. It is obvious that forms of storytelling, that are aligned to equal discourse or the qualification for participation, make special demands on information systems to be used in this kind of storytelling. These must be easily accessible, self-describing and conducive for learning; in short, ease of use is essential. An applicable information system that makes audio-based collaborative storytelling possible also for distributed use is based on a suitable process of collaboration. In the following we will determine the requirements to such an information system at first. Then we will regard related work. Subsequently, we present our solution and report on first experiences, in order to close with a view on future developments.

2   Requirements Analysis

In the following, we determine the requirements for supporting collaborative audio-based storytelling by describing a realistic scenario that could take place in a company or in an institution like a university. In this scenario, groups of employees interact with each other to report and comment on current events in their company or institution. This scenario serves perfectly for informal knowledge management processes and an initial requirement is:

R1: The audio-based collaborative storytelling support has to provide user and project management functionality.

To collaboratively create audio-based stories, users first have to record audio. However, these recordings are only available to one user. Thus, users have to be supported in sharing their recordings and thereby creating a shared audio database from which stories can be created collaboratively:

R2: The audio-based collaborative storytelling support has to offer a shared workspace for each story project, in which users can share and manage audio recordings.

Considering the scenario, it is obvious that users want to report on current events from different perspectives. They also want to have the possibility to comment existing stories and audio recordings. Such comments have to be placed directly at the point of interest and not at the end of an audio recording. To avoid cutting the audio recordings and thereby creating a huge bunch of smaller audio recordings, the following requirement has to be met to keep the information about the different points of interests in the audio recording:

R3: The audio-based collaborative storytelling support has to enable users to set marks in the audio recordings which can be used to link other audio recordings.

Such marks allow users to specify the exact point of time from where other audio recordings can be referenced. Thereby, users can collaboratively specify stories which consist of audio recordings, marks, and references between marks and audio recordings. The links between the different audio recordings allow the users to construct parallel threads in one story. Principally, it is possible to distinguish between linear and non-linear stories [14]. In a linear story there is exactly one thread. A non-linear story can have several parallel threads. In relation to our scenario, there could, e.g., be one original thread that reports on current events and a commented one which additionally includes the users' comments on the current events. To support users in creating such linked stories, the following requirement has to be met:

R4: The audio-based collaborative storytelling support has to facilitate the creation of links between marks and audio recordings for specifying non-linear stories.

These stories should not only be available to the group who has created it. Instead there should be a possibility to share the selected threads to a wider community. Podcasts are the current de-facto standard to share audio recordings and stories. For that purpose the following requirements have to be met:

R5: The audio-based collaborative storytelling support has to allow the selection of single threads in a non-linear story and its export as podcast.

The concept of non-linear stories allows users to create alternative stories and to add comments to existing stories. However, performing complex modifications on a shared story usually takes time and requires cognitive effort on the part of the user. In some cases, e.g. when users are collecting audio clips for a reportage, they might have no access to the Internet. Still, these users might want to add the collected audio clips to the shared story. This kind of nomadic work increases the probability of conflicting changes. To discard conflicting changes is inappropriate, since its originator has already expended much effort in performing the change. Hence, the following requirement must be met:

R6: The audio-based collaborative storytelling support has to allow users to work nomadically and support them in their synchronizing their results.

Finally, the community should have the possibility to discuss, comment, vote, and reuse the informally captured knowledge in the audio stories so that the following requirement emerges:

R7: The audio-based collaborative storytelling support has to include a community platform which allows to publish podcasts and discuss, comment, vote, and reuse the audio stories.

---

3   Related Work

There exist quite a few tools that support users in collaboratively creating stories. *StoryMapper* [15] supports groups in telling a story and structures the collaboration process by assigning different roles to the members of the group. A story is modeled as *conceptual map*. Each node in such a conceptual map can be linked to different media artefacts. *TellStory* [2] is a collaborative application that supports groups in creating text-based stories. *PhotoStory* [16] uses storytelling to increase the awareness in the group about its external presentation but also its social activities. For that purpose, the group can create stories that consist of a series a pictures with corresponding subtitles. Compared to our requirements, none of these tools focusses on audio-based stories for informal knowledge management. These tools violate R2, R3, and R4.

Apart from the above-mentioned tools that explicitly focus on the collaboratively creating a story, the following tools focus on the creation of stories in general. *MIST* [14] support the creation of non-linear multimedia stories. Users can integrate text, images, audio and video recordings in their story. Hence, it is also possible to create stories that solely consist of audio recordings. However, MIST does not focus on collaboration. Though different users can edit the same story, MIST does not support the users when synchronizing parallel changes or working nomadically. Hence, MIST violates R6.

iTell [17] makes use of a process that consists of four steps. This process allows users to create a text-based story which can be enriched with additional digital media items. Though Landry and Guzdial [18] consider collaboration as one of the fundamental activities when creating a story, iTell does not support it and thereby violates our requirements R2, R3, and specifically R6.

Rber et al. [19] introduce the concept of *Interactive Audiobooks*. Interactive audiobooks combine non-linear audio-based stories with interactive elements as known from computer games. However, such audiobooks are not created collaboratively and again R6 is violated.

StoryWriter [20] supports the creation of text-based stories that can be illustrated with images. For creating such a story, StoryWriter supports authors with a rule-based system, which, e.g., keeps track of the interaction among the different characters in the story or even generates text. But again, StoryWriter is not a collaborative application and does not support audio-based stories. Hence, R2, R3, and specifically R6 are violated again.

The above discussion shows that there is currently no sufficient support for collaboratively creating an audio-based story. In the following section, we will show how we address the requirements which we have identified for a tool and a process for collaborative audio-based storytelling.

4   Approach

Our approach is based on a web portal and a client application that supports the collaborative creation of audio-based stories. The process which is supported by this client application is a direct result of the first five requirements:

1. Creating a project team
2. Adding audio recordings
3. Segmenting audio recordings
4. Linking audio recordings
5. Publishing a story

These five steps may be executed in a nonsequential order. It is possible to skip steps or to return to steps in the process. Thereby, we support a free development of the story. When, e.g., linking audio recordings the need for an additional audio clip can become obvious. This clip can be added to the shared audio database, segmented, and then directly be linked.

Our approach consists of a storytelling client application which supports the different process steps, a web portal which offers the functionality to publish podcasts and discuss, comment, vote, and reuse audio stories, and a server which offers the functionality to manage and maintain the shared data. Fig. 1 highlights the main components of the resulting system architecture. The web portal as well as the server are based on Liferay [21]. The storytelling client application accesses the shared resources on the server via the Tunnel Servlet, whereas the web portal accesses these via a Struts [22] Servlet. At the server, both servlets can access the Liferay portal logic as well as the Storytelling Kernel which encapsulates the process functionality via an embedded Spring [23] layer. Finally, the Liferay portal logic and the Storytelling kernel provide access to the shared resources with are stored in a Java content repository and a MySQL database. The Java Content Repository is used to store and manage the shared files, i.e. audio clips and documents, whereas the MySQL database is used to manage the application-specific shared data.

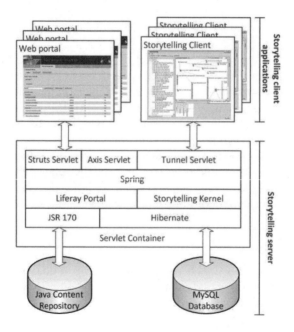

Fig. 1. Overall system architecture

In the following, we first describe the different process steps as supported by our client application, before we introduce our web portal in more detail.

4.1   Creating a Project Team

To collaboratively create a story, users must be able to build groups and define a common project (R1). As our web portal is based on Liferay [21], it reuses the provided functionality to request a login. Within the web portal users can create new projects, invite other users to join projects, or access the shared data within such a project. With the same login, users can connect from within the storytelling client application with the storytelling server to perform the same actions as well as to synchronize their local projects with the most recent versions at the server.

4.2   Adding Audio Recordings

In our second process step, users share their audio recordings for the story in the project workspace (R2). This can be done via the web portal as well as the client application. The client application has the advantage that it supports all process steps as well as the possibility to work nomadically and share the results when connected again.

For each audio recording, users can enter a textual description or even a transcription. This information can be used to search for audio recordings in the shared audio database. All audio recordings and its versions are managed by the server and stored within the JCR (cf. Fig. 1). Thereby, the server supports the creation of a shared audio repository which can be viewed via the web portal as well as the client application.

4.3   Segmenting Audio Recordings

In this process step, users can segment the audio recordings by setting marks (R3). Each mark is defined by a name and a exact time position in the audio recording. Apart from this information the client application also stores who created the mark, when the mark was created, and when and by whom it was last modified. To set a mark, users can listen to the audio recording while its wave form is visualized. By double-clicking on the wave form, users can add a mark to the audio recording.

The marks in an audio recording can later on be used to define links between audio recordings as they can serve as starting or end point for a link. By including marks in the audio recordings these have not to be split into separate parts when only a part is necessary for a story. Additionally, the marks are shared to all users that have access to the shared repository which supports a collaborative segmentation and adds additional value to the shared audio database.

Apart from the explicit marks which can be set by the users, each recording includes two implicit marks: the start and the end of the audio recording. These marks are set when the audio recording is imported into the shared workspace.

Fig. 2 shows the marking perspective, i.e. the user interface layout, of the storytelling client application and illustrates this process step. The left part of the screenshot shows the project explorer which shows the different projects in which the user participates and the project content, i.e. the corresponding audio recordings, documents, as well as the already proposed stories. The upper right part of the screenshot shows the list of marks for the selected audio recording.

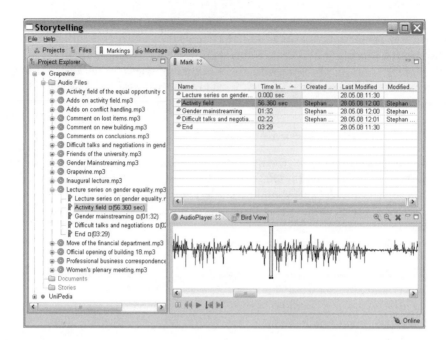

Fig. 2. The marking perspective

The lower right part shows the wave form of the selected audio recording. It allows users to listen to the recording and to define new marks by double-clicking on the point of time where they want to include a new mark. Existing marks can be dragged and dropped to a new position.

4.4   Linking Audio Recordings

After setting the explicit as well as the implicit marks, users can now specify different start and end points within an audio recording. However, users still need to be enabled to link different audio recordings. For that purpose, the storytelling client application has to offer appropriate support (R4).

Due to the marks that are associated with each audio recording it is possible to link the original audio recording without modifying them. In non-linear stories, the links between the different audio recordings represent a directed graph. The nodes in the story graph represent the parts of the shared audio recording which are in the story. The directed edges in the graph represent the links between these parts.

Each node is assigned to one audio recording. Users can select one starting mark for the node from the available marks. By allowing only one starting mark, the visualization of the story graph is simplified and more important the users possibility to follow the story flow is increased. However, as the use of an audio recording is not limited to one node in the story graph, users can use the same audio recording for other nodes with possibly different starting marks in the story graph again. Apart from the starting mark they can use each consecutive mark to add a link to another node. Though there is only one starting mark, there might be several links to other nodes.

Our client application supports the users in collaboratively creating such a story graph. When connected to the server, users can retrieve the most current version of the story graph or synchronize their local changes and thereby update the story graph at the server. By adding nodes and linking nodes, users can create a story graph in which each path between two nodes represents a new story. To ensure that users do not create cycles, the storytelling client application only allows links between two marks when this link does not lead to a cycle.

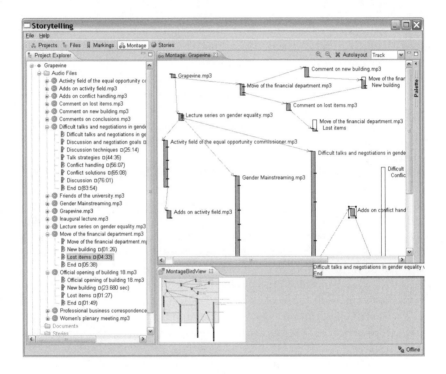

Fig. 3. The story graph perspective

Figure 3 shows the story graph perspective of our storytelling client application. The upper left corner again shows the project explorer with the shared audio recordings and their marks. The lower right corner shows a bird's view of the current story graph. The upper right part of the figure shows the current story graph. Each rectangle represents one audio recording. The filled part within the rectangle highlights the part of the audio recording which is used in the story. The mark which is used as starting mark are highlighted by a bar on the left side of the rectangle. Marks which are used to link to other audio recordings are shown as bar on the right side of the rectangle. New nodes can be included by dragging a selected mark on the canvas for the story graph. This mark serves as starting mark for the node. If this is not the implicit starting mark of the audio recording the rectangle is visualized as described above and the name of the starting mark is shown below the name of the audio recording right next to the rectangle, cf. node 'Move of the financial department' which starts at the mark 'Lost items'.

4.5   Publishing a Story

In this process step, users can select individual paths from the story graph and publish the path as a story in the group's shared workspace in the web portal(R5).

To publish a story, users have to select a starting node and an end node. Based on this specification, the client application calculates the shortest path between these two nodes. Users can adjust this path by adding new nodes to the story. Whenever such a node is added, the shortest path is recalculated to include this new intermediate node. The current path is always highlighted in a textual description on the left side of the publishing perspective as well as in the story graph by highlighting the included part of each node in a different colour. Users can also listen to the current path before it is finally published. When the user finishes the selection of the path, the client application uses the included information to create a continuous audio recording. This audio recording is then uploaded to the shared workspace in the web portal for further processing.

Once a mark or an audio recording with its representing node is included in a story, some restrictions concerning future changes apply. None of these can be deleted or changed. This only becomes possible again, when the story is deleted as well. Thereby, we ensure that the basic information for a story is available all the time. Additionally, this fosters users to create alternative stories when they want to comment on a story instead of deleting the disliked parts of a story.

Fig. 4 shows a screenshot of the publishing perspective. In this perspective the project explorer is replaced with a textual description of the currently defined story. The textual description starts with a name for the story and a textual description of the story. It contains all nodes in the order as they are included in the story. By selecting a node in the textual description, users can review additional information as the used starting mark or the textual description of the audio recording. It is also possible to listen to the part of the audio recording which is used in the story. The upper right corner contains the story graph in which the currently selected story components are highlighted in a different color. Currently, the local user has selected a story that starts at the 'Grapevine' audio recording and then passes the nodes, 'Move of the financial department', 'Comment on the new building', 'Move of the financial department, New building', 'Comment on lost items', and 'Move of the financial department, Lost items'. This story illustrates how different parts of one audio recoding can be used in one story to produce a commented version of an original recording, i.e. 'Move of the financial department'.

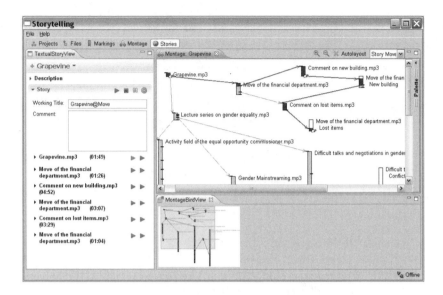

Fig. 4. The publishing perspective

4.6 Synchronization and Nomadic Work Support

Our client application stores most items (including audio recordings, marks and nodes and edges of a story graph) both locally and on the server, thereby allowing users to work nomadically and implementing the NOMADIC OBJECTS pattern [24]. However, after performing local changes, users need to synchronize their changes with the latest versions of the group's items and resolve any conflicts that might have occurred (R6).

Fig. 5 shows the part of the user interface which allows users to synchronize their local project data with the most recent version on the server. For this purpose, first users have to login to the server and provide their authentication data. This can be done via the *Login* button (1). The button to the right (2) allows users to download the most recent version from the server. Further The next button (3) allows to upload all local changes. Via the glasses button (4) users can highlight the most recent changes in the current perspective, i.e. new nodes, links, marks, files, etc. are equipped with the glasses. Apart from these options, the list (5) shows the local as well the remote changes since the last synchronization. By clicking on the text, the client changes the perspective and allows users to identify the most recent changes in detail. Thereby, the CHANGE INDICATOR pattern [24] is implemented to achieve group awareness.

After selecting a project and starting the synchronization mechanism, a user is presented with a merged version of local items and shared items. Our client application highlights the group's modifications and displays deleted items in the background, allowing the user to become aware of and comprehend the group's modifications. Furthermore, our client application points out conflicting changes. In these cases, only the latest modifications are shown, but access to

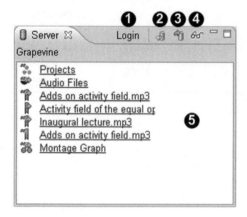

Fig. 5. Server synchronization and change awareness

the conflicting versions is also offered. The user has to review and resolve these conflicts before being allowed to commit local changes.

As stated previously, once a mark is used in a node, or an audio recording is used in a story, they may not be deleted. Therefore, in some cases our client application restores items deleted by the group. For example, if a user has utilized a mark (deleted by the group) to define a link in the graph, the client application restores the mark and only thereafter the graph may be shared.

4.7 Community Platform

As mentioned before, our web portal is based on Liferay [21]. It allows users to register, to create a project, to invite project members, to join ongoing projects, to upload and share audio recordings, to communicate via chat or message board, to view who else is currently, and to review all project-related information.

Apart from the above functionality, the web portal also supports users in starting story-related discussions, commenting stories, and voting on stories. These votes can be used by a project team to decide which of the alternative threads in the story graph is finally published as podcast to all members of the web portal and thus is available to the public. Such a published podcast can of course be included in a new story project, segmented by marks, linked with other audio recordings, and finally be published again.

Fig. 6 shows a screenshot of the web portal. The screenshot shows the audio library of the 'Grapevine' project. Here, users can access the different audio recordings, download and listen to the audio recordings, add new recordings, or search for recordings. The menu bar on top of the screenshot allows to access further functionality of the web portal, i.e. the project homepage with the corresponding user management functionality, the graphical view of the most recent story graph, as well as the voting functionality which allows to conduct voting on the published stories.

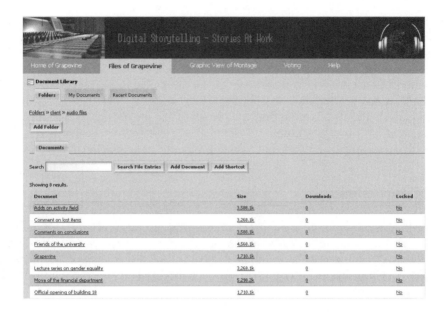

Fig. 6. The web portal

5   First Experiences

The evaluation of our support for audio-based collaborative storytelling is divided into three phases. In the first phase, we accomplished functional tests, in order to examine whether our requirements (R1) to (R6) are fulfilled. These tests were all together positive.

In a second phase, we firstly installed a test environment in our group. Secondly, we had the client application tested by different groups. The feedback received here confirmed that our solution allows to collaboratively produce audio-based stories. However, one feedback was that searching facilities have to be improved. Therefore, we currently work on social tagging functionality. In combination with a better support for transcribing audio clips, we thereby intend to improve the search functionality.

For a third phase, we plan to use and test our client application and web portal in real application scenarios. Our first application will be a seminar in the department of social sciences in which students have to retell, analyze, and discuss movies. The students will have to retell selected movies using our client application and afterwards add interpretation and discussion to the renarration. There, we are going to determine if and to what extend the provided support for the process enables collaborative editing and participative production of audio-based stories. We also want to evaluate the limitations of audio in dealing with e.g. ambiguity, anonymity, or inflection.

Since the design goal to enable the free and spontaneous development of collectively told stories is central, in particular those areas of application of the informal knowledge management are of interest in the third evaluation phase, which address equal discourse (*peer-to-peer*) or participation as well as qualification for participation (*bottom-up*). This comprises collaborative learning scenarios, both in school and in academics. In addition to this, forms of expert-lay communication take centre stage, e.g. in the appliance of participatory design methods in engineering and informatics, where users are included into product development. Parallels can be drawn to forms of the emancipatory use of media (as aspired by *grassroots media*), if the goal of participatory design is referred not only to technical artifacts, but also to social and organisational interactions.

6   Conclusion

According to our opinion, a comeback of listening can be determined on the basis of the risen demand for audio books and the unprompted rise of podcasting. Audio-based collaborative storytelling avails itself of this comeback and utilises the act of telling stories in groups. In this collaborative act everybody involved becomes a co-teller: by demands, additions, references, interpretation offers, contributions of individual perspectives and much more. The scope of possible areas of application is very broad and spans from open channels on local radio to the employment in collaborative learning and in knowledge management, leading to user oriented requirement analysis during product development.

All these scenarios are based on common principles, which require a technical support. For this technical support we identified seven requirements and presented a client application and a web portal for audio-based collaborative storytelling, which fulfils these requirements during a five-stage process. In comparison with other tools, our audio-based collaborative storytelling support is characterised by the fact that it enables users to collaboratively set marks and thus supports collaborative segmenting of shared audio recordings. Furthermore, our client application supports a collaborative assembly of single audio recordings, which results in a directed graph. Starting from a chosen starting mark, alternative paths, i.e. different versions of a story, can be selected from this graph for publication. Only by this collaborative assembly the alternative representation of events, knowledge, stories or requirements becomes possible as a genuine collaborative act of telling stories. This way informal knowledge management is explicitly supported, as a group of humans agrees stepwise on a common view of things.

Future research will show if other tools are needed, in order to reduce the barriers for the production of common stories, to assist collaborative creation or to improve navigation in the time-based audio clips. First experiences have shown that at a first step search facilities have to be improved. Here, we will focus on collaborative tagging and transcription rather than on speech-to-text technology, since the former is promising more success in the face of the expected diverse quality of audio recordings. Furthermore, future research will show which informal processes lead to a story and how a group as a whole arranges processes of evaluation and selection of knowledge in informal knowledge management.

Since this is not determined by our tool, the question arises, how consensus is reached on a relative stable version of a story as well as how long a stable version endures.

References

1. Jucks, R., Paechter, M.R., Tatar, D.G.: Learning and collaboration in online discourses. International Journal of Educational Policy, Research, & Practice 4, 117–146 (2003)
2. Perret, R., Borges, M.R., Santoro, F.M.: Applying group storytelling in knowledge management. In: de Vreede, G.-J., Guerrero, L.A., Marín Raventós, G. (eds.) CRIWG 2004. LNCS, vol. 3198, pp. 34–41. Springer, Heidelberg (2004)
3. Snowden, D.: The art and science of story or 'are you sitting uncomfortably?' part 1: Gathering and harvesting the raw material. Business Information Review 17, 147–156 (2000)
4. Snowden, D.: The art and science of story or 'are you sitting uncomfortably?' part 2: The weft and the warp of purposeful story. Business Information Review 17, 215–226 (2000)
5. Gillmor, D.: We the Media: Grassroots Journalism By The People, For the People. O'Reilly Media, Sebastopol (2006)
6. Fichtner, J., Günnel, T., Weber, S. (eds.): Handlungsorientierte Medienpädagogik im Bürgerradio. Forschungsergebnisse eines Modellporjekts mit ArbeitnehmerInnen und dessen Implikationen für die medienpädagogische Diskussion. KoPäd Verlag, Munich (2001)
7. Howley, K.: Radiocracy rulz! microoradio as electronic activism. International Journal of Cultural Studies 3(2), 256–267 (2000)
8. Beck, K.: eXtreme Programming Explained. Addison Wesley, Reading (1999)
9. Philips, D.: Talking books: The encounter of literature and technology in the audio book. Convergence 13(3), 293–306 (2007)
10. Friederichs, T., Hass, B.H.: Der Markt für Hörbücher. Eine Analyse klassischer und neuer Distributionsformen. MedienWirtschaft. Zeitschrift für Medienmanagement und Kommunikationsökonomie 3, 22–35 (2006)
11. Hein, G., Jakuska, R.: Podcast industry. iLabs – Center for Innovation Research, School of Management, University of Michigan-Dearborn (April 2007)
12. Martens, D., Amann, R.: Internetnutzung zwischen Pragmatismus und YouTube-Euphorie. Media Perspektiven 11, 538–551 (2007)
13. Ochs, E., Capps, L.: Living Narrative. Creating Lives in Everyday Storytelling. Harvard University Press (2001)
14. Spaniol, M., Klamma, R., Sharda, N., Jarke, M.: Web-based learning with non-linear multimedia stories. In: Liu, W., Li, Q., Lau, R. (eds.) ICWL 2006. LNCS, vol. 4181, pp. 249–263. Springer, Heidelberg (2006)
15. Acosta, C., Collazos, C., Guerrero, L., Pino, J., Neyem, H., Motele, O.: StoryMapper: A Multimedia Tool to Externalize Knowledge. In: Proceedings of the XXIV Conference of the Chilean Computer Science Society, pp. 133–140. IEEE CS Press, Los Alamitos (2004)
16. Schäfer, L., Valle, C., Prinz, W.: Group storytelling for team awareness and entertainment. In: NordiCHI 2004: Proceedings of the third Nordic conference on Human-computer interaction, pp. 441–444. ACM Press, New York (2004)

17. Landry, B.M., Guzdial, M.: iTell: supporting retrospective storytelling with digital photos. In: DIS 2006: Proceedings of the 6th ACM conference on Designing Interactive systems, pp. 160–168. ACM Press, New York (2006)
18. Landry, B., Guzdial, M.: Learning from human support: Informing the design of personal digital story-authoring tools. In: Proceedings of CODE 2006 (2006)
19. Röber, N., Huber, C., Hartmann, K., Feustel, M., Masuch, M.: Interactive audio-books: Combining narratives with game elements. In: Göbel, S., Malkewitz, R., Iurgel, I. (eds.) TIDSE 2006. LNCS, vol. 4326, pp. 358–369. Springer, Heidelberg (2006)
20. Steiner, K.E., Moher, T.G.: Graphic storywriter: an interactive environment for emergent storytelling. In: CHI 1992: Proceedings of the SIGCHI conference on Human factors in computing systems, pp. 357–364. ACM Press, New York (1992)
21. Liferay - Enterprise Open Source Portal (June 2008), http://www.liferay.com
22. Struts Framework (June 2008), http://struts.apache.org/
23. Spring Application Framework (June 2008), http://springframework.org/
24. Schümmer, T., Lukosch, S.: Patterns for Computer-Mediated Interaction. John Wiley & Sons, Ltd., Chichester (2007)

Toward a Taxonomy of Groupware Technologies

Daniel D. Mittleman[1], Robert O. Briggs[2], John Murphy[2], and Alanah Davis[2]

[1] School of Computer Science, Telecommunications and Information Systems, DePaul University
[2] Institute for Collaboration Science, University of Nebraska at Omaha
danny@cti.depaul.edu,
{rbriggs,jmurphy,alanahdavis}@mail.unomaha.edu

Abstract. The rise of the global marketplace and the advancing of the World Wide Web have given impetus to rapid advances in groupware. Hundreds of products now exist in the groupware marketplace, and more appear monthly. To ease the cognitive load of understanding what groupware technologies are, what capabilities they afford, and what can be done with them, we analyzed hundreds of computer-based collaboration-support products and distilled their attributes into two complementary schemas – a classification scheme and a comparison scheme. The classification scheme provides a way to organize the many products from the rapidly expanding groupware arena into a small set of relatively stable categories. The comparison scheme provides the means to compare and differentiate collaboration technologies within and across categories. Taken together, the classification and comparison schemas provide a basis for making sense of collaboration technologies and their potential benefits to the collaboration community.

Keywords: collaboration technology, computer-based collaboration-support products.

1 Introduction

We once worked with the headquarters staff of a large organization with world-wide presence. The organization's charter required that it develop creative, cross-disciplinary solutions to fast-breaking opportunities and challenges in its volatile operating environment. The best of these solutions seemed to emerge from a fluid, open exchange of the rich experiences and knowledge of the headquarters' 500+ knowledge workers. Yet the organization seemed bound in an outdated, multi-layered bureaucracy that was insufficiently responsive to the accelerating pace of change in their environment. We were chartered to help the organization develop a more effective work practice for conducting their weekly staff meetings. Hundreds of staff meetings took place every week to provide management status updates for current projects. Personnel we interviewed complained of the stultifying, painful nature of these meetings, which typically involved about ten people and lasted about two hours. The organization had a culturally ingrained practice of presenting from PowerPoint slides at these meetings. The meeting was usually the first time others saw the information in the slides. Knowledge workers briefed their managers; managers briefed executives;

R.O. Briggs et al. (Eds.): CRIWG 2008, LNCS 5411, pp. 305–317, 2008.
© Springer-Verlag Berlin Heidelberg 2008

and executives briefed senior executives up through the top levels of leadership. Management wanted a system that would allow people to preview status presentations before the meeting so that more attention could be paid to generating creative solutions that added value to the organization.

We worked with stakeholders in the organization to develop five iterations of a new work practice for status reporting and creative problem solving. Each iteration used a different collaboration technology. The first implementation was pen-and-paper. We subsequently prototyped the process with a group support system, a shared document editor, a wiki, and a web portal.. Each iteration revealed new problems and constraints. The paper-based system, for example did not provide shared access to critical information. The group support system provided shared access to information, but turned out to have unacceptable constraints on access controls. Its tools were well suited to in-meeting interactions, but were not as useful for between-meeting interactions. The shared editor provided just the right access controls, but lacked support for the variety of structures the group needed to create. The wiki provided for the variety of structures, but did not provide the requisite synchronicity for certain user actions. The web portal provided sufficient access controls, support for between-meeting coordination, variety of structures, and synchronicity of actions. Lacking a systematic way to analyze and articulate collaboration needs and to select among the many groupware possibilities, the evolution through these prototypes required more than a year.

The scope of this problem becomes more apparent when one considers that the rise of the global marketplace and the advancing of the World Wide Web have given impetus to rapid advances in groupware technologies and products. Today, hundreds of new groupware products exist to support collaboration by work teams, with more appearing monthly. This recent growth in the groupware marketplace is driven by the need for organizations to adapt to the rapidly-changing global business environment [1] and enabled by the maturation of Web2.0 development techniques (e.g. AJAX and Ruby on Rails). Online meeting collaboration and virtual project management are increasingly common modes of work as teams operate more and more frequently across departmental, organizational, and geographic boundaries to innovate and create value [see e.g. 2, 3, 4].

Commercial vendors tend to sculpt niches in the marketplace by bundling mixes of core technologies. For example, the most recent version of Yahoo Messenger has basic instant message functionality, along with file exchange, and audio conferencing, which are distinctly different core technologies and yet are packaged under a single "instant messaging" label. These "bundles of capabilities" – collaboration suites in a realm with no traditional or commonly accepted product categories or commonly accepted capability feature sets – make it difficult for practitioners to understand what capabilities they need, what capabilities a given product offers, and to select an appropriate groupware product for their mix of tasks [5, 6]. The rapid elaboration of the groupware space also raises complexity for groupware researchers to make sense of what value can be created with current technologies, how and why such value can be created, and what groupware challenges still remain unaddressed.

Journalists and bloggers offer several approaches to categorizing collaboration products, each of which is useful to its purposes.[1] More formal typologies have appeared over years of research, from the classic "time-place" classification of technologies [7, 8] to early schemes that grouped technologies according to whether they are communications- or process-oriented [9]. Later schemes further decomposed process-oriented technologies to accommodate information processing and overall group process perspectives [10]. Despite these incremental improvements, a simple, relatively stable, and comprehensive taxonomic model of elemental collaboration technologies has not yet emerged.

In this paper, we propose a starting place toward developing such a model. The purpose of this classification scheme is to provide a lens through which people can better understand the capabilities of, and relationships between, collaboration technologies. Such a framework might then help practitioners to select from among commercial collaboration software offerings, offer groupware designers a range of design choices for new systems, and reveal new challenges to the groupware research community.

This paper starts by describing the overall process used to gather and analyze the data necessary to develop the scheme. Next, it develops a comparison scheme that formed the basis by which collaboration products were grouped. An initial classification scheme for collaboration technologies is then presented. The paper concludes with examples of how to use the comparison and classification schemes, along with a discussion of potential directions for future research.

2 Scheme Development

Over the course of 5+ years a team of collaboration researchers has searched the commercial marketspace for new groupware products to develop the current listing of more than 250[2] products. Throughout that timeframe, the team has analyzed those products to derive factors that could be used to classify those products. As new products emerge, the team regularly re-examined the scheme to assess fit. This inductive process, driven by actual collaboration tools in use in the marketplace, drove the derivation of the original scheme and continues to shape it as collaboration products develop.

Initial analysis revealed that it would not be possible to create a taxonomy for groupware *products* because products tend to be heavily-overlapping bundles of *technologies* rather than distinguishable, classifiable entities. We define a groupware technology as a software solution which, if implemented, could help move a group toward its goals in some specific way. We define a collaboration product as an implementation of one or more collaboration technologies offered as an integrated package.

We therefore re-analyzed the products to distill from them a set of the elemental collaboration capabilities that comprised them. We next developed detailed functional descriptions of the products, and then reviewed those descriptions for accuracy and

---

[1] Examples include: http://www.collaboration-tools.com/, http://itre-dux.com/office-20/database/, and http://www.solutionwatch.com/515/ back-to-school-with-the-class-of-web-20-part-2/

[2] A list of the reviewed products is available upon request.

completeness. We then distilled from those descriptions a set of key themes that characterize the attributes by with which one could compare different implementations of groupware technologies, both within and across categories. We asked three groupware experts to apply the initial set of themes to the original functional descriptions to verify their completeness. We identified discrepancies, and refined the attribute set until the core set of attributes could be generally applied to any product in the set. The resultant comparison scheme is presented in Section Three.

The most critical attribute by which groupware products could be compared and contrasted was their fundamental technical capabilities. We therefore reexamined the groupware products to address the basic question, "What are the core underlying technologies represented in this constellation of products?" We organized the core technologies into a classification scheme with three main branches, which appears in Section Four.

3 Comparing Collaboration Technologies

The implementation of technologies and their details can vary from product to product. When comparing technologies, it is useful to sort out which affordances are present, which are absent, and which can be configured to adapt to variations of need.[3] In this section, we draw attention to key attributes by which one can compare and contrast collaboration technologies when designing or selecting among collaboration products. The comparison scheme consists of nine architectural constructs for collaboration technology affordances including: 1) core functionality, 2) content, 3) relationships, 4) supported actions, 5) action parameters, 6) access control, 7) session persistence, 8) alert mechanisms, and 9) awareness indicators. Table 1 lists these capabilities and briefly describes each attribute of the comparison scheme.

Core Functionality identifies the primary capability provided by a tool. The core capability of a *blog*, for example, is a page to which users may contribute shared text or hyperlinks to other pages (new posts). The core functionality of *voice conference* is a continuous audio stream. We discuss core capabilities in much more detail in Section Four, where we propose a classification scheme for collaboration technology.

Content describes the kinds of data that may be contributed to a particular collaboration [12]. These data structures include:

- *Text*: a block of textual information (e.g. a text message).
- *Links*: reference pointers with labels (e.g. a URL).
- *Graphic*: a pictorial image, object, or diagram (e.g. a jpg or gif picture)
- *Data stream*: a continuous data flow (e.g. a sound channel or desktop sharing).
- *Hypermedia*: Combinations of the content types above.

---

[3] A rich literature exists for mapping collaboration tool affordances to established group facilitation processes. Exploration of that literature, while important to a designer or user of collaboration products, is beyond the scope of this paper. However, an introduction into that literature can be found at 11. de Vreede, G.-J., and Briggs, R.O. Meetings of the Future: Enhancing Group Collaboration with Group Support Systems. *Journal of Creativity and Innovation Management*, 6, 2 (1997), 106-116.

Table 1. Summary of the Comparison Scheme Attributes

Capability	Affordances essential to the nature of the technology
Core Functionality	Primary functionality provided by the tool. This maps to the tool's location within the classification scheme (see Table 2).
Content	Possible content for contributions to a collaboration system are: text, links, graphic, and data-stream.
Relationships	Users can establish these kinds of relationships among contributions: collection, list, tree, and graph.
Supported Actions	Actions that users can take on structures or relations.
Add	Ability to create structures or relations.
Receive	Ability to receive, view, or read contributions to the system.
Associate	Establish relationships among contributions
Edit	Ability to modify content or relationships.
Move	Change relationships among contributions
Delete	Ability to eliminate content or relationships.
Judge	Render opinions on the relative merits of contributions
Action Parameters	Two key parameters that characterize or modify actions.
Synchronicity	Expected delay between the time one person executes an action and the time other users can perceive the effects of that action.
Identifiability	Degree to which users can determine who executed an action.
Access Controls	The granting or revoking of user ability to execute supported actions.
Session Persistence	The degree to which contributions are ephemeral or permanent.
Alert Mechanisms	The ways participants are notified that something or someone in the system requires their attention.
Awareness Indicators	The means by which users may know what other users have access to a session, the nature of their roles, and their current status.

Relationships are the associations users can establish among contributions. Four types of relationships are possible among contributions:

- *Collection*: connotes membership in a set of otherwise unrelated objects.
- *List*: a list an ordered set of objects (e.g. before/after, bigger/smaller).
- *Tree*: a set of objects in hierarchical relationships with each object (except the root) having only one parent, but having zero-to-many children (e.g. system, subsystem, component).
- *Graph*: an organization where each object can have zero to many links to other objects (e.g. parents, siblings, children, cousins…).

Relationships may be established among objects of same or differing content. Tools may support only a single kind of relationship or may support several types. This variety enables a vast array of information constructions. Some groupware may articulate the semantics of the relationships represented in the content. Other tools only represent the syntax of relationships, leaving it up to group members to agree on semantics.

Supported Actions indicate the things a system allows participant to do to content and relationships. These actions (already well established in the database and groupware literatures) are:

1) Add – contribute content to the group (e.g. add a new item to a blog; speak during an audio conference).
2) Receive – detect contributions made by self or others (e.g. view text contributions or hear an audio channel)
3) Associate - establish relationships among contributions (e.g. organize ideas into categories or arrange content into an outline)
4) Edit - modify content of a contribution (e.g. amend or change text already contributed to a session),
5) Move – Change relationships among contributions
6) Delete - remove a contribution from a session (e.g. delete text, erase audio)
7) Judge – render an opinion on the relative merits of contributions (e.g vote).

Action Parameters describe characteristics of actions that impact user's experience of contributions and of one another when using a collaboration tool.

- Synchronicity characterizes expected delay between the time that a user executes an action and the time other users respond to that action. For example, with audio conferences, participants expect a response to their contributions within a second or two, whereas with e-mail, users expect that responses may be delayed by hours or days. In some systems (e.g. audio conferencing) participants must wait their turn to contribute and in others (e.g. group support systems, wikis) participants may contribute simultaneously.
- Identifiability characterizes the degree to which users can determine who executed an action. Identifiability ranges from full anonymity, to subgroup identification, to pseudonym identification (so you may know which contributions came from the same person, but not who that person is), to full identification.

Access Control deals with the configuration of user's rights and privileges with respect to entering a session and executing supported actions. Some actions may be always available (e.g. in an instant messaging, all users may always add), or always blocked (e.g. in instant messaging, no users may edit or delete contributions. Still others may be configurable on the fly (e.g. in some group support systems, anonymity may be switched on or off as needed).

Session Persistence is the degree to which contributions are ephemeral or permanent. In some collaboration tools (e.g. video or audio conferencing) contributions may be ephemeral, disappearing as soon as they are made. In others, contributions persist only for the duration of a session and disappear when all users exit (e.g. instant messaging). Often time users may configure the degree to which their contributions persist. For example, in some system users may decide whether session contents will be saved. Other systems (e.g. e-mail) allow a user to delete contributions from their view, but the contributions remain in the views of others, or in a permanent record.

Alert Mechanisms are the way participants are interrupted or notified that something in the system demands their attention. For example, instant messenger systems typically signal an arriving contribution by making a sound and popping up a momentary visual cue. The interrupt is designed to attract immediate attention; however, it can be ignored or refused by the receiver. Alerts, like those from an RSS feed, for example, do not interrupt the user but rather require that the user deliberately seek them out.

Awareness Indicators are the ways users learn about the other people who have access to a session, the roles that they hold, and their current status. In some systems, the only indicator that others are present is arrival of new contributions. In others, people may see a list of participants who have been granted access. In some, users can learn who is currently active in a session, what they are doing, which tools they are using, which contributions they are manipulating, and in some cases, even their current state of mind (e.g. happy, confused, dissatisfied).

The comparison scheme presented here draws our attention to key attributes by which one can compare, contrast, optimize, and select among groupware technology implementation. The most important of these attributes is the core capability, which we elaborate more fully in the next section.

4 A Classification Scheme for Collaboration Technologies

In this section we provide a finer-grained examination of the core capabilities in the form of a classification scheme for collaboration technologies. The first level of the classification scheme divides all collaboration technologies into four main categories according to their most-fundamental capabilities: 1) jointly authored pages, 2) streaming tools, 3) information access tools, and 4) aggregated systems. The fourth category is for technologies that must integrate a mix of tools from the first three categories and optimize them to support work practices that that cannot be achieved with a single technology. The scheme further subdivides each of the top four categories into sub-categories by the functions they are optimized to support (see Table 2). This section summarizes these categories and their subcategories.

4.1 Jointly Authored Pages

The most fundamental capability for all technologies in the jointly authored pages category is a digital page, defined as a single window to which multiple collaborative participants can contribute, often simultaneously. The data structures of pages might include text, graphics, numbers, or other digital objects. However, regardless of content, any contribution made by a participant will generally appear on the screens of the other participants who view the same page. A given technology based on jointly authored pages may provide a single page or multiple pages. In some cases the contributions to one page serve as hyperlinks to other pages, allowing for the creation of hierarchies or networks of pages. Jointly authored pages are the basis for several subcategories of collaboration technology including: conversation tools, shared editors, group dynamics tools, and polling tools.

Conversation Tools are those primarily optimized to support dialog among group members. Email is a widely-used conversation tool as well as short message services (SMS) (i.e. cell phone text messaging) which is becoming increasingly common. According to Verizon Wireless, their customers sent and received more than 10 billion text messages in June 2007 [13]. Other conversation tools include instant messaging, chat rooms, and blogs or threaded discussions. Instant messaging and chat rooms

Table 2. Summary of the Classification Scheme for Collaboration Technology

Categories and sub-categories	Descriptions
Jointly Authored Pages	Technologies that provide one or more windows that multiple users may view, and to which multiple users may contribute, usually simultaneously.
Conversation Tools	Optimized to support dialog among group members.
Shared Editors	Optimized for the joint production of deliverables like documents, spreadsheets, or graphics.
Group Dynamics Tools	Optimized for creating, sustaining, or changing patterns of collaboration among people making joint effort toward a goal (e.g. idea generation, idea clarification, idea evaluation, idea organization, consensus-building).
Polling Tools	Optimized for gathering, aggregating, and understanding judgments, opinions, and information from multiple people.
Streaming Technologies	Technologies that provide a continuous feed of changing data.
Desktop / Application Sharing	Optimized for remote viewing and/or control of the computers of other group members.
Audio Conferencing	Optimized for transmission and receipt of sounds.
Video Conferencing	Optimized for transmission and receipt of dynamic images.
Information Access Tools	Technologies that provide group members with ways to store, share, find, and classify data objects.
Shared File Repositories	Provide group members with ways to store and share digital files.
Social Tagging Systems	Provide means to affix keyword tags to digital objects so that users can find objects of interest, and so they can find others with similar interests.
Search Engines	Provide means to retrieve relevant digital objects from among vast stores of objects based on search criteria.
Syndication Tools	Provide notification of when new contributions of interest have been added to pages or repositories.
Aggregated Systems	Technologies that combine of other technologies and tailor them to support a specific kind of task.

provide users with a single shared page to which they can contribute contributions to a chronologically ordered list. Participants may not move, edit, or delete their contributions. Instant messaging and chat rooms differ from one another only in their access and alert mechanisms. With instant messaging an individual receives a pop-up invitation that another individual wishes to hold a conversation, while with chat rooms an individual browses to a web site to find and join a conversation. Blogs (otherwise known as Web Logs) and threaded discussion tools are optimized for less-synchronous conversations. Users make a contribution, then come back later to see how others may have responded. Blogs and threaded discussions are typically persistent (i.e. their content remains even when users are not contributing) whereas chat rooms and instant messaging are usually ephemeral (i.e. when the last person exits a session, the session content disappears).

Shared Editor tools are typically a jointly authored page optimized for the creation of a certain kind of deliverable by multiple authors. The content and affordances of these tools often match those of single-user office suite tools (e.g. .word processing, spreadsheet); however they are enhanced to accept contributions and editing by multiple simultaneous users. A wiki (the Hawaiian word for 'fast') is another example of joint document authoring. Wikis are simple web pages that can be create directly through a web browser by any authorized user without the use of off-line web development tools.

Group Dynamics Tools are optimized for creating, sustaining, or changing patterns of collaboration among individuals making a joint effort toward a goal. The patterns these tools support include generating ideas, establishing shared understanding of them, converging on those worth more attention, organizing and evaluating ideas, and building consensus [14]. These tools are often implemented as multiple layers of jointly authored pages such that each contribution on a given page may serve as a hyperlink to a sub-page. The affordances of such tools are typically easily configurable, so at any given moment a group leader can provide team members with the features they need (e.g. view, add, move) while blocking features they should not be using (e.g. edit, delete).

Polling Tools are a special class of jointly authored pages, optimized for gathering, aggregating, and understanding judgments, or opinions from multiple people. At a minimum, the shared pages of a polling tool must offer a structure of one or more ballot items, a way for users to record votes, and a way to display results. Polling tools may offer rating, ranking, allocating, or categorizing evaluation methods and may also support the gathering of text based responses to ballot items.

4.2 Streaming Technologies

The core capability of all tools in the streaming technologies category is a continuous feed of dynamic data. Desktop sharing, application sharing, and audio/video conferencing are common examples of streaming technologies.

Desktop and Application Sharing Tools allow the events displayed on one computer to be seen on the screens of other computers. With some application sharing tools, members may use their own mouse and keyboard to control the remotely viewed computer.

Audio Conferencing Tools provide a continuous channel for multiple users to send and receive sound while Video Conferencing Tools allow users to send and receive sound or moving images. Typically all users may receive contributions in both types of tools, however systems may vary in the mechanisms they provide for alerts and access control as well as by the degree to which affordances can be configured and controlled by a leader.

4.3 Information Access Technologies

Information access technologies provide ways to store, share, classify, and find data and information objects. Key examples from this category are shared file repositories, social tagging, search engines, and syndication tools.

Shared File Repositories provide mechanisms for group members to store digital files where others in the group can access them. Some such systems also provide version control mechanisms such as check-out, check-in capabilities, and version back-ups.

Social Tagging allows users to affix keyword tags to digital objects in a shared repository. For example, the web site, del.icio.us allows users to store and tag their favorite web links (i.e. bookmarks) online so they can access them from any computer. Users are not only able to access their own bookmarks by keyword, but bookmarks posted and tagged by others as well. More significantly, users can find other users who share an interest in the same content. Social tagging systems allow for the rapid formation of communities of interest and communities of practice around the content of the data repository. The data in a social tagging repository are said be organized in a folksonomy, an organization scheme that emerges organically from the many ways that users think of and tag contributions, rather than a taxonomy, organized by experts.

Search Engines use search criteria provided by users to retrieve digital objects from among vast stores of such objects (e.g. the Worldwide Web, the blogosphere, digital libraries). Search criteria may include content, tags, and other attributes of the objects in the search space. Some search engines interpret the semantic content of the search request to find related content that is not an exact match for the search criteria.

Syndication tools allow a user to receive a notification when new contributions to pages or repositories they deem to be of interest (e.g. blogs, wikis, and social networks). Users subscribe to receive update alerts from a feed on a syndicated site. Every time the site changes, the feed broadcasts an alert message to all its subscribers. Users view alerts using software called an aggregator. Any time a user opens their aggregator, they see which of their subscription sites has new contributions. Therefore, users do not need to scan all contents to discover new contributions.

4.4 Aggregated Technologies

Aggregated technologies integrate several technologies from the other three categories and optimize them to support a task that cannot be executed using a single technology [5]. Aggregated technologies deliver value which could be achieved with a collection of stand-alone tools. There are many examples of aggregated technologies, among them virtual workspaces, group support systems, and social networking systems. Virtual workspaces often combine document repositories, team calendars, conversation tools and other technologies that make it easier for team members to execute coordinated efforts (e.g. Groove or SharePoint). Remote presentation or web conferencing systems often combine application sharing and audio streams with document repositories and polling tools optimized to support one-to-many broadcast of presentations, with some ability for the audience to provide feedback to the presenter (e.g. Webex or SameTime). Group support systems integrate collections of group dynamics tools to move groups seamlessly through a series of activities toward a goal, for example, by generating ideas in one tool, organize them in another, and evaluating them in yet another (e.g. GroupSystems or WebIQ). Social networking systems (e.g. MySpace or Flickr) combine social tagging with elements of wikis, blogs, other shared page tools, and a search engine so users can find and communicate with their acquaintances as well as establish new relationships based on mutual friends or mutual interests. Thus, aggregated technologies may combine any mix shared-page, streaming, and information access technologies to support a particular purpose.

5 Conclusion

We present these comparison and classification schemes as a starting place toward a taxonomy of collaboration technology. One can use the classification scheme to analyze, compare, and contrast the capabilities offered by groupware products.

Additionally, one can use the comparison scheme to compare and contrast important implementation choices in technologies within the same category or across categories.

5.1 Example Uses of the Comparison and Classification Schemes

Table 3 illustrates how the comparison scheme can be used to weight the capabilities of two collaboration technologies (i.e. instant messaging and video conferencing) against each other.

Table 3. Example of the Comparison Scheme Attributes

Attribute	Instant messaging	Video Conferencing
Core Functionality	Creation and exchange of single text pages	Single video stream, usually paired with a single audio stream
Content	Text	A/V streams
Relationships	Time-ordered list of text contributions	Time-ordered sequence of synchronized sounds and images
Supported Actions		
Add/	Text	Audio and Video in parallel;
Receive/	Yes	Yes
Associate	No	No
Edit	No	No
Move	No	No
Delete	No	No
Judge	No	No
Action Parameters		
Synchronicity	Immediate display of contributions to all participants; users may add content in parallel	Immediate presentation of all contributions to all participants; users will add content in parallel
Identifiability	Identification of contributor by login-name is automatic and mandatory	Identification of contributor typically only by cues embedded into the stream (e.g. sound of voice, face recognition)
Access Control	Receive by invitation only. Once invitation is accepted, all users have both add and receive rights	Varies by system. Access ranges from browsing to dial-up access. Control ranges from open to password or access code. Once in, all users have view rights. Add rights may be under the control of a moderator
Session Persistence	For duration of session by default; manual or automatic saving optional	For the duration of session by default; manual or automatic saving optional
Alert Mechanisms	Interrupt by invitation with sound, pop up visual cue	Vary by system, ranging e-mail invitations, to audio and visual interrupts

Note that any given implementation of IM or video conferencing system could differ from these configurations in many ways and still be essentially the same class of technology. The differences, though, could have significant impact on the degree to which the implementation serves the needs of the users. To better illustrate this point, consider the following scenario. One of this paper's authors was consulted about a technology selection by a systems integrator who operates out of the Midwest United States to service customers nationwide. They had selected MS-LiveMeeting® as the technology platform for conducting a distributed requirements elicitation for a product under development. Without knowing anything more about the situation, the author predicted that the chosen solution would be ineffective. He was able to make that judgment based on the fact that he knew that though LiveMeeting was quite capable as a streaming technology, it lacked key technical features found in jointly-authored page and information access categories that would also be important for the success of the project. This scenario demonstrated how basic knowledge of the classification scheme can help assess potential technical solutions.

5.2 Implications and Future Research

The goal in presenting these comparison and classification schemes was to address the challenge of understanding and selecting among various collaboration technologies. Additionally, the scheme can be used to identify new opportunities for collaboration technologies by identifying gaps or holes where technologies are not offered.

We have found that even this first version significantly reduced cognitive load for understanding the broad groupware space. It has its limitations, particularly in terms of accommodating the aggregated products. However, the classification and comparison schemes represent a credible step forward that may help groupware researchers, designers, and users to analyze and understand the sometimes complex "bundles of capabilities" found in collaboration products. The schemes may also help researchers and designers to understand the range of implementation choices available to them, and may help researchers to discover a) which of the many technological interventions account for effects observed in the field, and b) what groupware challenges remain unaddressed. We anticipate that further research will be required making additions and revisions to the schemes to bring them to a state where they can account for all elemental collaboration technologies and all design and configuration choices for those technologies.

References

1. Watson-Manheim, M.B., Chudoba, K.M., Crowston, K.: Discontinuities and continuities: a new way to understand virtual work. Information Technology & People 15(3), 191–209 (2002)
2. Evans, P., Wolf, B.: Collaboration rules. Harvard Business Review 83(7/8), 96–104 (2005)
3. Nunamaker, J.F., Briggs, R.O., de Vreede, G.-J.: From information technology to value creation technology. In: Dickson, G.W., DeSanctis, G. (eds.) Information technology and the new enterprise: New models for managers. Prentice Hall, Englewood Cliffs (2001)
4. Munkvold, B.E., Zigurs, I.: Integration of e-collaboration technologies: Research opportunities and challenges. International Journal of E-Collaboration 1(2), 1–24 (2005)
5. DeSanctis, G., Poole, M.S.: Capturing the complexity in advanced technology use: Adaptive structuration theory. Organization Science 5(2), 121–147 (1994)

6. Zigurs, I., Khazanchi, D.: From profiles to patterns: A new view of task-technology fit. Information Systems Management (forthcoming)
7. Grudin, J., Poltrock, S.E.: Computer-supported cooperative work and groupware. Advances in Computers 45, 269–320 (1997)
8. Nunamaker, J.F.J., Dennis, A.R., Valacich, J.S., Vogel, D.R., George, J.F.: Electronic Meeting Systems to Support Group Work. Communications of the ACM 34(7), 40–61 (1991)
9. Pinsonneault, A., Kraemer, K.L.: The impact of technological support on groups: an assessment of the empirical research. Decision Support Systems 5, 197–216 (1989)
10. Zigurs, I., Buckland, B.D.: A theory of task/technology fit and group support systems effectiveness. MIS Quarterly 22(3), 313–334 (1998)
11. de Vreede, G.-J., Briggs, R.O.: Meetings of the Future: Enhancing Group Collaboration with Group Support Systems. Journal of Creativity and Innovation Management 6(2), 106–116 (1997)
12. Briggs, R., Mittleman, D., Santanen, E., Gillman, D.: Collaborative Molecules: A Component-Based Architecture for GSS. In: Americas Conference on Information Systems (AMCIS), Indianapolis, IN, pp. 182–184 (1997)
13. Zeman, E. Verizon Wireless Subscribers Send 10 Billion SMSs In June, Probably Have Carpal Tunnel. Information Week (2007)
14. Briggs, R.O., Kolfschoten, G.L., de Vreede, G.-J., Dean, D.L.: Defining key concepts for collaboration engineering. In: 12th Americas Conference on Information Systems (AMCIS-12), Acapulco, Mexico (2006)

PaperFlow: A Platform for Cooperative Editing of Scientific Publications

Guilherme Saraiva[1], Orlando Carvalho[1], Benjamim Fonseca[2], and Hugo Paredes[1]

[1] UTAD, Quinta de Prados, Apartado 1013, 5001-801 Vila Real, Portugal
`pimentelsaraiva@hotmail.com, orlandocroccia@gmail.com,`
`hparedes@utad.pt`
[2] UTAD/CITAB, Quinta de Prados, Apartado 1013, 5001-801 Vila Real, Portugal
`benjaf@utad.pt`

Abstract. The production of scientific publications requires usually the participation of several authors that contribute to the final result according to their role
in the work being described. Nevertheless, this is obviously a cooperative activity and requires the simultaneous presence of the collaborators or the exchange
of documents and annotations through email. Cooperative editors introduced a
further step in the cooperation, but the current solutions do not accommodate
issues like referencing and publishing. This gap motivated us to build a platform that integrates the three main functionalities required to effectively produce scientific publications: a cooperative text editor, a cooperative reference
manager and a connector to scientific digital repositories. This paper presents
this solution, which we called PaperFlow and was specified with the aid of a
study conducted with Portuguese and Spanish researchers, which results are
also presented, with the aim of evaluating the platform requirements.

Keywords: Groupware, Group Editors, Scientific Digital Repositories, Scientific publications.

1 Introduction

The advances in communication technologies lead to the era of information society,
where the access to information is a key factor to human interactions. In this context one
of the most important activities is the production and publication of digital content. The
increasing availability of digital content brought about challenging problems concerning
its storage, manipulation and access. A particular case among these digital contents is
scientific publications, where digital repositories play an important role. Scientific digital repositories were developed to respond to this increasing use of digital support to
disseminate institutional and scientific work, facilitating and accelerating the associated
processes of publication and dissemination. However, scientific digital repositories do
not cover the production of digital content and act as isolated tools to publish and retrieve scientific work.

The creation of scientific publications is usually a cooperative activity, since it requires the intervention of more than one author. Creating scientific publications includes, among other tasks, the writing and revision of specific sections or the entire

R.O. Briggs et al. (Eds.): CRIWG 2008, LNCS 5411, pp. 318–323, 2008.

document, involving either synchronous or asynchronous cooperation. Usual cooperative applications used to support this activity are email, text editors with revision tools, Content Versioning Systems (CVS) and more recently specific tools like Google Docs. The unavailability of cooperative bibliography management tools, which are also important, leads us to specify an integrated platform for the complete cooperative process of creating scientific papers, from planning to its effective publication on scientific digital repositories. In this paper we present a study to evaluate the requirements for developing a platform for the creation of scientific publications and based on this study we present the specification of PaperFlow, a platform that integrates a group editor for the creation of scientific papers, a cooperative bibliography manager and annotator, and a connector for scientific digital repositories.

2 Related Work

Scientific publications are a set of various processes that require the participation of many actors [1]. SciX (open, self-organising repository for scientific information exchange) project [2] identifies seven different actors that participate in the publication process [2]:

- Researchers/research group - perform the research and write the publications;
- Publishers - manage and carry out the actual publication process;
- Academics - participate in the process as editors and reviewers;
- Libraries - archive the publications and provide access to them;
- Bibliographic services - facilitate the identification and retrieval of publications;
- Readers - search for, retrieve and read publications;
- Practitioners - implement the research results directly or indirectly.

According to this model, the scientific creation process is delegated to researchers, who, based on their research, write scientific publications. This process is mainly characterized by multiple interactions between a group of researchers and can be divided in four main phases: planning, writing, reviewing and publishing.

In the first phase (planning) the research group defines the key concepts of the publication, based on their collective knowledge and sustained on existing publications. To facilitate the activities of this phase, researchers normally use software tools to search, retrieve, annotate and catalogue publications. This set of tools is commonly known as bibliographic references managers. Among the current features of bibliographic references management tools are the creation of bibliography databases and its management as well as the integration with text editors and multiple bibliography format options. Examples of solutions for bibliographic references management are software tools like EndNote [3] and Procite [5].

In the following phases, writing and reviewing, the main activities of the researchers are editing, reviewing and annotating the publication text. These activities are usually performed cooperatively. However the tools used in this phase do not reflect the cooperative nature of these tasks. Frequently, the tools used are text processors and communication tools like email or messaging services. Occasionally cooperative tools such as group editors are used. Cooperative editors were developed to provide functionalities based on communication and information technologies, enabling group work and aiming to increase its productivity and efficiency. The main characteristic of cooperative editors resides in the ability to edit, review and annotate documents in a cooperative environment, where each element of the group can participate in the activities synchronously or asynchronously depending on the features of the tools used.

The final phase of the process is publication, usually performed by one of the authors, which also informs other group members of the action performed. Nowadays, a common place to publish scientific publications is scientific digital repositories. These repositories enable the storage of digital scientific publications, with the advantage of preserving and managing all of its content for long periods of time and providing the correct access. The core of scientific digital repositories resides on the data and metadata that is stored. By associating scientific digital publication to a set of standard metadata allows powerful searches on the data stored on them. The metadata can contain, among other things, information about the author, year of publication, article subject and the publication content. Moreover, through sharing metadata, authors who want to make their work available, will fulfill their objective by sharing metadata between various institutions. This aspect is used by universities to disseminate their work, creating mechanisms to legitimate and stimulate publication[4].

3 Requirements Analysis

Currently, there are several different solutions for bibliographic reference management and cooperative editors. In addition the use of digital repositories is increasing and each year more institutions make available their work in Open Access repositories. Each of these tools supports a specific stage in the cooperative edition of scientific documents but none supports the overall process. This situation motivated us to propose an integration of these three resources in order to support the overall process with a single application. To help us in the specification of the application a previous requirements analysis was made, beginning with the study of existent solutions for both cooperative editors and bibliographic reference managers. The objective of this study was to understand the similarities and also the differences between the existing solutions, namely concerning the functionalities they provide to its users towards the platform specification of our solution.

The second phase of the requirements analysis stage comprised the elaboration of a survey in order to gather information of several aspects concerning the field of cooperative work [6] [7]. The aspects addressed in the survey were the time and space distribution of the group elements, user activity coordination issues, functionality concerns based on user's experience with collaborative tools and the importance given to the integration of groupware tools.

The survey had the participation of 47 academic members from several Portuguese and Spanish universities. Based on the information collected with this survey, we can present some substantiating conclusions that support the solution we proposed. The first conclusion, which arises from the collected information, is that the synchronous option is residual, because different place/asynchronous distribution gathered 85.71%, and same place/synchronous 14.29%.

Regarding user activity coordination, the results obtained shows a balance between the 3 approaches. Indeed, a sequential coordination, allowing only one user in each moment to work directly in the project passing on to the next upon his task completion, received 35%. A parallel coordination, where group members work simultaneously although in an independent manner received 30% of the choices, exactly the same percentage gathered by a reciprocal coordination, where all group members work together in predetermined times and parts of the project. From these results the balance between the different approaches is evident, despite the fact that 65% of the results fell for options (Sequential and Parallel) where group members work independently.

The results obtained for the desired functionalities based on user's experience with cooperative tools are shown in Figure 1.

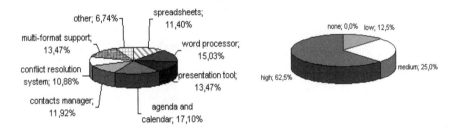

Fig. 1. Cooperative editor functionalities   **Fig. 2.** Cooperative editor/reference manager integration

Figure 2 shows the importance assigned to the integration of a cooperative editor with a bibliographic reference manager software tool, in order to increase the functionality of both tools. The results reveal a high interest of the academic members inquired in the integration of a cooperative editor with a reference manager tool (62,5%).

One last issue addressed in the survey concerned the possible integration of bibliographic reference management with scientific digital repositories. This possibility was largely classified with a 'high' importance level (75%), which leads us to conclude that the various possibilities that this integration open, combined with a cooperative editor are very interesting and that academic members are aware of these possibilities, although groupware tools for cooperative work support are not as widely spread and used as they could.

4 PaperFlow Specification

The requirements analysis provided important information for the specification of the PaperFlow platform. The corresponding architecture is shown in Figure 3 and emphasizes the integration aspects discussed in the previous section.

First of all, it is important to mention that presently the application deals with the time variable asynchronously, allowing only one group member in each moment to submit or edit his contributions to the shared document. The next version of Paper-Flow is planned to introduce synchronous operation.

PaperFlow is based in the integration of three main components: a cooperative editor, a reference manager and scientific digital repositories. Based on the conclusions from the requirements analysis and beginning with the cooperative editor, like other editors of the same nature, it provides several functionalities: an article document management component to enable users to manage, edit, move or upload their articles much like emails management functions of ordinary inbox mails, as Google Docs [8] provides. A list of all collaborators within a project will also be available, presenting all the collaborators that are working with the user in identified shared projects. With this information, users can manage and also visualize his collaborators and shared content in a simple way. An edition area is also available for the creation, development and revision of the documents. These functionalities are already available in other existing solutions, beginning the differentiation of our platform in the use of a framework that provides services to the application and consequently to the users. In the core of PaperFlow is Web Services Architecture for Groupware Applications (SAGA) [9], a generic cooperative framework that provides the platform a set of functionalities in the form of web services which add cooperative features to the application.

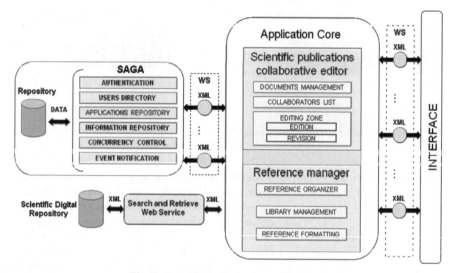

Fig. 3. PaperFlow platform specification

A reference manager is integrated in the cooperative editor. The reference manager module is an essential part of the application because it can provide support to the reference harvesting that the cooperative development process requires and also to manage and reference all the information sources. The reference manager module comprises several functions that support the creation of scientific articles in all its phases. Among others, one of the main components of the reference manager, which were chosen based in the survey conclusions, are a reference organizer, which enables users to manage and organize their bibliographic references in libraries, simplifying their organization and providing a fast access to them and to the detailed information comprised in each one. Since bibliographic references are going to be stored in libraries, it also has a library management component. This component enables users to create, edit, share and manage their libraries. This is very important since it organizes references by categories defined by the user, which simplifies their storage and facilitates the search process. Essential to every reference manager, is the references formatting capabilities. Due to this fact, there is also a reference formatting component for users to freely use the different reference formats they need. Another interesting feature of the reference manager is the ability to publicly annotate the references, enabling collaborators to share their opinions and knowledge about the stored scientific publications and the subjects they encompass.

5   Final Remarks

Throughout the study of several cooperative tools and considering the survey results, we concluded that the solution we propose is viable and we assumed that we could gather the best functionalities from the three technologies: the cooperative editor of scientific publications: the bibliographic references management for best organization and gained time for development of bibliography itself; and the integration with digital repositories. Nowadays, scientific digital repositories act like isolated technologies, without direct interaction with other applications supporting the scientific community work. As has already been described, these repositories are good storage facilities for scientific content, and its integration with other technologies can bring many advantages for researchers and institutions, by providing a participation increase of all community on these spaces. This way, institutions and researchers can improve their scientific publications through collaborative work, with the best functionalities of bibliographic references management, storing directly into the repository or harvesting digital content from it.

Several improvements can be made to the current version of PaperFlow. Besides fine-grained/details improvements, the most important evolution will be the addition of the synchronous operation to the application, enabling collaborators to conduct sessions of simultaneous work that may empower the cooperation effectiveness by enriching the discussion of ideas among group members.

References

1. Bo-Christer Bjork, T.H.: A formalised model of the scientific publication process. Online Information Review 28(1), 8–21 (2004)
2. Bo-Christer Bjork, Z.T.: Bob Martens. Re-Engineering the Scientific Publishing Process for the "Internetworked" Global Academic Community. In: elpub 2002. VWF Berlin, Karlovy Vary (2002)
3. EndNote (cited February 2007), http://www.endnote.com/eninfo.asp
4. Lynch, C.A.: Institutional Repositories: essential infrastructure for scholarship in the digital age, ARL. pp. 1–7
5. Procite (cited February 2008), http://www.procite.com/pcinfo.asp
6. Jonathan, D., Fouss, K.H.C.: Classifying groupware. In: ACM Southeast Regional Conference Proceedings of the 38th annual on Southeast regional conference. ACM, Clemson (2000)
7. Jiten Rama, J.B.: A survey and comparison of CSCW groupware applications. In: ACM International Conference. South African Institute for Computer Scientists and Information Technologists, Somerset West (2006)
8. Stijn Dekeyser, R.W.: Extending Google Docs to Collaborate on Research Papers, p. 11. The University of Southern Queensland, Queensland (2007)
9. Fonseca, B., Carrapatoso, E.: SAGA: A Web Services Architecture for Groupware Applications. In: Dimitriadis, Y.A., Zigurs, I., Gómez-Sánchez, E. (eds.) CRIWG 2006. LNCS, vol. 4154, pp. 246–261. Springer, Heidelberg (2006)

A Model Based Approach for GUI Development in Groupware Systems

William J. Giraldo[1], Ana I. Molina[2], Cesar A. Collazos[3], Manuel Ortega[2], and Miguel A. Redondo[2]

[1] System and Computer Engineering, University of Quindío, Quindío, Colombia
wjgiraldo@uniquindio.edu.co
[2] Department of Information Technologies and Systems.
University of Castilla – La Mancha.Ciudad Real. Spain
{AnaIsabel.Molina,Manuel.Ortega,Miguel.Redondo}@uclm.es
[3] IDIS Research Group, University of Cauca, Popayán, Colombia
ccollazo@unicauca.edu.co

Abstract. This paper proposes a methodological approach for Model Based User Interface Development of Collaborative Applications. This proposal is based on the use of several models for representing collaborative and interactive issues. Therefore, several techniques and notations are used. We describe the integration process of two notations: CIAN, which involves collaboration and human-computer interaction aspects; and UML, which specifies groupware systems functionality. In addition, we describe how this model is integrated into the Software Engineering Process. Both integration processes are developed by using software tools like CIAT and EPFC.

Keywords: Software Engineering, Groupware design, Interaction design, Model Based Design and Development.

1 Introduction

In this paper we propose a methodological approach for Model Based User Interface Development of Collaborative Applications. We propose a systematic modeling framework that relates technologies such as Enterprise Architecture, Model Driven Architecture (MDA) [1], meta-modeling approach, domain specific methodology (DSM), model transformation and framework-based development, etc. It supports the interface design of groupware applications, enabling integration with software development processes through UML notation and the Unified Method Architecture (UMA) [2].

The Software Engineering is a discipline focused mainly on the functionality, efficiency and reliability of the system in execution. From the software engineering point of view, several authors have proposed valid process models for the design of user interface [3]. Our proposal allows developers to implement their applications taking into account usability parameters. The user centred design (UCD) refers to the process that focuses on the usability during the whole project development. However, these methodologies do not guarantee usable developments [4]. The design of an interactive

R.O. Briggs et al. (Eds.): CRIWG 2008, LNCS 5411, pp. 324–339, 2008.

groupware design involves disciplines such as Software Engineering (SE), CSCW, and Usability Engineering (UE), therefore, it requires the interaction of multiple stakeholders by using their own specific workspaces [5, 6]. These workspaces should provide support for modeling diagrams by using different notations. The specified information on each workspace could serve as a complement for the modeling on other workspaces, both in the same perspective and in other one for the same abstraction level. Therefore, each developer represents the system in a more effective manner by using adequate, readable, comprehensible and expressive notations that support their job. For example: UML activity diagrams are adequate and provide good expressive power to describe activities. However, task models are more adequate to design usable interfaces [7].

Our aim is to integrate the information specified with a specific notation for modeling interactive and group work issues called CIAN[1] (*Collaborative Interactive Applications Notation*) [5] with the information gathered in the UML models, and so, try to reduce the gap between the development of the interface and the software development process, as well as the mapping between the two types of notations.

This paper is organized in the following way: section 2 introduces our methodological approach for designing interactive groupware applications, presenting a brief explanation of principles and foundation as the integration basis. Section 3 introduces the artifacts structure of this integration approach, putting emphasis on interchange points. Section 4 explains in a more detailed way our integration proposal, especially the integration layer that supports it. Section 5 presents an example of the application of our integration framework. Finally, the conclusions and further work are presented.

2 Methodological Proposal

In this section we present a methodological proposal for designing interactive collaborative systems. So, we took different approaches to address different system aspect. Our proposal is drawn from our previous research, i.e. CIAM (Collaborative Interactive Applications Methodology); related works regarding with these same issues, i.e. Usability Design [8]; and Industry Tools, i.e. RUP (OpenUP). We explain the interest in each proposal and why we decide to adopt OpenUP [9]. Our interest is centred mainly on information related to user interfaces. Our approach focuses on integrate processes and notations for supporting collaboration and usability issues into the Software Development Process. This proposal provides usability throughout the entire development process.

Our goal is to establish a more user-centred attitude in the development of groupware systems. Thus, individuals and developers should be more focused not only on their personal goals but also on the needs of users. Usability designers do not use UML to specify user interface information; it is gathered in more adequate notations by using story boards, tasks models, and prototypes, etc [7]. Therefore, different approaches should be integrated, such as: Object-oriented, User-Centred Design, Software Engineering, and data domain, etc. In software development, i.e., the object-oriented approach emphasizes objects and operations; and it is effective to model

---

[1] CIAN Notation is commented in the whole text.

internal system aspects in terms of object behaviour. The Flowcharts, Entity-relationship diagrams, and Relational databases are specialized for a different purpose. These techniques loose sight of the interactive system and how it relates to whole information and user needs. In interactive applications development, it is more important to begin by identifying tasks and then the related objects. Task decomposition into subtasks of lower difficulty is the usual way in which human beings work. Once a task model is defined, the objects manipulated by the User Interface must be shown.

Integration foundation: Our proposal is based on the assumption that an interactive groupware system can be classified and, therefore, modelled through one or more layers, or sets of specifications families. Likewise, the development processes can be related to each other through a common core as UMA. This idea is depicted in Figure 1. Our proposal is aimed at modeling and integration of layers having in mind different abstractions of a system. A layer is a set of diagrams organized according to a particular criterion, for example: diagrams modelled with the same notation, some representing a particular abstraction level, some representing a certain quality indicator, etc. Our goal is to integrate some models from CIAN and UML Notation; however, our integration proposal can be applied to a large number of notations, where each notation is used in a specific aspect of the system.

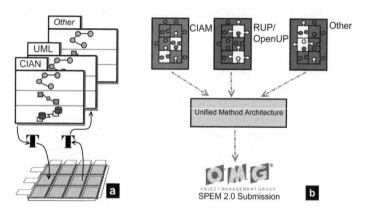

Fig. 1. Layers of an interactive groupware system. Processes Structure.

The integration or separation is carried out by using one or more integration layers, whose purpose is to store the useful and relevant information in each notation that is used for those purposes. A way to combine information from UML and CIAN models directly by using a layer of integration is showed in Figure 1(a). The common information of model elements on modeling notations is classified and organized in this layer into different perspectives and views. Each development process focuses on any aspect of the system; however, integration processes are required for coordinating additional iterations where more than one layer is involved, for example, usability iterations. Our goal is to find interchange points both on notations and processes. The use of EPF[2] and CIAT[3] [10] allows stakeholders to focus on their own notations and processes. The usability aspect is not only important for the final application, but also for the representations used in the design process [8].

RUP Adoption: By concentrating on the overall information system, how each element relates to each other, and how everything fits together, we need adopt a consolidated development process. The Rational Unified Process (RUP) is adopted for the definition of the activities and artifacts that support qualities as FURPS[4]. Likewise, RUP defines system structure by means of the business and software architectures. RUP is a process focused on architecture and guided by use cases [11]. Although RUP do not consider usability as a crucial aspect, through its ability to customize, there are proposals to provide it with usability [12]. Göransson et al [8] add a new discipline to the RUP, which is called "Usability Design", in order to make this process more user centred. Philips et al [13] incorporate two artifacts that support prototyping and modeling of interfaces, creating a bridge between them. Souza et al [14] propose the adaptation of the RUP towards having HCI aspects integrated into its main workflows, called RUPi. There are some approximations in order to include usability aspects in the software development processes; however, they are not integrated in the business modeling. This research is important because it incorporates advances not carried out until now in RUP.

Usability approach: In relation to usability issues, we do not propose a new discipline; instead, we use activities that arise from best practices for user-centred design and have been carefully tested in the usability design discipline presented by Göransson et al [8] proposal and on CIAM. We have distributed the activities in different disciplines. These activities help projects focus on usability and the users throughout the system development lifecycle from business modeling until system modeling –not only in Requirements discipline. Its Usability Designer role is adopted; Göransson et al [8] explain how it covers the other roles necessary in their discipline. In addition, we propose an outline for organizing information related to usability at all stages of development. This is done by using a metamodel that allows us to define where to locate the usability notations and techniques into the overall process, and to control the traceability in a more effective manner. Our proposal is the integration of models between UML and other notations, in this case specifically CIAN. UML is effective by supporting software development but not by supporting user interface design and development.

OpenUP adoption: RUP is not only a standard software engineering process, but also a process framework for tailoring iterative processes and a process framework product by IBM integrated with different IBM tools [2]. Our proposals are more focused on free distribution software based on Eclipse Foundation. We adopted a customization of RUP named OpenUP [9]. This subset of RUP through an Eclipse project will allow all interested parties to adopt the concepts of RUP as an open-source process framework.

---

[2] The Eclipse Process Framework (EPF) Project provides an open and collaborative ecosystem for evolving software development processes.
[3] CIAT is an Eclipse-based tool that helps developers specify models CIAN and UML.
[4] Functionality, Usability, Reliability, Performance, Supportability + others.

[2]. Process integration of CIAM and OpenUP and their respective tool supporting facility will be done by using Jazz [15]. It is a scalable, extensible team collaboration platform for seamlessly integrating tasks across the software lifecycle. Jazz is a team collaboration platform for the full software lifecycle, designed to support seamless integration of tasks across all phases of the software lifecycle [15]. Jazz will be used to provide effective collaboration environments to motivate and encourage stakeholders to participate on cross-functional areas5 and cross-aspect6 collaboration, integrating workflows across different qualities. Figure 2 depicts the inception workflow into OpenUP. OpenUP is a minimally sufficient and complete process in the sense it can be manifested as an entire process to build a system. It is based on use cases and scenarios, and an architecture-centric approach to drive development [9]. It is extensible, therefore, that we used its foundation to add content and tailor it as needed.

CIAM proposal: CIAM [5, 16] is our methodological approach to deal with the conceptual design of applications for supporting work groups. It is based on the use of specific notations, for the design of interactive workgroup applications. This methodology intends to connect high-level requirements models with low-level interaction models with the aim of deriving the final UI more directly. In this paper we relate CIAM and OpenUP. For more information about CIAM refer to [17]. The main activities of CIAM are depicted in Figure 2(left).

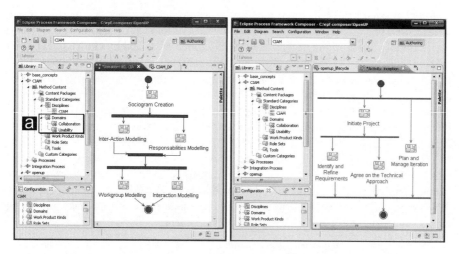

Fig. 2. (Left) CIAM WorkFlows, (Right) Inception WorkFlow OpenUP

Process Foundation: The UMA meta-model provides a language for describing method content and processes. UMA is an architecture to conceive, specify, and store method and process metadata [2]. Basic elements of UMA are work products, roles, tasks, and categories. We define CIAM and the Integration Process by using these elements, depicted in Figure 2. Categories are used to group content elements in disciplines, aspects,

---

[5] I.e. Business Modeling, Requirements, Analysis and Design, etc.
[6] Aspect is related to different layers or qualities. i.e., usability, functionality, collaboration.

and qualities, etc. The categories allow us to classify artifacts and activities that involve different system qualities or system aspects. The process CIAM is responsible for the artifacts that involve collaboration and usability. This is shown in Figure 2(a).

Best Practices: We adopt the best practices from RUP, OpenUP, and CIAM. Likewise, we have taken some of the practices proposed by Gulliksen et al [18 189]. They propose a series of Key Principles for User-Centred Design Systems. We adopted the following:

Table 1. Best usability practices

Practice	Use
User focus—	We provide specific work spaces for stakeholders and specific support for the goals modeling relate of the activities. The integration process adds usability iterations, each iteration differs from software engineering iteration in the fact that usability iteration is user oriented, it is not use case oriented.
Active user involvement—	We involve the representative users who should actively participate on the entire development process.
Evolutionary systems development—	Both OpenUp and CIAM are iterative and incremental
Simple design representations—	The artifacts and notations can be easily understood by users and all other stakeholders.
Prototyping—	Prototypes are used both in business modeling and system modeling to complement use cases
Evaluate use in context—.	Usability iterations are specified in integration process to evaluate usability goals
Explicit and conscious design activities—	Both OpenUp and CIAM contain design activities
A professional attitude—.	Both OpenUp and CIAM are conducted by multidisciplinary teams
Holistic design—.	Each system aspect can be developed by using their specific process. It can be in parallel
Process customisation—	UMA and EPF provide the foundation for process modeling.

3 System Artifacts

In order to reach our objectives, we decided to adopt the OpenUP development process, however, we continue considering the same model structure from RUP. Our approach is to provide the same RUP philosophy but with low-ceremony level through an agile process. Agile methods have drawn our attention back to the importance of coordinating understanding, benefiting stakeholders over unproductive artifacts and formality [9]. While RUP focuses too much on artifacts, our proposal focuses on providing minimum work-products (artifacts) to gather information related to various aspects of the system, i.e., functionality, collaboration, interaction, etc. Our purpose is to focus on the "big picture" and that each role is encouraged to get his work done in multi-disciplinary teams. The notation for each work-product should be the most consistent with each stakeholder profile. Provided several aspects are considered in software developing, such as user interaction, collaboration and functionality, it is not easy to identify and separate classes and objects involved in all these aspects, as this information is conceptualized in a different way among different stakeholders. For example, while for an ethnographer a "data object" is an attribute of an activity into an Inter-Action diagram in CIAN, the same "data object" is considered as a business entity into a diagram business objects diagram in UML by an analyst.

Our goal is to divide the information about system in different abstraction levels, or architectural levels. As we have explained, OpenUP is a use cases driving process and their design activities are centred on the notion of *architecture*. The use cases provide guidance for starting the design of both *business architecture* and *software architecture*. Although use cases are a technique that nowadays has become de facto standard for software developers, in user-centred development has been discussed its effectiveness [7]; however, they are very useful to direct integration with the user interface and collaboration design. The *use cases* are used to capture "functional" requirements or specify detailed behaviours. In addition, we use CIAN for modeling interactive and collaborative aspects. The overview about artifacts is depicted in Figure 3.

Model-based approaches for interactive applications have paid particular attention to task models. To model the user interaction, a notation exists broadly diffused in the community of the Computer Human-Interaction. This language is CTT [19, 20]. By using CTT we can reach high levels of detail in the interaction model. This facilitates the obtaining of the final design of the user interfaces. CTT allows to specify cooperative tasks but it cannot be used to model collaborative tasks and shared context. We proposed CIAN notation with the aim of covering this lack and allowing joint modeling of interactive and collaborative issues. Paternò [7] presents a proposal for integrating task models represented in CTT and UML. He proposes building a new UML for interactive systems. It is done by inserting CTT in the set of available notations still creating semantic mapping of CTT concepts into UML metamodel. As it was mentioned before, our proposal is aimed at modeling and integration of layers for having in mind different abstractions of a system. The model integration is explained in next section.

Our proposal focuses on the integration of use cases models and tasks models primarily. But, when use cases are associated with task models, certain discrepancies between software developers and usability designers can arise, as they have different requirements about the granularity of the use cases. So, if we separate the business model and the system model, these requirements of both software developers and Usability Designers can be considered in each of these specification levels (business and system). In the business model, the use cases correspond to the tasks to be performed by users, being modelled usually by means of activity diagrams. However, Usability Designers can specify tasks using the CIAN notation, mainly, with the aim to model collaborative interfaces. In the system model, the use cases are small to express functionality; however, Usability Designers can use CTT diagrams (used in the last stage of the CIAM methodology) to express interactive tasks that allow to access to the functionality through the user interface. That is why we see a real benefit in the separation of business modeling from system modeling.

We classify use cases into three categories, *core*, *supporting* and *Management*. Supporting use cases as core use cases are responsible for the running and maintenance of a company's infrastructure. From the modeling perspective, there are no real differences between core use cases and supporting. Both types of use cases should have the same requirements of usability and effectiveness [21]. Supporting use cases provides common functionality to core use cases. Management use cases are responsible for managing both the business and system; that is, for running in order to provide internal services or functionality that is not offered directly to users.

Core use cases are used to the integration with tasks models. They are services exposed for the application for supporting the users in their work. These use cases should be modeling in participation with the end-users, while the management and some support use cases should be modeling focusing on the software system. Specifically core system[7] use cases should be specified as essential use cases. These must be complemented with the CTT task models obtained in the last stages of CIAM. We have to point out that the CTT obtained in this last stage is enriched with additional information related with group work issues. See Figure 3(e). The business use cases are semantically related to tasks into inter-action diagrams in CIAN. The inter-action diagrams provide a logistic model to process orchestration. It is explained in detail in the section 5. See Table 2.

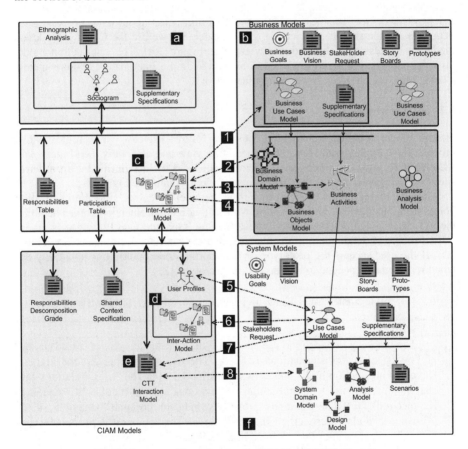

Fig. 3. System artifacts. Integrating CIAM and UML Models.

We are focused on use cases, their interaction with actors, the interface design, and the interaction between actors. So, we need distinguish between human and non-human actors and the possibility for modeling interactions outside the system scope. This is done by combining capabilities of UML and CIAN, for example, the CIAN Sociogram.

---

[7] There are core business use cases (business model) and core system use cases (system model).

3.1 Business Model

The business model defines how real people interact between them and with resources to reach a common objective. See Figure 1(b). Therefore, a business model should be a human oriented model that facilitates the understanding of the organization domain knowledge and its business [22]. Business objectives represent the intended position of a company. The objectives achieve business goals to generate value. This value should consider the customer value. The intended value helps us focus on aspects of context that can serve as a guide towards a systematic approach to HCI [23]. Early on the process, business use cases must be complemented with storyboards and proto-types, as mentioned above, to gain insight into user ideas and functionality. Here, the focus is to gather user intentions rather than concreting user interaction, making them independent device. Business model contains information about data, process, goal, people and network, but it lacks of logistic model. So, we use the CIAN inter-action model to express this abstraction o view. See Figure 3(c). The inter-action model provides information about postconditions, precondition, interchange of information, events, time, etc., used to process coordination.

3.2 System Model

System model contains architectural views as follow: use-case view, logical view, process view, deployment view, implementation view and data view. See Figure 3(f). Identically to business model, the system model lacks of logistic model. In order to specify this logistic, we use the CIAM Inter-Action Model. See Figure 3(d).

Software requirements arise from business model when we find automatization points from business activities. Dealing with software requirements involves detailing the use cases and supplemental specification. The User-Interface Prototype is an excellent source of detailed requirements elicited during initial requirements. See Figure 3(f). If the user accepts the prototype, it is important to explicitly document any detailed requirements needed to implement the prototype [21].

Table 2. Interchange Points for integrating CIAM and UML Models

#	Description
1	The mapping between the use cases and the task models can be based on the following basic transformations [24]: (a) The use cases represent the highest levels of abstraction in the hierarchical task models. (b) The "uses" relations can be interpreted as temporal order expressions (in particular a sequence connection). (c) The "extends" relations indicate optional behaviors. This situation can also be specified in a task model. (d) Temporal dependencies are related to post conditions and preconditions in activities diagram.
2	Business data provides domain information for activities. An Inter_Action model consists of a set of tasks carried out in a certain order and considering certain data or temporal restrictions among them. For each task, the roles involved, the data manipulated, and the product obtained as a result of the task are specified. From the data specified in the context of a task, we can specify the access modifiers to the objects, which can be *reading, writing* or *creation*

Table 2. (*continued*)

3	The tasks in the Model are interconnected by means of several kinds of relationships that can be interpreted as dependencies: *temporal dependencies* (order relationship), *data dependencies* (when tasks need data manipulated by previous tasks) and *notification dependencies* (when it is necessary for a certain event to occur so that the work flow continues). The dependencies acts like preconditions and postconditions into activities diagram. It allows designers to define task attributes, such as the category, the type, the objects manipulated, frequency, and time requested for performance. Inter-Action Model is more expressive than Activity diagrams and use cases in order to design logistic models
4	Business Object Models define all actor interaction with domain objects. UML do not provide semantic for storing object access information. We capture into activities information about object access; next, this information is stored into responsibilities table.
5	The information gathered within the use cases model about actors should complement the information gathered about user profiles. This information is useful to complement prototypes and to prioritize the user interface development when we use attributes for any user, i.e. priority, value, importance, etc.
6	This step of integration is as important as in Business use cases and can be used similarly.
7	The core use cases should be transformed into CTT tasks diagrams. A CTT diagram should be generated for each use case. It is assumed that the interdependencies between use cases have been modelled with inter-Action diagrams. The inter-action diagrams kept the same semantics like CTT. The modeling of activities and use cases should be complemented with scenarios. The scenarios are diagrams for validating each one of the flows of user interaction, and can be easily interpreted by them.
8	Information about data is very important in order to design the user interface. We have enriched the CTT notation with three new icons that represent three visualization areas. These icons are used separately as roots of the subtrees in the interaction tree in CTT notation: (a) the subtree that represents the interaction with the shared context, common for all group members involved in a multiuser task (collaborative visualization); (b) the individual interactions of each member in the group (individual visualization); and (c) the subtree that specifies the dialogue with the area of the shared context that can be accessed exclusively by one member of the group at a time. By using our extension of CTT we can identify additional information about the areas that comprise the collaborative user interface. Thus, this extension has higher-level semantics, which better organize and express specific interactive issues for collaborative applications.

In this stage, we integrate core system use cases with CTT task models. Task models allow designers to obtain an integrated view of functional and interactive aspects [7]. See Figure 3(e).

Our approach of integration consists of finding the *semantic mapping* between different notations. Subsequently, we must identify the correlation existing between different model elements. By means of the use of taxonomy, one or more layers of integration and models transformation, the integration is achieved.

The Table 2 presents the exchanged information between OpenUP process and CIAM. The model integration process is explained in the next section.

4   Integrating Software Engineering and Groupware Design

As it was mentioned before, we use an integration layer for integrating different layers. The integration layer which we propose is based on the Zachman Framework [25]. This Layer is defined in two dimensions organized in perspectives and views. The intersection of views and perspectives leads to 12 Modeling cells, (Figure 4 left). Each cell provides a container for models that address a particular perspective and view. A *perspective* is an architectural representation at a specific abstraction level. This classification enables designers to establish independence between different levels of abstraction by using perspectives; however, it is necessary to have a solid architecture that allows its subsequent integration. MDA (*Model Driven Architecture*) [26] is an architecture that promotes design guided by models and there is a relationship between the perspectives and levels of MDA. Frankel et al [1] describe the mapping between Zachman Framework and MDA. The concept of view, or abstraction, is a mechanism used by designers to understand a specific system aspect. So, we focus first on abstractions, and later in implementations that are derived from these abstractions[27]. For capturing all software system requirements is necessary to provide multiple views, i.e. the data, function, network and time view.

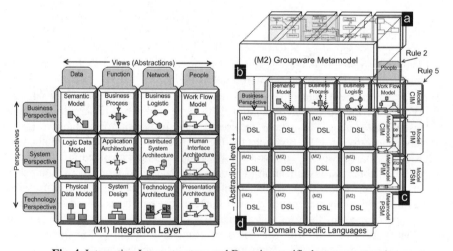

Fig. 4. Integration Layer structure and Domain specific languages structure

Rules are defined by taxonomy to obtain integrity, uniqueness, consistency and recursion of the specified information. So, the seven rules of the Zachman Framework have been adopted and refined [28]. Examples of these rules are: (R2) all of the cells in each column-view-is guided by a single metamodel. (R5) The composition or integration of all models of the cells in a row is a complete model from this perspective. (R7) The logic is recursive.

MDA provides the ***conceptual structure for specifying the notations*** or domain specific languages (DSL) used in every cell in the integration layer. Therefore, each one of these models of the cells is related to their respective metamodel (DSL). Figure 4(b). All models into MDA are related because they are based on a more abstract metamodel called MOF (Meta Object Facility) [26]. MOF facilitates the definition of the necessary transformations to integrate models.

The information into integration layer cells must be related to each other in two directions, views and perspectives. Therefore, a base metamodel should be specified (Figure 4(a)). This metamodel control the models cells consistency into the same view -rule 2- and it is necessary for the integration or composition of the models into cells of the same row -rule 5 - performing an integration role at perspective level. It is possible to specify a base metamodel for each integration layer, which depends on the nature of the family of languages (DSL) that is specifying. For example, a single base metamodel can be used to define common information useful for integration of models in UML and CIAN.

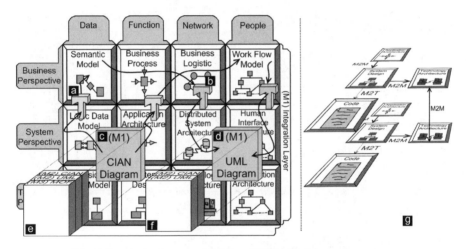

Fig. 5. Integration between CIAM and UML. Model transformations.

MDD proposes model transformations to reduce the complexity of software design [29, 30]. ***The integration of models in UML and CIAN*** is done through an integration layer; see Figure 5 (left). The integration layer is populated by using transformations applied to CIAN models; see Figure 5(a). The structure of notations is represented by some boxes containing metamodels at M2 and M3 levels. Figure 5(e, f). The cell which contains the CIAN diagram –Inter_Action- lies in the level M1 (Model); in addition, the notation CIAN which is defined as a UML Profile lies in the level M2 (metamodel). The transformations have as input metamodel CIAN and as output metamodel the DSL defined for these cells. In Figure 5(b) the process to transform models from the integration layer to generate UML diagrams is shown. It is not always possible to obtain complete UML diagrams; therefore, the generated information serves as a starting point for the subsequent modeling in UML.

The transformation and integration process are controlled through the integration layer metamodel. The first transformation uses the CIAN metamodel as the input metamodel and the integration layer metamodel as the output metamodel. The second transformation uses the integration layer metamodel as the input metamodel and the UML metamodel as the output metamodel. CIAT recognizes these three metamodels and it is possible to edit models by using editors for each one of these.

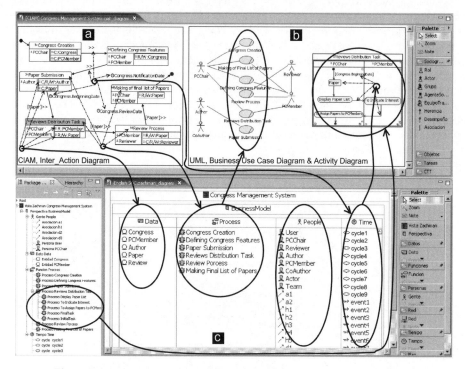

Fig. 6. Integration example between CIAN and UML by using the CIAT tool

The ATLAS Transformation Language (ATL) is used to implement transformations between models. We used the ATL plug-in for eclipse.

5 Case Study (the Congresses Management System)

We have chosen the congress management system as a case study. It is done because of, the nature of its cooperative and interactive requirements. This problem have been studied in literature by using several approaches[31, 32]. The modeling process follows the stages shown in the section 2. In this section a brief example of the application of this method for integrating CIAN and UML using CIAT is presented.

The first integration is done in business modeling. This process is shown in the Figure 6. It integrates business use cases and inter-action tasks, as it is explained in section 3. This integration is done in the same way as depicted in Figure 5. In this example, we only use the business model perspective in the integration layer; it is presented in Figure 6(c), which has complete information for data views, function, network and people. This information is generated from several diagrams in CIAN. The Inter_action Diagram is shown in Figure 6(a).

The transformation of model elements between models UML and CIAN is done at various interchange points, as follows:

1) ***Define the Sociogram:*** Although this paper does not show the sociogram, we have the following roles: PC-Chair, PCMember, Reviewer, Author and Co-Author. The information regarding the roles and relationships among organization members is processed through the transformations to generate partial information of Business Model and System Model perspectives. This information is classified into these two perspectives by people view mainly. See column people in Figure 6(b).

2) ***Group-Work Tasks Modeling Stage***: In this phase, we identify group task (collaborative or cooperative) and the relationships in order to specify group work.

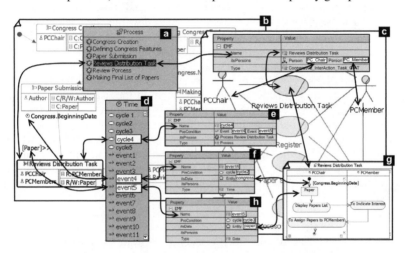

Fig. 7. Detailed integration example between CIAN and UML

The Inter_Action diagram, see Figure 6(a), illustrates the system macro activities and their interdependencies. The Figure 6(c) illustrates the information extracted from this diagram. The transformations separate information as follows: (a) The Inter-Action activities are associated with business use cases. The cooperative activities are transformed into diagrams activity. (b) The interdependencies are associated with preconditions, post conditions and events among various activity diagrams. (c) The domain objects are associated with business entities. A business object diagram is derived from the information in each activity, which is related with roles and objects.

3) ***Business Modeling:*** In the business modeling, the use cases business model and analysis model should be designed. Designers follow activities specified on OpenUP process. Core business use cases are defined to interact with business users.

4) ***Business Model integration detailed description***: The Figure 7 shows a possible integration scenario between CIAN diagrams and UML diagrams. In this scenario we need define the business use case diagram that is related with the inter_action diagram. A transformation generates the business use cases diagram -Figure 7(c)- and the activity diagram -Figure 7(g)- from Inter_action diagram mainly -Figure 7(b). The integration is based on information from the column process (function) -Figure 7(a)- and the column time -Figure 7(d)- into the integration layer. The variables cycle4, event4 and event5 have the information needed to build these diagrams in UML. See Figure 6(e,f,h), respectively. The structure of these variables is defined in the integration

layer metamodel. The event type variables become preconditions or postconditions of business use cases. In Figure 7(g) is observed how the event4 and event5 are transformed into the guard [Congress.Beginning.Date] and the object node "Paper". Similarly, the variable "Reviews Distribution task", Figure 7(a), stores the information required to relate the business use case with their respective Actors - Figure 7(i).

6 Conclusions and Further Work

In this paper, we submitted a proposal to integrate processes and notations for the purpose of addressing several aspects found in an interactive groupware system. To this end we have integrated several technologies, proposals and standards. We have adopted OpenUP as basic development process and we have shown its integration with a methodology for specifying interactive and collaborative issues, called CIAM., and the RUP. We have reached the aim of integrating the information specified with a specific notation for modeling interactive and group work issues called CIAN (Collaborative Interactive Applications Notation) [5] with the information gathered in the UML models, and so, trying to reduce the gap between the development of the interface and the software development process, as well as the mapping between the two types of notations. We have showed that it is possible to integrate different aspects about groupware system through the use of a layer of integration. We have presented a case study in which diagrams from CIAN and UML are integrated. We have implemented tools to support the process, modeling and integration. We have used free tools and standards found in the eclipse foundation. In this way, we can bridge the gap between software engineering and usability.

We plan to integrate this proposal in the Jazz [15] framework to improve team collaboration and integrate tools necessary for generating the final interface. In addition, this proposal must be tested with other aspects of software or other domains such us mobile and ubiquitous computing.

Acknowledgments. This work has been supported by Universidad del Quindío, the Castilla-La Mancha University and the Junta de Comunidades de Castilla-La Mancha in the projects AULA-T (PBI08-0069), M-CUIDE (TC20080552) and mGUIDE (PBC08-0006-512), and Microsoft Research Project Fund, Grant R0308LAC001 and CICYT TIN2008-06596-C02-2.

References

1. Frankel, D.S., et al.: The Zachman Framework and the OMG's Model Driven Architecture. MDA Journal (2003)
2. Shuja, A.K., Krebs, J.: IBM Rational Unified Process Reference and Certification Guide— Solution Designer, p. 284. Ibm press, Pearson Education, London (2008)
3. Granollers, T., et al.: Integración de la IPO y la ingeniería del software: MPIu+a. In: III Taller en Sistemas Hipermedia Colaborativos y Adaptativos, Granada España (2005)
4. Gulliksen, J., et al.: Key Principles for User-Centred Systems Design. BIT (2003)
5. Molina, A.I., Redondo, M.A., Ortega, M.: A conceptual and methodological framework for modeling interactive groupware applications. In: Dimitriadis, Y.A., Zigurs, I., Gómez-Sánchez, E. (eds.) CRIWG 2006. LNCS, vol. 4154, pp. 413–420. Springer, Heidelberg (2006)
6. Gutwin, C., Greenberg, S.: Design for Individuals, Design for Groups: Tradeoffs between power and workspace awareness. In: ACM CSCW 1998. ACM Press, Seattle (1998)

7. Paternò, F.: Towards a UML for Interactive Systems. In: 8th International Conference on Engineering for Human-Computer Interaction (2001)
8. Göransson, B., Lif, M., Gulliksen, J.: Usability Design—Extending Rational Unified Process With A New Discipline. LNCS. Springer, Heidelberg (2003)
9. Balduino, R.: Introduction to OpenUP (Open Unified Process). Eclipse site (2007)
10. Giraldo, W.J., et al.: CIAT, A Model-Based Tool for designing Groupware User Interfaces using CIAM. In: Computer-Aided Design of User Interfaces VI, CADUI 2008. Springer, Heidelberg (2008)
11. IBM_Rational, Too Navigator (Rational Unified Process) (2003)
12. Kapor, M.: A Software Design Manifesto. ACM Press, New York (1996)
13. Phillips, C., Kemp, E.: In Support of User Interface Design in the Rational Unified Process. In: Third Australasian User Interface Conference, Australian Computer Soc. (2002)
14. Souza, K.S., Furtado, E.: RUPi -A Unified Process that Integrates Human-Computer Interaction and Software Engineering. In: ICSE 2003 (2003)
15. Rational, I.: Jazz Platform Technical Overview. IBM Systems Journal (2007)
16. Molina, A.I., et al.: A proposal of integration of the GUI development of groupware applications into the Software Development Process. In: Haake, J.M., Ochoa, S.F., Cechich, A. (eds.) CRIWG 2007. LNCS, vol. 4715, pp. 111–126. Springer, Heidelberg (2007)
17. Molina, A.I., et al.: CIAM: A methodology for the development of groupware user interfaces. Journal of Universal Computer Science (JUCS) (2007)
18. Gulliksen, J., et al.: Human-Centered Software Engineering — Integrating Usability in the Software Development Lifecycle. Human-Computer Interaction Series, ed. S. Netherlands, vol. 8, pp. 17–36. Springer, Heidelberg (2005)
19. Paternò, F.: ConcurTaskTrees: An Engineered Notation for Task Models. In: The Handbook Of Task Analysis For HCI (2004)
20. Paternò, F., Mancini, C., Meniconi: ConcurTaskTree: A diagrammatic notation for specifying task models. In: IFIP TC 13 International Conference on Human-Computer Interaction Interact 1997. Kluwer Academic Publishers, Sydney (1997)
21. IBM_Rational, Too Navigator (Rational Unified Process), Concepts: User-Centered Design (2003)
22. Couprie, D., et al.: Soft Systems Methodology, A report, University of Calgary (2004)
23. Cockton, G.: Value-Centred HCI. In: Proceedings of the third Nordic conference on Human-computer interaction. ACM Press, New York (2004)
24. Lu, S., Paris, C., Vander Linden, K.: Towards the automatic generation of task models from object oriented diagrams. In: Chatty, S., Dewan, P. (eds.) Engineering for Human-Computer Interaction. Kluwer academic publishers, Boston (1999)
25. Zachman, J.A.: A Framework For Information Systems Architecture. IBM Ssystems Journal 26(3) (1987)
26. Miller, J., Mukerji, J.: MDA Guide Version 1.0.1 (2003) (cited 08-07-2007), http://www.appdevadvisor.co.uk/express/vendor/domain.html
27. Kaisler, S.H.: Software Paradigms. John Wiley & Sons, Inc., Chichester (2005)
28. Sowa, J.F., Zachman, J.A.: Extending and formalizing the framework for information systems architecture. IBM Syst. J., 590–616 (1992)
29. Frankel, D.S.: An MDA Manifesto. MDA Journal (2004)
30. Jouault, F., Kurtev, I.: On the architectural alignment of ATL and QVT. In: Proceedings of the 2006 ACM symposium on Applied computing. ACM, Dijon (2006)
31. Carlsen, S.: Action Port Model: A Mixed Paradigm Conceptual Workflow Modeling Language. In: Proceedings of the 3rd IFCIS (1998)
32. Trætteberg, H.: Model-based User Interface Design, in Department of Computer and Information Sciences, p. 211. Norwegian University of Science and Technology (2002)

Comparative Study of Tools for Collaborative Task Modelling: An Empirical and Heuristic-Based Evaluation

Jesús Gallardo[1], Ana I. Molina[1], Crescencio Bravo[1], Miguel Á. Redondo[1], and César A. Collazos[2]

[1] Escuela Superior de Informática
Departamento de Tecnologías y Sistemas de Información
Universidad de Castilla-La Mancha, Spain
{jesus.gallardo,anaisabel.molina,
crescencio.bravo,miguel.redondo}@uclm.es
[2] Grupo IDIS
Departamento de Sistemas- FIET
Universidad del Cauca, Colombia
ccollazo@unicauca.edu.co

Abstract. Within groupware systems, collaborative modelling systems play an important role. They are useful and promising tools for many fields of application. One of these fields in which collaborative modelling tools can be useful is the design and systematic development of usable User Interfaces (UI) using task models. In this paper, the use of a generic modelling groupware system, SPACE-DESIGN, is proposed for task modelling using CTT, one of the most used notations in this area. In order to evaluate the utility of the approach, a comparative study with two groups of experienced users and a heuristic evaluation using some well-known frameworks have been made. Results and conclusions of the evaluation are discussed.

Keywords: Groupware, Collaborative modelling tools, Usability and awareness evaluation, User Interface Design, Task Modelling.

1 Introduction

Due to the high demand of new technologies and the increase in information access by users, the design and systematic development of usable User Interfaces (UI) is becoming increasingly important. The UI is a fundamental part in the development of any application, and it is therefore essential for its design to be in accordance with the needs of the final user [1]. One of the techniques applied for obtaining a more user centred application is the use of task models. Task modelling for the specification of the users' interaction in a complex application may require the participation of several designers/engineers. For example, it might be interesting for the UI engineer to carry out the modelling work together with the customer/user. Therefore, the user is implied in the modelling process, which could result in an improvement in the usability of the interactive application to be developed. This engineer-user collaboration could be useful at several stages of the design/modelling process: when making the first design,

R.O. Briggs et al. (Eds.): CRIWG 2008, LNCS 5411, pp. 340–355, 2008.

when evaluating the final version, or at intermediate stages in the design of the user interface. It would also be interesting for several members of the same team (e.g., UI designers) to be able to carry out the modelling in a collaborative way. Indeed, in the context of computer programming, the *pair programming* paradigm is already gaining importance as it has been shown to significantly improve performance in design tasks [2]. In such situations it is useful to have a support for collaborative modelling, which in some situations can be distributed when the stakeholders involved in the design cannot work in a co-localized way. Taking into account this setting, the need to use a collaborative application or *groupware* system [3] becomes evident as this enables multiple users to interact in the joint development of the task model.

For this situation we propose the use of a groupware tool called SPACE-DESIGN [4], a collaborative modelling system able to adapt itself to several design domains by means of a procedure of domain specification in XML. One such domain is the CTT (ConcurTaskTrees) notation [5], one of the most widespread task modelling notations in the Computer-Human Interaction (CHI) community. The SPACE-DESIGN tool is framed in the context of a model-driven architecture for groupware system generation. Another way to work collaboratively for creating task models in CTT is by sharing a single user application, i.e., the CTTE [6] tool, by means of a shared windows system (e.g., NetMeeting). However, this paper is based on the hypothesis that the use of a groupware system such as SPACE-DESIGN provides advantages over this second possibility. Testing such hypothesis requires evaluate in a proper manner each one of these alternatives. The analysis will be centred on usability and awareness support. In order to perform that comparison it is necessary to use adequate evaluation techniques of the so called *groupware usability*. [7]. The evaluations must be done in such a way by following some kind of procedure so that the results make sense and can be used to enhance the research. There are many approaches trying to evaluate different aspects within CSCW and CSCL areas. The demand for groupware evaluation can be observed by the number of papers and research reports addressing this issue and by the recent workshops totally devoted to this theme [8] [9]. As a result of this work we aim to show the suitability of SPACE-DESIGN against the other option considered for collaborative modelling with CTT. In addition, this work will suppose an example of how to apply evaluation techniques that focus on the interactive aspects and usability of groupware systems. Accordingly, different evaluation techniques and procedures are mixed to produce an evaluation framework. The primary idea is to compare the groupware system with an existent mono user tool shared by means of a shared windows system. The techniques and procedures used are: to apply an awareness evaluation framework, to apply a heuristic-based usability evaluation and to measure the users' opinion and satisfaction by using, for instance, questionnaires with structured and open questions.

The next section explains in more detail the two ways of making a CTT model collaboratively that have been considered: the use of CTTE in conjunction with a shared windows system, in particular NetMeeting, and the use of the collaborative system SPACE-DESIGN. In the third section, a revision of some approaches for CSCW evaluation is discussed. Next, the study carried out for validating the initial hypothesis is described in detail. Finally, we present the conclusions drawn as a result of this study and the future lines of work arising from it.

2   Collaborative Task Modelling

In order to validate the hypothesis enunciated in the previous section, two different ways of carrying out a modelling task working in groups with CTT notation are considered. Each of these forms is characterized by the architectures and tools that are used. On the one hand, the use of the single-user CTTE application together with a shared windows application like Microsoft NetMeeting is studied. In this case, CTTE is used simultaneously by several users who carry out a single modelling task in a collaborative way. On the other hand, the SPACE-DESIGN tool has been used as an infrastructure to create the necessary support for collaborative task modelling with CTT. SPACE-DESIGN is a synchronous collaborative modelling tool which adapts itself to any domain that can be modelled by means of an XML specification that the tool processes.

In order to compare the modelling processes, a study in which users with experience in modelling with CTT work together will be made. The participants in the study will form two groups. One of them will create the models using the SPACE-DESIGN system, whereas the other group will make use of CTTE+NetMeeting. Later, the users will fill out a questionnaire in which they will express their opinion about the system used and about the development of the study. The results of the questionnaires will be used to make the comparative analysis of both systems. Also, a heuristic analysis based on some well-known evaluation frameworks will be done in order to reinforce the results of the empirical study.

Both possibilities for collaborative task modelling will then be explained in detail as well as how each of them has been prepared in order to ensure its correct functioning and consequently to make an evaluation of its efficiency, ease of use and utility when carrying out modelling tasks by means of CTT notation.

2.1   CTTE+NetMeeting

There is not any collaborative tool for working with CTT, so the second collaborative environment used in the study is made up by a mono user tool and a shared windows system. CTT notation includes a mono user tool for editing, validating and simulating the task models. This application is called CTTE (*ConcurTaskTrees Environment*) [6]. CTTE allows the flexible edition of the interactive tasks hierarchy and the specification of the temporal operators that connect them.

On the other hand, there are several shared windows environments that enable an application that is executed by one user on his/her computer to be viewed and manipulated by other users connected to a certain design session. In our case, we decided to use one of the most commercially known systems: Microsoft NetMeeting. NetMeeting integrates several tools for supporting collaboration, such as the shared windows system, a videoconference client, a chat, an electronic whiteboard, remote desktop sharing and file transfer. Although it is already known that awareness is not an important feature in NetMeeting, we have chosen to work with it since it is the corporative shared windows tool in the institution where the study has been made and because it is one of the most used software tool by users, so its functionality is known. However, other tools such as VNC could have been used.

2.2 SPACE-DESIGN

The SPACE-DESIGN tool (Figure 1) is a system that is included within a methodo-logical approach for the model-driven development of synchronous collaborative modelling systems [10]. In particular, SPACE-DESIGN is a tool with support for distributed synchronous work that allows users to carry out modelling tasks. It is domain-independent since the system processes the specification of the domain ex-pressed by means of an XML-based language and spawns the user interface and the necessary functionality to support that specific modelling.

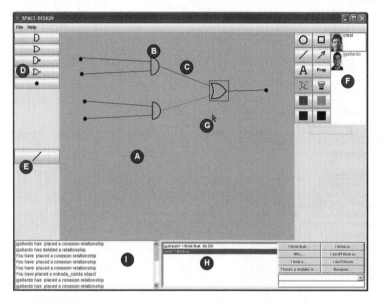

Fig. 1. The SPACE-DESIGN tool working with the digital circuits' domain

As shown in figure 1, SPACE-DESIGN has a shared whiteboard (A) on which the users can work with the different elements that form the application domain. These elements can be of two types: objects (B) and relationships (C). Both types are instan-tiated from the toolbars that are located on the left side of the user interface (D, E). These toolbars will vary according to the domain on which the system is working, and the objects and relationships will be those appearing in the domain specification.

An important characteristic of SPACE-DESIGN is the elements for awareness [11] and collaboration support that it includes by default. These elements are: a session panel that shows the users participating in the design session and identifies them by means of a specific colour (F), the identification of the elements that the users select by means of colours, the telepointers that indicate where the other users are pointing out (G), a structured chat for communication between the participants (H) and a list of interactions that indicates what actions have taken place and who has carried them out (I).

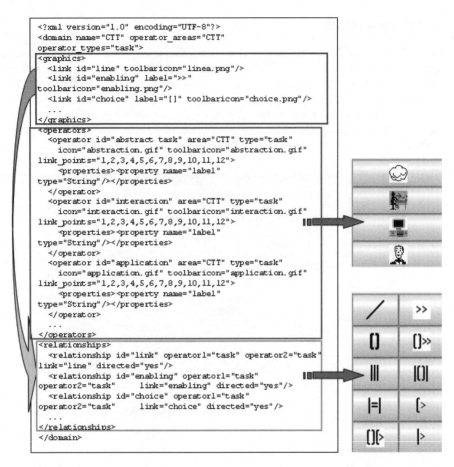

```
<?xml version="1.0" encoding="UTF-8"?>
<domain name="CTT" operator_areas="CTT"
operator_types="task">
<graphics>
    <link id="line" toolbaricon="linea.png"/>
    <link id="enabling" label=">>"
toolbaricon="enabling.png"/>
    <link id="choice" label="[]" toolbaricon="choice.png"/>
    ...
</graphics>
<operators>
    <operator id="abstract task" area="CTT" type="task"
      icon="abstraction.gif" toolbaricon="abstraction.gif"
link_points="1,2,3,4,5,6,7,8,9,10,11,12">
        <properties><property name="label"
type="String"/></properties>
    </operator>
    <operator id="interaction" area="CTT" type="task"
      icon="interaction.gif" toolbaricon="interaction.gif"
link_points="1,2,3,4,5,6,7,8,9,10,11,12">
        <properties><property name="label"
type="String"/></properties>
    </operator>
    <operator id="application" area="CTT" type="task"
      icon="application.gif" toolbaricon="application.gif"
link_points="1,2,3,4,5,6,7,8,9,10,11,12">
        <properties><property name="label"
type="String"/></properties>
    </operator>
    ...
</operators>
<relationships>
    <relationship id="link" operator1="task" operator2="task"
link="line" directed="yes"/>
    <relationship id="enabling" operator1="task"
operator2="task"      link="enabling" directed="yes"/>
    <relationship id="choice" operator1="task"
operator2="task"      link="choice" directed="yes"/>
    ...
</relationships>
</domain>
```

Fig. 2. Excerpt of the specification of the CTT domain and its translation to the user interface

The presence of these elements for awareness and collaboration support is one of the features that differentiate SPACE-DESIGN from other similar systems, like Synergo [12] or CoolModes [13]. But the main differences between SPACE-DESIGN and these systems are that the former adapts itself in a flexible way to new domains, incorporates awareness mechanisms, and stores the developed models in XML files (Figure 2); whereas the other systems have difficulty incorporating new domains, have fewer awareness mechanisms and, in the case of CoolModes, store the models in a proprietary format. As regards the supported domains, the mentioned systems allow the modelling of several domains from a series of specifications programmed in the system itself, whereas SPACE-DESIGN defines the domains in a way which is external to the system by means of specifications which can be built by end users. This means that any domain made up of objects and the relationships between them as well as actions to manipulate them can be modelled in this way and SPACE-DESIGN can be used to work in a collaborative way with it. In particular, SPACE-DESIGN has already been tested with domains such as digital circuits, use case diagrams, conceptual maps, bayesian networks, etc.

In order to adapt SPACE-DESIGN to the characteristics of CTT notation and thus to evaluate the initial hypothesis, the specification of CTT has been made following the DTD that describes the domain specification language recognized by SPACE-DESIGN. In so doing, a specification is obtained that can be processed by SPACE-DESIGN to enable the development of collaborative task modelling sessions with CTT (Figure 2).

The specification is divided in three sections. The first one, delimited by the *graphics* label, represents some visual aspects of the domain in which definitions of types of line (*line*) and definitions of connections (*link*) that can use the previously defined types of line are included. The second section represents the domain objects (*operators* label). Each object belongs to an area (which allows object grouping and classifying), is represented on the whiteboard (*icon*) and also in the toolbar (*toolbaricon*), specifies a list of linking points (corresponding with the hours in the clock sphere) and has a set of properties. Finally, the last section represents the relationships (*relationships* label). The relationships are binary. Each relationship has the following attributes: an identifier, the types of the two domain objects implied, the type of graphical connection of the relationship and a property that indicates whether the relationship is directed (i.e., the order in which the domain objects are linked is relevant) or not.

Starting off from the specification of CTT, SPACE-DESIGN adapts its user interface to give support to modelling with this notation. This consequently provides a collaborative tool prototype that enables task modelling with the CTT notation, which will serve to validate the investigation hypothesis by means of the comparative study that will be explained in detail below.

3   Comparative Study

Evaluation is a very broad concept and that is why it is a must to state explicitly what elements or aspects we must consider and under what focus. This paper is centred on the evaluation of aspects related with usability and awareness support. Firstly, we are going to present some related works that deal with these issues and then we will describe our comparative study.

3.1   Related Work: CSCW Evaluation

CSCW evaluation requires questions of multiple theories, methods, perspectives and stakeholders to be considered. Araujo et al. [14] have proposed a CSCW Lab that is a laboratory for conducting groupware pilot evaluations. Besides the physical space, it includes guidelines and instruments for executing groupware evaluations. Groupware evaluation involves a great amount of effort. The planning, design, accomplishment and replication of an evaluation are costly activities. The design of the experiment is an activity that should be carefully performed in order to guarantee that the results and measures obtained are relevant for interpretation [14]. By introducing instruments such as questionnaires or by incorporating direct observation, evaluators can be aware of participants' satisfaction and have an indication about the collaboration that occurs among group members.

There are different methods that have been used for evaluating usability on single user systems. This is the case of consistency inspection techniques or techniques for inspecting standards [15], the use of cognitive walkthroughs [16] and evaluation heuristics [17]. However, these techniques that have been useful and validated in their application have not been appropriate to evaluate groupware systems. In that way, there are some other techniques to evaluate usability of this kind of groupware systems. The work developed by Pinelle and Gutwin [7] proposes a *groupware usability* definition as the "extent to which a groupware system allows teamwork to occur – effectively, efficiently and satisfactorily– for a particular group and a particular group activity". Among the new techniques for groupware usability evaluation we find basic inspection methods [8], cognitive walkthrough adapted to collaborative systems [7] [18] and an adaptation of the Nielsen heuristics to apply at groupware systems [19].

Among the methodologies proposed for evaluating collaborative systems are worth to be mentioned the ones by Cugini [20] and Ramage [21]. Besides, a number of conceptual frameworks have been proposed that outline the major factors relevant to analyzing CSCW [22] [23]. They have several properties in common: group characteristics, situation factors (context), individual characteristics, group process, task properties, and task and group outcomes. Each one of these factors can have a number of aspects associated with them. Much of these frameworks stem from early research on group behaviour. The factors in these frameworks correspond generally to situation, task and human considerations in any type of applied research endeavour. However, there are other frameworks with different approaches.

Taking into account another groupware evaluation perspective, there are some works related with not only the evaluation of the process (mainly for the analysis of the discussion process), but also of the products obtained as a result of these kind of work group process. In CSCL, there are some frameworks to evaluate collaboration, as the proposal by Muhlenbrock and Hoppe [24] or Constantino-González et al. [25]. Collazos et al have developed a framework for evaluating collaborative process using digital games, taking into account aspects like: strategies definition, intragroup cooperation, success criteria review and monitoring [26].

Next, we are going to present the study we have developed in order to evaluate the *groupware usability* issues taking as study domain a task of collaborative modelling. Once the two options for collaborative task modelling with CTT explained in the previous section were working, a comparative study of both alternatives with students from the Computer Science School of the University of Castilla-La Mancha (UCLM) was carried out. The study consisted of the accomplishment of a task modelling exercise with CTT. The development of the study is commented below in detail, starting with the actions followed, and continuing with the results obtained and their discussion and interpretation. The study does not follow an exhaustive quantitative and qualitative evaluation, as this is just a previous study that will be followed by some other rigorous ones, with a larger sample and a wider scope.

Besides this study, also a heuristic analysis has been made in order to verify to what extent both systems fulfil the heuristics proposed by some of the aforementioned evaluation frameworks. This heuristic analysis will be made analyzing on the one hand the aspects regarding usability and on the other hand the ones related to awareness, as it has been made in the study with users.

3.2 An Experiment with Users

Thirty-eight students who attend a course on Computer-Human Interaction in the Computer Science School at UCLM participated in the experiment. They had previous knowledge of CTT notation and of CTTE application, both of which they had already worked with. The 38 students were organized randomly into pairs whose members were physically separated. All the pairs had to do the same task modelling exercise, but using different technologies. From the 19 groups, 11 had to work with SPACE-DESIGN and 8 with CTTE+NetMeeting. The study was recorded to identify later possible usability problems.

Once the modelling was completed, the students filled in a survey with a series of questions whose answers would be used to compare different aspects of both systems. These aspects are divided into *aspects related to the usability* of both systems, on the one hand, and with its *awareness elements*, on the other. As for the questions referring to the usability, the survey begins with some questions about general aspects regarding the functioning of both systems, such as the ease of use or the facility of configuration. The aspects were given a value ranging from 1 to 5, 1 being the lowest value and 5 the highest. Then, the users were asked about their personal impressions of other usability aspects of the collaborative modelling tasks and they could freely answer two questions about the main problems arising in the interaction with the system and about their final impressions of the study. Regarding awareness aspects, the users answered some questions related to the awareness mechanisms of both systems. Also, users that worked with SPACE-DESIGN answered some additional questions about the awareness mechanisms of the collaborative tool.

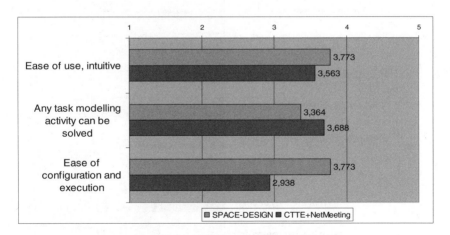

Fig. 3. Results of the questions about usability aspects

The results obtained in the evaluation survey filled in by the users are described below. As far as the usability aspects of both systems are concerned (Figure 3), it can be seen that the students working with SPACE-DESIGN have rated much more highly its facility of configuration and execution than those using the CTTE+NetMeeting combination. In the other two questions however the difference between values also exists, but is not so significant.

Later, the users were asked about their personal impressions regarding other usability aspects of the collaborative modelling tasks. In this case the questions were yes-no questions. When asking the users whether they had felt comfortable when carrying out the collaborative CTT modelling, the difference is remarkable. Whereas 77% of the users who worked with SPACE-DESIGN answered affirmatively, only 38% of those who worked with CTTE+NetMeeting gave a positive answer. The answers to the question about the existence of conflict situations during the modelling process are also important. 94% of the users that worked with CTTE+NetMeeting confirmed having encountered conflict situations whilst in the case of those who used SPACE-DESIGN it was only 68%.

Finally, the users could freely answer two open questions about the main problems arising in the interaction with the system and about their final impressions about the experiment. Among SPACE-DESIGN users, the main problems that arose were the conflicts when trying to work on the same objects and the difficulty in reaching an

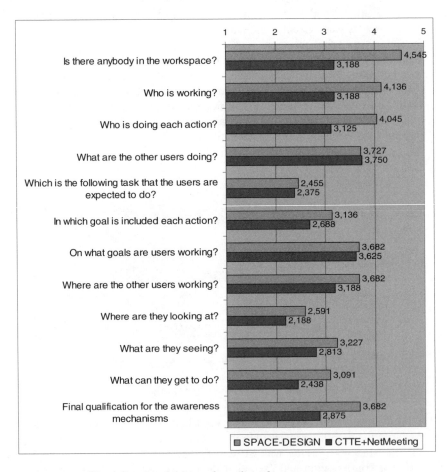

Fig. 4. Results of the questions about the awareness aspects

agreement between the participants in the session. Approximately 30% the users made such comments. The users also criticized the appearance of the application. As far as possible improvements of the system are concerned, although most users valued very positively the structured chat, they also suggested that the use of an audio tool or of a video-conference system would improve communication between the members of the design session. In relation to the users who had worked with CTTE+NetMeeting, more than 50% criticized the way in which turn taking is implemented in NetMeeting. Some of them even preferred that only one of the members worked and that the other simply commented his/her opinions through the chat tool.

Awareness support in the two modelling alternatives has been also compared according to the awareness dimensions defined by Gutwin and Greenberg in their evaluation framework [27]. In particular, for each system the specific elements that support each one of the ten dimensions of the framework have been identified. In this analysis it can be seen how SPACE-DESIGN has several awareness mechanisms that cover all the dimensions defined by Gutwin and Greenberg, whereas CTTE+NetMeeting only covers a few dimensions by means of a few elements. Thus, the utility of SPACE-DESIGN as far as the support of awareness is concerned is also validated.

The participant users were also asked about the awareness mechanisms for each dimension in the framework (Figure 4): the advantage of SPACE-DESIGN with respect to CTTE+NetMeeting is noticeable, as almost all the indicators give a greater value to SPACE-DESIGN than to the union of CTTE and the shared windows system. The difference in the values given to the first three aspects, referring to the perception of the other users and their work in the shared workspace, is especially important. Finally, the global evaluation given by the users to the awareness mechanisms is also quite significant.

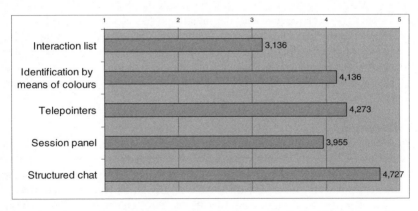

Fig. 5. Information quality evaluation for the awareness and collaboration support elements in SPACE-DESIGN

In addition to the aforementioned comparative questions, users who worked with SPACE-DESIGN were also asked about its awareness mechanisms. Thus, in Figure 5 the evaluation made by the users on the quality of information provided by the awareness components of SPACE-DESIGN can be seen. In this comparison several elements were analyzed: from simple awareness mechanisms as the identification by means of colours of objects and relationships to the more complex components for collaboration support as, for example, the structured chat. We can see that the structured chat is the component better valued by students, although most of them received values around 4, that is, a good evaluation. The component least valued by users was the interaction list.

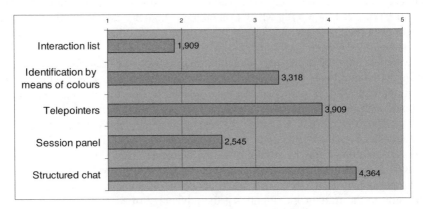

Fig. 6. Use statistics of the awareness and collaboration support elements in SPACE-DESIGN

Another activity that the users who worked with SPACE-DESIGN made was to order the awareness elements depending on their use during the design session. Thus, the elements were evaluated using values between 1 and 5, where 1 corresponded to the element less used and 5 to the most used. The results of this question are reflected in Figure 6. We can observe that the chat, which was the most valued by users, is also the most used. The same occurs with the interaction list, which was the least valued and is now the least used. The most remarkable aspect about the remaining elements is that the session panel, which received a good rating in the previous question, now is considered the second less used element. This may be due to the fact that the session panel offers more specific information that may not be constantly consulted during the design session.

3.2.1 Discussion

The most outstanding conclusions drawn from the analysis of the results are presented in this section. A first conclusion is that we have validated the utility of the approach of using a modelling groupware system for task modelling with CTT. Most users have evaluated very positively the utility of SPACE-DESIGN, its facility of configuration and its awareness mechanisms in contrast with the same characteristics in the combination of CTTE+NetMeeting. Similarly, SPACE-DESIGN has caused fewer situations of conflict than CTTE+NetMeeting, mainly due to the relatively inflexible way in which turn taking is implemented in the latter approach.

In addition to this, although it is not a conclusion that can be drawn from the analysis of the surveys, direct observation of the students during the work enabled us to detect the advantages of the use of the chat as a planning tool. The groups that decided to distribute the responsibilities (the modelling of different parts of the system) obtained better results and reduced both the time dedicated to solving the activity and the number of conflicts arising during the development of the solution.

One point, however, that received a low evaluation by the users is the question asking whether any task of CTT modelling can really be carried out with SPACE-DESIGN. The reason for this may be the evident limitations of a generic modelling system with respect to a system that has been developed specifically for a certain notation. However, the difference in the evaluation with respect to CTTE is very small and compensated when taking into account the additional power provided by a tool that can deal with any modelling notation that can be expressed by means of a previously defined XML-based language.

Apart from this, a possible improvement in the system that would facilitate communication would be, as previously mentioned, the inclusion of an audio tool or of a desktop video-conference system. As far as the awareness systems are concerned, although they have been generally well valued by the users of the system, those referring to the identification of the place in which the users carry out the different actions could be improved. That would mean implementing another kind of telepointers or increasing the functionality of the current ones. It would also be useful to provide some information about the following actions to take and about how they are framed within greater goals. In order to provide the system with these mechanisms, a work distribution tool or a similar one could be developed and integrated with SPACE-DESIGN. This would correspond with the proposals made in the methodology for the development of groupware systems on which SPACE-DESIGN is based [10].

Finally, as regards turn taking, the comments of the users indicate that the use of a tool for turn taking is useful, but only if it is implemented in a different way to the one used by NetMeeting. Thus, it would be possible to implement a simple tool that, by means of a given low-level protocol of actions, would regulate the use of the collaborative whiteboard.

3.3 Heuristic Evaluation

Besides the evaluation carried out by the users, another analysis has been made based in some heuristic proposals and evaluation frameworks that were mentioned in the related work section. In particular, and following the division between usability evaluation and awareness evaluation, the heuristics proposed by Baker et al. [19] will be used to analyze usability aspects.

Regarding usability aspects, the support given by both SPACE-DESIGN and CTTE+NetMeeting for each characteristic mentioned by the heuristic proposal by Baker et al. [19] has been verified. In table 1 it can be seen how for most aspects SPACE-DESIGN provides a greater support than CTTE+NetMeeting. It is especially remarkable that CTTE+NetMeeting does not provide any support for aspects of gestural communication and people perception, aspects that SPACE-DESIGN does cover. Other aspects in which SPACE-DESIGN shows a greater difference with respect to CTTE+NetMeeting are the communication resulting from the manipulation of artefacts or the search of collaborators and establishment of contact.

Table 1. Collaborative support of SPACE-DESIGN and CTTE+NetMeeting following the heuristic evaluation by Baker et al.

Heuristic	Support in SPACE-DESIGN	Support in CTTE+NetMeeting
Provide the means for intentional and appropriate verbal communication	Structured chat	NetMeeting chat
Provide the means for intentional and appropriate gestural communication	Telepointers, session panel	
Provide consequential communication of an individual's embodiment	Telepointers, session panel	
Provide consequential communication of shared artefacts	Telepointers, list of interactions, identification using colours, visual feedback	Visual feedback
Provide Protection	Identification using colours	Turn taking.
Management of tightly and loosely-coupled collaboration	WYSIWIS, telepointers, visual feedback	Shared windows, visual feedback
Allow people to coordinate their actions	Structured chat	Turn taking, NetMeeting chat
Facilitate finding collaborators and establishing contact	List of sessions, session panel	Contact by IP

4 Conclusions and Future Work

In this study the utility of an approach based on a generic collaborative modelling tool for the design of task models with CTT has been presented and analyzed. It has been demonstrated by means of a study using expert users, who can be considered as legitimate judges of the experience, that this approach has numerous advantages in comparison to the use of a single user system combined with a shared windows system. So, the initial hypothesis has been successfully tested.

Among the main advantages of our approach that have been identified through the study the following ones stand out: the facility of installation and configuration, the utility and versatility of the awareness mechanisms (mainly in relation to knowing who is working and with what objects), and the lack of conflict situations during the work sessions.

On the other hand, some features of the proposed approach that could be improved have also been identified, such as the identification of the following activity to be carried out by the users, the implementation of a structured system for turn taking or the improvement of some graphical aspects of the system.

Also a heuristic analysis has been realised to strengthen the conclusions drawn from the study with users. This way, it has been verified the fact that SPACE-DESIGN is a more usable tool and has a greater support for awareness than CTTE+ NetMeeting when doing task modelling in workgroup situations.

Taking all the aforementioned into account, we can affirm that SPACE-DESIGN can contribute to support collaborative design of task models with CTT and, therefore, to the design of fundamental models for the development of user interfaces. In addition, it provides a more suitable support (from various points of view) than other choices such as, for instance, the sharing of the CTTE application.

From the evaluation carried out it can be derived an evaluation framework following a mixed methods approach, with techniques such as the users' evaluation or the application of usability and awareness frameworks. The idea behind this is to apply such techniques in the frame of a comparative study between a groupware system and a mono user application shared by means of a shared windows system.

As a result of the analysis carried out, the inclusion of the improvements identified in the study will be considered in the future development of the SPACE-DESIGN tool. In addition, the utility for more domains of the approach described will be validated by carrying out further studies similar to the one described in this paper. More specifically, the next studies will be undertaken on domains related to Software Engineering. In so doing it will be possible to verify whether the system is really valid for a wide set of collaborative modelling tasks.

Acknowledgments

This work has been partially supported by the Ministerio de Educación y Ciencia (Spain) in the TIN2008-06596-C02-01 and TIN2008-06596-C02-02 projects, by the Junta de Comunidades de Castilla-La Mancha in the PBC08-0006-5212 project and by the Microsoft Research Project Fund, Grant R0308LAC001.

References

1. Myers, B.A., Rosson, N.B.: Survey on user interface programming. In: CHI 1992 Conference Proceeding on Human Factors in Computing Systems. ACM Press, Nueva York (1992)
2. Lui, K.M., Chan, K.C.C., Nosek, J.: The Effect of Pairs in Program Design Tasks. IEEE Transactions on Software Engineering 34(1) (2008)
3. Ellis, C.A., Gibbs, S.J., Rein, G.L.: Groupware: Some issues and experiences. Communications of the ACM 34(1) (1991)
4. Bravo, C., Gallardo, J., García-Minguillan, B., Redondo, M.A.: Using specifications to build domain-independent collaborative design environments. In: Luo, Y. (ed.) CDVE 2004. LNCS, vol. 3190, pp. 104–114. Springer, Heidelberg (2004)
5. Paternò, F.: ConcurTaskTrees: An Engineered Notation for Task Models. In: The Handbook Of Task Analysis For HCI, pp. 483–501. D. Diaper and N. A. Stanton, LEA (2004)
6. Paternò, F.: CTTE: Support for Developing and Analyzing Task Models for Interactive System Design. IEEE Transanctions on Software Engineering 28(9) (2002)

7. Pinelle, D., Gutwin, C.: Groupware walkthrough: Adding context to groupware usability evaluation. In: Proceedings of the 2002 SIGCHI Conference on Human Factors in Computing Systems, Minneapolis, pp. 455–462. ACM Press, New York (2002)

8. Steves, M.P., Allen, R.H.: Evaluating Collaborative Enterprises – A Workshop Report. In: Proceedings of the 10th Int. Workshops on Enabling Technologies: Infrastructure for Collaborative Enterprises. MIT, Cambridge (2001)

9. Pinelle, D., Gutwin, C.: A Review of Groupware Evaluations. In: Proceedings of the 9th Int. Workshops on Enabling Technologies: Infrastructure for Collaborative Enterprises, Gaithersburg, Maryland (2000)

10. Gallardo, J., Bravo, C., Redondo, M.A.: An ontological approach for developing domain-independent groupware. In: Proceedings of the 16th IEEE International Workshops on Enabling Technologies: Infrastructures for Collaborative Enterprises (WETICE 2007), pp. 206–207. IEEE Computer Society, Los Alamitos (2007)

11. Dourish, P., Bellotti, V.: Awareness and Coordination in Shared Workspaces. In: Proceedings of the Conference on Computer Supported Cooperative Work CSCW 1992, Toronto, Canada. ACM Press, New York (1992)

12. Avouris, N., Margaritis, M., Komis, V.: Modelling interaction during small-groups synchronous problem-solving activities: The Synergo approach. In: Proceedings of the 2nd International Workshop on Designing Computational Models of Collaborative Learning Interaction, Maceio, Brazil (2004)

13. Pinkwart, N., Hoppe, U., Bollen, L., Fuhlrott, E.: Group-Oriented Modelling Tools with Heterogeneous Semantics. In: Cerri, S., Gouardéres, G., Paraguaçu, F. (eds.) ITS 2002. LNCS, vol. 2363, Springer, Heidelberg (2002)

14. Araujo, R.M., Santoro, F.M., Borges, M.R.S.: The CSCW Lab for Groupware Evaluation. In: Haake, J.M., Pino, J.A. (eds.) CRIWG 2002. LNCS, vol. 2440. Springer, Heidelberg (2002)

15. Wixon, D., Jones, S., Tse, L., Casaday, G.: Inspections and design reviews: Framework, history, and reflection. In: Nielsen, J., Mack, R. (eds.) Usability Inspection Methods, pp. 79–104. John Wiley & Sons, Chichester (1994)

16. Polson, P., Lewis, C., Rieman, J., Wharton, C.: Cognitive walkthroughs: A method for theory-based evaluation of user interfaces. Int. J. Man-Machine Studies 36, 741–773 (1992)

17. Nielsen, J., Molich, R.: Heuristic evaluation of user interfaces. In: Proceedings of the 1990 SIGCHI Conference on Human Factors in Computing Systems, Seattle, pp. 249–256. ACM Press, New York (1990)

18. Ereback, A.L., Hook, K.: Using Cognitive Walkthrough for Evaluating a CSCW Application. In: Proc. ACM CHI 1994, pp. 91–92 (1994)

19. Baker, K., Greenberg, S., Gutwin, C.: Heuristic Evaluation of Groupware Based on the Mechanics of Collaboration. In: Nigay, L., Little, M.R. (eds.) EHCI 2001. LNCS, vol. 2254. Springer, Heidelberg (2001)

20. Cugini, J., Damianos, L., Hirschman, L., Kozierok, R., Kurtz, J., Laskowski, S., Scholtz, J.: Methodology for Evaluation of Collaboration Systems. The evaluation working group of the DARPA intelligent collaboration and visualization program, Rev. 3.0 (1997)

21. Ramage, M.: Developing a methodology for the evaluation of cooperative systems. In: Proceedings of IRIS 20 (Information Systems Research in Scandinavia), Hankø Fjordhotel, Norway, August 9-12 (1997)

22. Olson, G.M., Olson, J.S.: Technology support for collaborative workgroups. In: Olson, G.M., Malone, T.W., Smith, J.B. (eds.) Coordination Theory and Collaboration Technology, pp. 559–584. Lawrence Erlbaum, Mahwah (2001)

23. Pinsonneault, A., Kraemer, K.L.: The impact of technological support on groups: An assessment of the empirical research. Decision Support Systems 5(2), 197–211 (1989)
24. Muhlenbrock, M., Hoppe, U.: Computer Supported Interaction Analysis of Group Problem Solving. In: Hosadley, R. (ed.) Proc. of CSCL 1999, pp. 398–405 (1999)
25. Constantino-González, M., Suthers, D.: Coaching Web-based Collaborative Learning based on Problem Solution Differences and Participation. In: Moore, J.D., Redfield, C.L., Lewis Johnson, W. (eds.) Proceedings of the Int. Conf. AI-ED 2001, pp. 176–187. IOS Press, Amsterdam (2001)
26. Collazos, C., Guerrero, L., Pino, J., Renzi, S., Klobas, J., Ortega, M., Redondo, M., Bravo, C.: Evaluating Collaborative Learning Processes using System-based Measurement. Educational Technology & Society 10(3), 257–274 (2007) ISSN 1436-4522
27. Gutwin, C., Greenberg, S., Roseman, M.: Workspace Awareness in Real-Time Distributed Groupware: Framework, Widgets, and Evaluation. In: People and Computers XI, Proceedings of HCI 1996 (1996)

Using Interactive Whiteboard Technology to Support Collaborative Modeling

Michiel Renger[1], Gwendolyn L. Kolfschoten[1], and Gert-Jan de Vreede[1,2]

[1] Department of Systems Engineering, Faculty of Technology Policy and Management,
Delft University of Technology, The Netherlands
d.r.m.renger@tudelft.nl, g.l.kolfschoten@tudelft.nl
[2] Institute for Collaboration Science, University of Nebraska at Omaha, USA
gdevreede@mail.unomaha.edu

Abstract. Modeling is a key activity in conceptual design and system design. Users as well as stakeholders, experts and entrepreneurs need to be able to create shared understanding about a system representation. Technology like interactive whiteboards may provide new opportunities in the support of collaborative modeling. We conduct an exploratory research on experiences in using interactive whiteboards in collaborative modeling, based on semi-structured interviews. This paper offers a first overview of advantages and disadvantages of interactive whiteboards and creates a research agenda to explore how process support can help in harvesting the efficiency gain that we believe can be achieved using this technology.

Keywords: collaborative modeling, interactive whiteboards, system and design, groups, technology.

1 Introduction

Modeling is a key activity in conceptual design and system design. There is broad agreement that it is important to involve various experts, stakeholders and users in a development cycle [1-3]. While these parties are often interviewed or in other ways heard, they often lack the skills to actively participate in the modeling effort. If users are not involved in systems analysis tasks, their problems, solutions, and ideas are difficult to communicate to the analyst. Further, analysts and entrepreneurs might have mental models, visions of a solution or system design, but might lack the adequate means of articulating these in terms familiar to all stakeholders involved [4]. While there are means to verbally explain models, such as metaphors, a graphical representation is often more effective. ("A picture tells more than a thousand words" [5]). In order to use models as a basis for discussion, it would be useful if the all stakeholders can be actively engaged in the construction and modification of such models.

However, building models in groups can be challenging [6]. There is an on-going research to develop new ways to support model building groups using facilitation techniques and technology, see for example [7-9]. Research in technological support for collaborative modeling has mainly focused on group support systems [10, 11]. Less is known about alternative technologies for collaborative modeling.

R.O. Briggs et al. (Eds.): CRIWG 2008, LNCS 5411, pp. 356–363, 2008.

The goal of this paper is to provide first insights in the various settings in which interactive whiteboards can be used to support collaborative modeling. To this end we first define collaborative modeling and describe the functionality of interactive whiteboards. Next we describe the interviews that we conducted to elicit experiences with interactive whiteboards for collaborative modeling. Subsequently, we discuss the lessons learned that we derived from the interviews. We end with conclusions and directions for further research.

2 Collaborative Modeling and Interactive Whiteboards

For the purpose of the research presented in this paper, we define collaborative modeling as the joint creation of a shared graphical representation of a system. In [12], we identified three major schools of thought in collaborative modeling research:

- Problem Structuring Methods refers to a broad variety of methods and tools developed in the UK to cope with complexity, uncertainty, and conflict [13-15].
- Group Model Building is considered a special case of Problem Structuring Methods for system dynamics modeling and simulation [16, 17].
- Enterprise Analysis has a stronger focus on the development of software tools and facilitation techniques to support collaborative modeling efforts [10, 18, 19].

While in Enterprise Analysis, producing high quality models is the primary goal, in Problem Structuring Methods model quality is subservient to the group's shared understanding and buy-in. This shows that different purposes of modeling efforts exist and that each requires different modeling approaches.

An interactive whiteboard (IWB) (also known as electronic whiteboard) is an interface device which has a large display that is accessible for a group, and the possibility to manipulate content on the display by the use of styluses, fingers or other devices as a mouse pointer. Specific modeling software for IWB's often enables text or line recognition and transformation into digital text boxes and straight connection lines, which enhances the intuitiveness of the interface, see for example [20].

Although specific group support systems to support collaborative modeling have been developed (e.g. [21, 22]), Aytes suggests that traditional whiteboards are more suitable for collaborative modeling tasks that require considerable group interaction [23]. Interactive whiteboards can have the benefits of digital technology without sacrificing the 'live' visual interaction within a group. Moreover, IWB's are expected to be of value for collaborative modeling because they allow group members to manipulate the model directly. Therefore, they could stimulate participation, feelings of data ownership and buy-in.

Research in the use of IWB's to support collaborative modeling is mainly directed at the design and development of software tools [20, 24]. However, as is stressed by group support system researchers, the effectiveness of technological group support depends on its use in practice [25, 26]. In this paper we explore how IWB's can be used effectively to support collaborative modeling.

3   Method

Due to the exploratory nature of this research, we based our findings on the experiences collaborative modelers had with IWB's. We used in-depth semi-structured interviews to gather our data, which allowed us to explore and elicit the findings and opinions of the interviewees with a flexible approach [27]. In total we interviewed nine persons about their experiences with collaborative modeling efforts using IWB's in an educational or research setting, usually in the role of supervisor during workshops. In order to obtain a broad picture of the experiences, the interview protocol, which can be found in the appendix, covered human factors, technological factors and factors with regard to the modeling approach.

In most modeling sessions that were discussed during the interviews, several groups worked in the same room on separate IWB's. The groups created one or two models on the IWB in two to four hours. Most of the sessions were part of modeling courses in the bachelor or master curriculum of the Faculty of Technology, Policy and Management at Delft University of Technology in the Netherlands. The purpose of these sessions was to learn the modeling approach and language. For these assignments the students obtained a instruction of the modeling approach, the syntactical rules of the model and a case description of the process or system they had to model. During the session supervisors walked around to give students feedback on their modeling syntax and validity. In one session, the participants were colleagues, and the purpose of the session was to exchange knowledge and explore possible synergy between participants. Some interviewees had been involved in the same sessions.

All IWB's discussed in the remainder of this article are Smart Boards for Flat Panel Display px346 with a 46 inch touch screen display, combined with MS Visio or Smart Ideas, which enables users to drag and drop blocks and arrows on a page, and to edit these by typing or writing on the IWB.

4   Results

The results and lessons learned from the interviews are discussed below on three different topics: group composition, technology and modeling approach.

4.1   Group Composition

From the interviews we identified several lessons learned about the group composition, concerning the group size, level of participation and role assignment.

1. **Group size.** In the different sessions mentioned during the interviews, the group sizes varied from 2 to 8 persons. Most interviewees perceived 4-5 persons as the ideal group size to model on the IWB's. As the size of the display allows for a limited number of people to interact directly with the screen, non-participating group members or free-riding behavior was observed at groups of 5 or more.
2. **Participation.** Larger groups make communication and engagement more difficult, while very small groups share less ideas and criticism, which is required to produce rich and complete models. Other cases were mentioned where one or two

persons operated the IWB while leading the discussion in a larger group. IWB's are expected to stimulate participation as digitally storable input is perceived to be more permanent than writings on ordinary whiteboards.

3. **Role assignment.** One interviewee stressed that someone operating the IWB for the group, requires that such person is not only a skilled IWB user, but also at least to some extent a skilled modeler, so (s)he can reflect on syntax and representation issues before making changes to the model. In terms of the roles in collaborative modeling as described by Richardson and Andersen [28], the interviewee thus recommends that the role of recorder should be combined with the role of modeler/reflector.

4.2 Technology

We studied the use of the IWB interface rather than specific software tools. Therefore, we only focus on very general features of modeling tools. Some interviewees felt that the available software does not yet exploit the full potential of the IWB's. One reason is that the available tools are still based on a 'traditional' mouse-based interface. The intuitiveness of the IWB would benefit from an interface that corresponds more to the use of non-digital whiteboards, see for example the gestured-based interface described in [20]. Moreover, most available tools do not provide explicit process structure for groups that use IWB's to collaboratively build a model.

There is some disagreement among the interviewees about skills required to operate the IWB. Experiences ranged from intuitive use to 15-60 minute training to support using manuals. Some felt that learning to operate the IWB's can be easy for hands-on learners and young students, but can be problematic for older people and non-academic professionals. The learning curve can reduce the efficiency of the technology in the short term. Summarized the following functionalities affect collaborative modeling with IWB's:

1. **Manipulation and access rights.** The interactive element of easy manipulating content directly on the display makes it a suitable medium for discussion. Most interviewees felt that the ease of model manipulation could increase efficiency. Access right to manipulate the model is achieved using styluses. Given the learning curve to use the IWB, some suggested that one skilled user should manipulate the model based on the group discussions. We also identified positive experiences with a person who specifically provided the group with technological support. Although the technology allows for more persons to work on one board at the same time, the software that we used does not support parallel work, unless multiple boards are used for one model. Access rights affect the possibilities of process support discussed below.

2. **Text and structure recognition.** The clarity of the model improves because no handwriting and drawing is involved, and the model can be easily changed to become more readable, e.g. rearranging and aligning blocks.

3. **Storage and versioning.** Revisions of the model or different versions can be saved separately for later comparison. This possibility allows users to explore and evaluate different versions of a model recursively.

4.3 Modeling Approach

A couple of sessions were meant to teach formal modeling techniques: IDEF0 and UML. Therefore, the semantic and syntactic quality of the models is very important. In some sessions the goal was to learn policy analysis from a multi-actor perspective, which involved multiple techniques like mind mapping and causal diagramming. In another session system dynamics simulation models were manipulated on the IWB. During the exploratory sessions with colleagues mind maps were created, where no syntactic rules applied. Regarding the support for collaborative modeling with IWB's, we identified three alternative approaches:

1. **No process support**. In this setting, every participant can manipulate the model. In this setting, participants tend to hold on to the IWB stylus, and therewith to the access rights. Therefore, it is important to facilitate active and equal participation, for instance using a turn-taking rule. Such protocols can be based on the passing of the styluses. Multiple interviewees observed the emergence of process structure and roles. In all cases, the observed approach taken by participants can be identified as top-down, starting with a very coarse model and working toward more details. Interviewees believed that this complies with the standard modeling techniques that participants are taught, and that the use of an IWB did not affect the modeling approach of the groups. Further research is required to see if this effect also manifests in organizational settings.
2. **Chauffeured.** In this setting one or two persons operate the IWB based on the group discussion. One interviewee felt that this could be advantageous because the IWB functions as a center of attention, so no subgroups can emerge in the group discussion. However, like interrupting speaking, interrupting drawing can be experienced as disruptive social behavior, and thus a turn-taking solution might work better. Because there needs to be agreement before changes can be made to the model, more discussion is encouraged. Because changes made to the model are more 'final' in this set-up, the recorder operating the IWB should also have a modeler/ reflector role. This setting allows for 'free-riding', i.e. observing without participating.
3. **Facilitated.** A process facilitator leads the group to create the model in several steps; e.g. first creating a list of elements, and subsequently identifying relations between elements. Although we did not experience this setting, we expect that it can result in richer and more complete models, but that it might conflict with the individual cognitive modeling process. One interviewee suggested that such a separation could avoid a tunnel vision, meaning that no alternative modeling perspectives are considered by the group. Furthermore, separating generation and organizing tasks have in-built model completeness checks, which is less apparent if the tasks are combined.

5   Conclusion and Further Research

In this paper, we conducted an exploratory interview-based research about the experiences in supporting collaborative modeling with interactive whiteboards. We identified different ways of using IWB's. We stress that the way the IWB's are used depends on the primary goal of the modeling effort, e.g. learning, creating shared understanding, and creating consensus about a system representation.

Given these advantages we identify a research challenge in exploring which group size and role allocation, approach, and tool set is more efficient and effective to support collaborative modeling. Further research is required to confirm the advantages and disadvantages of the different options we discussed, and to understand their effect on efficiency and effectiveness with respect to the different purposes of collaborative modeling. Furthermore, research is required to understand cognitive implications of the integration of individual system representations and its relation to the efficiency of different approaches to support collaborative modeling. In terms of technology, the intuitiveness of the interface could benefit from a design that resembles traditional whiteboards. Moreover, the IWB environment could be extended to enable flexible allocation of (parallel) access and manipulation rights in order to enable process facilitation while keeping the ability for each participant to interact with the model.

Acknowledgement

We would like to thank the interviewees for their time and effort.

References

1. Boehm, B., Gruenbacher, P., Briggs, R.O.: Developing Groupware for Requirements Negotiation: Lessons Learned. IEEE Software 18 (2001)
2. Fruhling, A., de Vreede, G.J.: Collaborative Usability Testing to Facilitate Stakeholder Involvement. In: Biffl, S., Aurum, A., Boehm, B., Erdogmus, H., Grünbacher, P. (eds.) Value Based Software Engineering, pp. 201–223. Springer, Berlin (2005)
3. Standish Group: CHAOS Report: Application Project and Failure (1995)
4. Hill, R.C., Levenhagen, M.: Methaphors and Mental Models: Sensemaking and Sensegiving in Innovative and Entrepreneurial Activities. Journal of Management 21, 1057–1074 (1995)
5. Larkin, J.H., Simon, H.A.: Why a Diagram is (Sometimes) Worth Ten Thousand Words. Cognitive Science 11, 65–100 (1987)
6. Renger, M., Kolfschoten, G.L., de Vreede, G.J.: Challenges in Collaborative Modeling: A Literature Review. Advances in Enterprise Engineering I, 61–77 (2008)
7. Eden, C., Ackermann, F.: Cognitive mapping expert views for policy analysis in the public sector. European Journal of Operational Research 127, 615–630 (2004)
8. Orwig, R., Dean, D.: A Method for Building a Referent Business Activity Model for Evaluating Information Systems: Results from a Case Study. Communications of the AIS 20, article 53 (2007)
9. Rouwette, E.A.J.A., Vennix, J.A.M., Thijssen, C.M.: Group Model Building: A Decision Room Approach. Simulation & Gaming 31, 359–379 (2000)
10. Dean, D.L., Lee, J.D., Orwig, R.E., Vogel, D.R.: Technological Support for Group Process Modeling. Journal of Management Information Systems 11, 43–63 (1994)
11. Lee, J.D., Dean, D.L., Vogel, D.R.: Tools and methods for group data modeling: a key enabler of enterprise modeling. SIGGROUP Bulletin 18, 59–63 (1997)
12. Renger, M., Kolfschoten, G.L., de Vreede, G.J.: Patterns in Collaborative Modeling: A Literature Review. Group Decision & Negotiation Proceedings (2008)

13. Rosenhead, J.: Rational analysis for a problematic world: problem structuring methods for complexity, uncertainty and conflict (1993)
14. Eden, C., Ackermann, F.: Where next for Problem Structuring Methods. Journal of the Operational Research Society 57, 766–768 (2006)
15. Shaw, D., Ackermann, F., Eden, C.: Approaches to sharing knowledge in group problem structuring. Journal of the Operational Research Society 54, 936–948 (2003)
16. Andersen, D.F., Vennix, J.A.M., Richardson, G.P., Rouwette, E.A.J.A.: Group model building: problem structuring, policy simulation and decision support. Journal of the Operational Research Society 58, 691–694 (2007)
17. Andersen, D.F., Richardson, G.P.: Scripts for Group Model Building. System Dynamics Review 13, 107–129 (1997)
18. Morton, A., Ackermann, F., Belton, V.: Technology-driven and model-driven approaches to group decision support: focus, research philosophy, and key concepts. European Journal of Information Systems 12, 110–126 (2003)
19. Dean, D.L., Orwig, R.E., Vogel, D.R.: Facilitation Methods for Collaborative Modeling Tools. Group Decision and Negotiation 9, 109–127 (2000)
20. Damm, C.H., Hansen, K.M., Thomsen, M.: Tool support for cooperative object-oriented design: gesture based modelling on an electronic whiteboard. In: Proceedings of the SIGCHI conference on Human factors in computing systems, pp. 518–525. ACM, New York (2000)
21. Dean, D.L., Lee, J.D., Pendergast, M.O., Hickey, A.M., Nunamaker Jr., J.F.: Enabling the effective involvement of multiple users: methods and tools for collaborative software engineering. Journal of Management Information Systems 14, 179–222 (1997)
22. Ram, S., Ramesh, V.: Collaborative conceptual schema design: a process model and prototype system. ACM Transactions on Information Systems 16, 347–371 (1998)
23. Aytes, K.: Comparing Collaborative Drawing Tools and Whiteboards: An Analysis of the Group Process. Computer Supported Cooperative Work, vol. 4, pp. 51–71. Kluwer Academic Publishers, Dordrecht (1995)
24. Qi, C., Grundy, J., Hosking, J.: An e-whiteboard application to support early design-stage sketching of UML diagrams. In: Proceedings. 2003 IEEE Symposium on Human Centric Computing Languages and Environments, pp. 219–226 (2003)
25. Bostrum, R.P., Anson, R., Clawson, V.K.: Group Facilitation and Group Support Systems. In: Jessup, L.M., Valacich, J.S. (eds.) Group Support: New Perspectives. Macmillan Publishing Company, Basingstoke (1993)
26. Fjermestad, J., Hiltz, S.R.: An assessment of group support systems experimental research: methodology and results. Journal of Management Iformation Systems 15, 7–149 (1998)
27. Berry, R.S.Y.: Collecting Data by In-depth Interviewing. In: British Educational Research Association Anual Conference (1999)
28. Richardson, G.P., Andersen, D.F.: Teamwork in Group Model Building. System Dynamics Review 11, 113–137 (1995)

Appendix: Semi-structured Interview Protocol

1. Can you tell something about your background and your specialization?
2. How often have you been involved in a collaborative modeling effort, and in what context?
3. Since when and how did you use interactive whiteboards in collaborative modeling?
4. What was the primary goal of these sessions, and what were the deliverables?

5. What are your experiences with the available time for a session, and the efficiency of using interactive whiteboards?
6. What are your experiences with the group size and background of group members?
7. How much and how are group members stimulated to participate in the process?
8. Can you identify steps in the approach taken by groups when they model with interactive whiteboards?
9. To what extend would groups have behaved differently without an interactive whiteboard?
10. Do participants need special skills to operate interactive whiteboards?
11. To what extend and how did you or someone else have a steering or guiding role in the process?
12. How do participants themselves experience working with interactive whiteboards?
13. How do interactive whiteboards provide advantages in collaborative modeling?
14. Do you see limitations in using interactive whiteboards in collaborative modeling?

Supporting the System Requirements Elicitation through Collaborative Observations

Renata Guanaes Machado, Marcos R.S. Borges, and José Orlando Gomes

Graduate Program in Informatics
Federal University of Rio de Janeiro – Brazil
rguanaes@ufrj.br, mborges@nce.ufrj.br, joseorlando@nce.ufrj.br

Abstract. Many approaches to work analysis have been proposed to enhance the requirements elicitation for systems design. However, systems delivered at dynamic, complex and socio-technical workplaces have still failed at satisfying the users' real needs, mainly because they are unable to support users' activities entirely, especially those related with cognition and collaboration aspects. We argue that the use of a combination cognitive and observation techniques can contribute to enhance the requirements elicitation activity, particularly if a collaborative approach is also adopted. This paper describes a collaborative observation model and a collaborative observation method aimed at improving the quality of the requirements elicitation process. We also include the description of a groupware prototype that supports our approach.

Keywords: Requirements Elicitation, Cognition, Collaborative Observation.

1 Introduction

Requirements engineering methodology is widely recognized as one of the most important steps of system development [21]. The requirements elicitation activity is the first step in bringing about the stakeholders' needs and expectations from the proposed system which is intended to better support their strategic and operational goals in organizations. Consequently, requirements elicitation is often regarded as the most critical phase for successful system project [11] since it deals with identifying system's goals, tasks, properties and constraints. During requirements elicitation, a basic understanding about the organization domain, people's roles, work activities, problems and opportunities for system improvements is required.

However, requirements elicitation is neither free from ambiguity nor a straightforward task. Lack of user input and incomplete requirements had been the leading factors that resulted in challenged or impaired system projects [23]. Furthermore, the rate of projects that did not fully meet the user's needs accounted for 46 percent in 2006, despite of being reduced from 52.7 percent in 1994 [20].

Many factors contribute to requirements inadequacy. One of them is the requirements elicitation technique applied. Most tools, methods and techniques, rely on traditional elicitation approaches, such as interviews, questionnaires, surveys and analysis of existing documentation [4]. Some of them use group elicitation techniques, such as brainstorming sessions, JAD or RAD structured workshops [18]. In case of high

R.O. Briggs et al. (Eds.): CRIWG 2008, LNCS 5411, pp. 364–379, 2008.
© Springer-Verlag Berlin Heidelberg 2008

uncertainty, prototyping methods may be included. However, despite of how long, extensive and efficient these techniques could be, it is evident that they do not provide all information required for achieving high quality, consistent and complete system requirements. In general, there are limits to what participants can tell about their work, particularly in reporting the role of tacit knowledge needed to their activities' accomplishment [10]. A noteworthy absence in their reports is the need for interaction with other stakeholders in order to accomplish their tasks.

In this paper, we assert that a deeper understanding about the real activities is an essential necessity in order to develop a more precise set of system requirements. Observation methods can help in uncovering work practices, so employees must be observed in their natural workplace environment, engaging in natural activities [17]. Moreover, cognitive approaches are equally important, as they enable understanding of people's knowledge: how their minds work, what they struggle with, and how they manage to perform complex work adeptly [3].

We propose the use of cognitive approaches and the observation methodology in system design so as to complement, rather than substitute, the most traditional techniques of requirements elicitation. We believe this approach will result in a realistic, better modelling and analysis of problem domain, while discovering additional users' system requirements.

The remaining of the paper is organized as follows. Section 2 discusses the limitations of the most traditional requirements elicitation. Section 3 shows how ethnography and cognitive, as emerging approaches, can improve the requirements elicitation. Then, in Section 4 we propose an Observation Conceptual Model and a Collaborative Observation Method for executing ethnographic studies. Once our approach is discussed, we describe an experiment carried out in a real setting aimed at comparing the set of requirements obtained by interviews in contrast to that obtained by our approach. Finally, Section 6 concludes the paper.

2 Drawbacks Requirements Elicitation Traditional Approaches

Each requirements elicitation method has its strengths and weakness aspects [4]. In general, the most traditional requirements elicitation approaches do not afford complete information about current work practices along with its actual difficulties and adversities. A sort of the potential limitations and problems can be exemplified as follow, but are not limited to these.

First, it is a commonplace that stakeholders often do not know exactly what their real needs are, so the traditional interviewing approaches do not help. Stakeholders also may not describe their activities entirely during interviews or focus groups, either due to lack of time, low recall capability, fear, omission, or because it is burdensome to articulate their routines, skills, abilities and tacit knowledge. As a major and critical concern, sometimes stakeholders say what they should do instead of what they really do. Stakeholders at managerial level are usually interviewed, even though the agreed requirements do not always reflect the final users' real needs.

Stakeholders are not the only issue. System developers usually collect requirements data from very limited perspectives, as most system design models have proven too narrow to adequately assess users' needs. They often focus on systems goals and its technical aspects, such as its data, functions, operations and restrictions [27]. In doing so, they may either miss or neglect the human, cognitive, social, political and cultural aspects [4], such as: users' knowledge and beliefs; group interactions and communications pattern; subtle organizational power relationships; and informal working practices. Without taking in account these non technical aspects, proposed requirements could be incomplete, unrealistic and impractical.

Often, system developers try exposing requirements formally rather than to analyze its appropriateness. Besides, considerable communication obstacles are common due to distinct mental models: stakeholders have domain specific knowledge, whereas system developers are familiar with formal requirement methodologies [25].

In other words, there is a gap between information obtained from stakeholders during interviews or focus groups, and the rich, dynamic and complex reality of workplaces. Therefore, instead of asking stakeholders to describe what they need (user-centered approaches) or what they do (task-centered approaches), it may be more effective to apply a work-centered perspective which aims to capture the actual activities being carried out in the context of work settings.

Consequently, it could be possible to understand: the overall organizational culture and context; difficulties users face; mistakes occurred; current systems vulnerabilities and usability problems; and opportunities for system improvements. This set of information, once uncovered, identified, interpreted, analyzed and confirmed, can contribute to derive more authentic, detailed and complete system requirements.

Another type of requirement, usually neglected in elicitation methods are that related to communication and collaboration among system users and between them and other stakeholders, clients and management. Users and system analysts tend to overlook these interactions unless they are formally required. Only by following closely the workers routine these interactions can be revealed.

3   Emerging Approaches to Requirements Elicitation

There have been some undertakings toward the development of more adequate methods and techniques to address the requirements elicitation problems discussed in the previous section, such as HCI; Ergonomics and Human Factors; Observation methodology; and Cognitive approaches. The last two are introduced and discussed in more detail, as they are adapted for use in our proposed approach.

3.1   Cognitive Methods

Contemporary factors such as the increasing use of information technology, the growth of automation and the intense division of labor lead to arise of dynamic, complex and distributed organizations worldwide, and equally, to the larger recognition of cognitive tasks. Consequently, any analysis of work, no matter if at management or operational level, usually involves both physical and cognitive elements, the latter being more and more predominant.

Thus, cognitive approaches are concerned with identifying and understanding topics related to human thinking and knowledge, such as reasoning and judgment processes, sense making, situation awareness, decision making and the like, fairly exposed in activities of planning, execution, problem detection, diagnosis and resolution. Social aspects of performance, as relationship, collaboration, cooperation and communication are also considered significant cognitive aspects [14].

There have been a large number of different frameworks and methods aimed at capturing, analyzing and assessing the cognitive aspects collected, each one for specific purposes of the investigation and depending on which cognitive variables are more relevant. Besides, there are many reports about applying these frameworks in studies of field settings [8]. This set of approaches together constitutes the Cognitive Systems Engineering discipline.

3.2   Observation Methodology

Observation methodology is a social research that comes from the Social Anthropology discipline that provides detailed descriptions of human activity and behaviour in primitive and unknown societies, along with its social interactions and cultural practices. It results from an observer situated in the natural environment for very prolonged periods, sometimes several years, getting close to where the action is.

Once accepted as a generic qualitative data collection and analysis approach, ethnography has been a methodological orientation independently of a specific subject matter. In fact, it has been widely used and extended throughout various research areas well beyond Anthropology [6], [10], [22].

Observation refers to a set of methods and techniques used within the field of qualitative research. Examples of data collection methods of ethnographic work are participant or shadowing observations, unstructured or semi-unstructured interviews [17]. However, ethnography should not be seen merely as fieldwork-based studies with data collecting and organizing methods, but includes its interpretive and analytic components [1]. Similarly, requirements for system development are not the primary output of ethnographic investigations. In some cases, its most effective outcome might be to recommend what should not be built rather that to recommend what should be [5]. That is the case of a study of air traffic controllers, in which paper strips were maintained instead of replacing them by their electronic replacement [13].

3.3   Ethnography in System Design

Ethnographic research is thus well suited in providing information systems researchers with a deep understanding of the people, the organization, and the broader context within which they work; allowing an intimate familiarity with the dilemmas, frustrations, routines, relationships, and risks that are part of everyday life [16]. Bringing real aspects of workplace through naturalistic field studies is a means of discovering and understanding the following aspects in a holistic way:

- What the actual working knowledge and practices are and what the adaptive strategies developed by users in response to the organization's demands are;
- How people learn, interact and use artefacts that are part of their activities;
- How people interact with systems, as well as how they cooperate with others;
- How people use cognition to cope with complex or unanticipated situations that arise during their activities, detecting and solving problems as they happen;
- How work situations and characteristics may be strongly context-dependent.

This set of knowledge about what happens in the workplace, and the considerable acquaintance with organizational setting can challenge system developers team assumptions, value judgements and pre-conceived ideas about stakeholders' activities and needs, as "things are not what they seem" [1]. Moreover, communication problems between developers and users may be reduced, since observing what they do allow better acknowledging of their point of view. As a result, a growing number of system researchers have been recognizing the value of ethnographic method for system requirements elicitation [9], [15], [22], [26].

Given the researcher immersion within a specific work setting, ethnography is regarded as the most intensive, in-depth research methodology, which enables opportunities for exploration, discovery and surprising insights. So, a hallmark of doing ethnography is to get surprised [1].

There are many possibilities to be surprised. A good example is to uncover the sources of collaboration in group work, as the understanding of human cooperation in air traffic control field studies [22]. These ethnographic studies put in question some widely-held design assumptions, which are appropriate for one controller at a human machine interface, but not adequate in actually cooperative work environments.

4 Our Proposal

The Observation Conceptual Model (OCM) is a generic framework that provides organized field studies for complex organizations. It combines cognitive and ethnographic methods already discussed and some ideas of CSCW research to support effectively the ethnographic analysis in multidisciplinary groups.

This model is primarily concerned with supporting the practical efforts of observing and analyzing people's real activities and performance in the context of workplaces to capture the missing social, collaborative and cognitive aspects. As shown in Figure 1, OCM is composed of a multi-level Dimension of Analysis, and implements a systematic Process of Analysis, in which collected and analyzed field data may deliver work representations and system requirements.

From Dimensions of Analysis, OCM specifies "what to observe" and "how to observe". The former refers to the most relevant work domain variables that must be collected from the field, so as to focus observation as suggested in [15]. The latter describes possible ways to capture those variables, such as through audio and video equipment or by multiple observers situated in the workplace, supported by manual or technological appliances as field notes, portable computers and observation templates (prompted forms for checking data instead of making transcription manually).

Once activities data are stored on a central repository, the model describes "how to analyze", driving the teamwork endeavours for collective elaboration of sketches, models and other representations about the work setting. More significantly, it can be done through controlled and collaborative virtual sessions of ideas brainstorming, debating, deliberation, and negotiation of conflicting issues. People from workplaces investigated, being observed or not, are strongly recommended to participate in the session, confirming the analysis or sharing domain knowledge not yet captured.

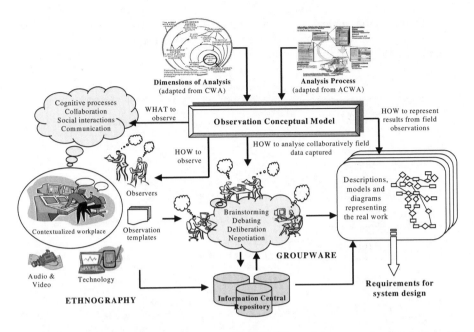

Fig. 1. Components of Observation Conceptual Model. The model supports the observation and analytical activities by describing repositories to store the results of these activities. It also supports the data produced by the interaction process.

The Process of Analysis involves creating successive representations until system requirements, and for each of its steps the collaborative processes happens repeatedly. Observation is not limited to initial analysis, but may occur in all phases of systems development whenever an ongoing confirmation or a deeper analysis is needed.

Finally, after ending the collaborative sessions, "how to represent" refers to viewing the analysis results at different viewpoints through computational interfaces. Based on [24], each viewpoint addresses a particular aspect of the work setting, so as to allow navigation through the most relevant information, according to the user preferences, and also traceability to requirements. Both activities of analyzing and representing results can be supported by a groupware.

4.1 The Collaborative Observation Method

The second part of our proposal is the Collaborative Observation Method that provides a methodological dynamic for observation work that enables delivering system requirements in a more systematic and structured way.

First, we establish some main premises for using this method. We are considering field studies at organizations in which there is clearly a demand for existing systems improvements. Moreover, we are assuming that ethical or methodological concerns are already addressed: after initial contacts and objectives clarification with users, they allow and consent the observations about their working activities.

The Observation Method is composed into iterative and cyclical phases, since field studies preparation to system requirements representation. Figure 2 outlines each of its phases with their principal activities.

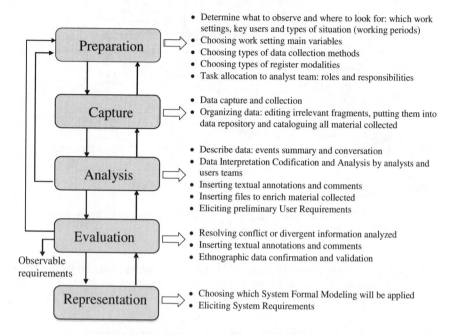

Fig. 2. Overview of the Observation Method Phases

The Preparation, Capture and Analysis phases are presented as if they occur in sequence, but actually they will be repeated iteratively until team feels comfortable to confirm and validate the captured, interpreted and analyzed ethnographic data with users once observed. Also, these phases repeat continuously for refinement of initial understanding or even to overcome limitations of previous observations. Each of its phases is detailed as follow.

Preparation

Rarely does a requirements analyst enter a workplace with an open brief on what to observe in the work setting [10]. Thus, this phase is concerned with doing the first informal social contacts, that aims to communicate to users what the main purposes of observations are. Furthermore, this phase is also related to initial groundwork that provides for a basic familiarization with the organization domain.

If possible, informal, open and exploratory observations should be carried out to uncover what seem to be the most critical activities in a given work setting. Afterwards, based on the initial facts confronted, possible discoveries and according to actual users' demands, the scope of next systematic observations should be established. Some questions, as exemplified below, needed to be answered:

- Which set of work settings, types of situations or the scope of activities will be the focus of next systematic observations;
- What key users will be observed;
- What working periods will be most interesting to observe;
- What types of data collection methods and strategies will be applied, according to work settings characteristics;
- What types of register modalities will be used – audio and video equipment, microphones, field notes, questionnaires;
- What roles and responsibilities will be allocated to ethnographic team – moderator, observers and analysts.

Therefore, a set of preliminary and tentative hypothesis is formed after exploratory observations in a specific work context, in order to be confirmed on the next systematic observations to be performed. Briefly, as ethnography is atheorethic, the analyses of previous observations guide the next set of observations, confirming, adjusting or changing hypothesis, also refining all supplementary information collected and arising new ones.

Table 1. Examples of different video orientations in an ethnographic study

Video Orientation	Descriptions of possible perspectives
The overall workplace	Description of the physical layout and workplace conditions. Description of user's overall communications, postures, gestures and physical moves.
Team cooperation	Description of activities performed simultaneously by teams. Description of social and collaborative aspects. Description of informal conversations, speech and information exchange, stories, sayings, jargon.
The user's own workplace or the scope of human machine interaction	More detailed description of user's communications, gaze direction and gestures. Activities performed by users and their actions on working artifacts – documents, manuals.
System interfaces	Visualization of systems functions that are actually used. Visualization of possible usability problems.

Ethnographic studies are usually undertaken using audio and video equipments. Video is considered as a richer and detailed source of data. It has several advantages, since it is possible to: allow events review; register of hard discrimination working variables such as system interfaces; register of many variables simultaneously that it is difficult for just an observer to register. In setting-oriented record, one or more video camera is positioned to cover as much as possible of the physical workspace. In person-oriented record, it has its focus on the work activities of a particular person, so as to understand his/her work. In object-oriented record, it traces a particular artefact or technology. Finally, a task-oriented record has focus on one task over the particular period of time. This can involve several persons at different physical locations. As an example, Table 1 shows possible workplace perspectives to be captured and its information to be captured, from a broader context to a more specific one.

Capture

After establishing the scope of observation studies in the previous phase, the capture phase is characterized by going to the field to collect data about workplace environment, contextual information and user's activities, interactions and roles [19]. Besides video recordings, observers should also make textual annotations about activities being observed and informal conversations, as well as their impressions, feelings, and questions which emerge during field observations. Furthermore, they should use observation templates with possible probes of what to look for [3].

Observations can be obtrusive, so a good recommendation is to have a user as a key informant while others are being observed without interruptions. His role is to accompany the observers and answers their questions and doubts during or after observations. It is sometimes referred to contextual inquires, where observers ask the key informant to detail or to explain activities while they are being executed.

Any material collected, such as audio and video recordings, photographs, field annotations, sketches, drawings, and artefacts given by users (documents, manuals, and procedures) should be organized, catalogued and stored into the groupware tool. This material should be submitted for subsequently collaborative analysis and interpretation by team in next phase [3]. Figure 3 outlines the steps in this phase.

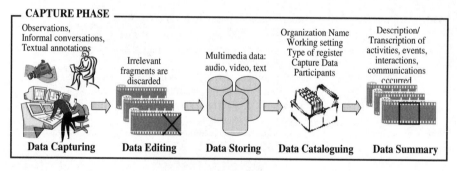

Fig. 3. A more detailed view of Capture Phase. Note that many documents and data are requested as a result of the observation process.

Analysis

At every observation, data should be accessed, summarized, interpreted and confirmed as appropriate [16], before conducting the next set of observations. Collection of field notes, audio and video records from observations are turned into a set of coherent, meaningful findings about activities with its central issues. Therefore, after organizing and cataloguing the observation data into the repository, this phase is mainly focused on data interpretation and analysis.

Figure 4 outlines the steps in this phase: the analyst team accesses and visualizes the observation data so as to proceed on further data codification and interpretation. In fact, at the outset of this analysis phase, the team moderator must settle on an explicit analysis approach. Patterns of activities, users' behaviour, cognitive tasks and collaboration are identified and described. From these, descriptions about problems,

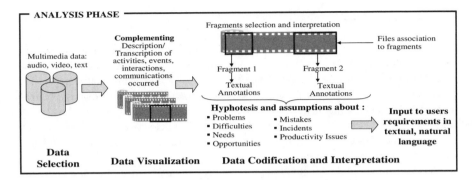

Fig. 4. A more detailed view of Analysis Phase. This phase analyzes the material generated from the observations and the data supplied by the users. It does not negotiate the requirements, yet.

difficulties, needs, opportunities for systems improvements, mistakes, incidents accounts and the like, are identified and classified.

Collaborative virtual sessions of data analysis should be made on the groupware tool that supports the observation method. During analysis and multiple data passes, emergent ideas are identified, reflected and revised by team. This continues until all agree on the interpretation, or they have found questions that ought to be observed in the workplace field again or to be asked to reveal more about the activity. Finally, after data interpretation, team synthesize this into a model of a domain, as activity diagram with its issues and constraints correlated. In this model, user requirements are proposed in natural language for later confirmation in the next phase.

Evaluation

Our approach to evaluation consists of setting up interviews meetings with the users – especially those who were observed – and presenting the final set of data observation, so as to obtain their confirmation and validation about data interpretation, as well its accuracy and completeness. It is important to present fragments of video with its descriptions and issues, since it can facilitate communication and interaction between analysts and users, fostering participation in the confirmation and validation of all information presented.

Representation

Finally, based on observation data and preliminary requirements confirmed by the users in the previous phase, more formal system requirements are developed. As mentioned elsewhere, our proposal does not imply in letting formal system design methods aside. Rather, the initial phases of requirements elicitation should include observation studies before the use of the most traditional requirements elicitation techniques. However, the study about the systems requirements representation is not part of our proposal.

5 The Case Study

We state at the outset that traditional requirements elicitation definitely does not provide complete information about current work practices along with its actual difficulties and adversities, so it could not afford complete and consistent system requirements. The main reason for this requirements inadequacy is the gap between information obtained from stakeholders during interviews, and the rich, dynamic and complex reality of workplaces. Furthermore, we also assert that the use of an observation methodology can contribute to enhance the organizational domain and work analysis, resulting in better system requirements elicitation.

Thus, our initial goal is to elaborate a set of preliminary field studies in real organizations to evaluate the efficacy of interviews and observation approaches. Our hypothesis is that observations about current users' work activities in their contextual workplace provides richer, detailed and consistency information compared to those obtained through users' interviews.

Two sets of experiments have been carried out in a real organization that sells health care plans. These experiments were conducted by two different groups, as outlined in Table 2. Each group used templates that are aimed to assist them in what kind of information should be obtained from interviews (possible questions to be made) and observations (possible contextual question probes and thinks to look for).

Table 2. Types of experiment

Type	Experiment Description	Group Size	Group Composition
Observation approach	Observations about current users' work aimed to uncover users' activities, difficulties and system usability problems.	3 members	Two members act as observers and one interprets the data collected.
Interview approach	Meeting with users so as to get descriptions of their activities, actual problems and users needs.	3 members	Two members act as interviewers and one checks interview notes.

Some initial premises have been established: both experimental groups cannot access the collected material from each other so as not to be influenced by newer information not yet captured; interviews and observations should be performed with users that actually make use of systems; and observations must occur before interviews. The reason for this decision was because after the interviews, users can reflect about their work and may change their behaviour during the observations. In Table 3, we present the schedule of the experiment planning. As mentioned before, they spent the same time in both observing and interviewing.

Results from each approach have been evaluated in quantitative (Table 4) and qualitative way (Table 5). For example, we checked which group presented the greater number of users' expectations and needs, as well as problems, difficulties and opportunities for systems improvements. Furthermore, we also verified which group

Table 3. Distribution of events along the experimental period

Experiment Planning: Methodology and Visits			4 Days	
Observations and Interviews				
Event	**Begin**	**End**	**Begin**	**End**
1st Visit	Day 1	Day 1	Day 1	Day 1
Preliminary Analysis	Day 2	Day 4	No material available	
2nd Visit	Day 5	Day 5	Day 8	Day 8
Preliminary Report	Day 6	Day 12	Day 9	Day 24
Verification	Day 13	Day 22	Day 25	Day 27
Report Revision	Day 22	Day 22	Day 27	Day 27
Final Report	Day 23	Day 26	Day 28	Day 52
	Group G1 - Interviews		**Group G2 - Observations**	

presented higher level of detail, completeness and accuracy through quality of their elaboration of glossary of terms, activities diagram or models. The full results of the experiments are reported in [12].

From Table 4, we have some indication that the observation method generated much more information than the interviewing method. The difference was quite significant for all measures. If the observation method identified more problems and difficulties, then it is expected it will generated more system requirements.

Table 4. Summary of the quantitative evaluation

Information	G1 - Interviews	G2 - Observations
Activity Flows	**2** flows	**11** flows
Total of Activities	19 activities of *underwriting* e 2 activities of another area	148 activities
Total of Problems or Difficulties reported	**11**	**24**
Total of Requirements	6	18

In Table 5 we present some qualitative results, comparing the reports generated by each group. As the observation method included a video, then it is expected that a number of additional problems and situations emerged. The interviews were recorded, but, of course, the video is much richer for the analysis purpose. An interesting result was the association between the identified problem depicted in the video and the requirement proposed. The interviewing group did not do this, perhaps because the audio association was not easy.

These results are not conclusive because it is very dependent on the domain where the experiment has been carried out, the participants of each team, the lack of appropriate tools (a groupware tool is under development) and several other factors. We can read these results as an indication that the observation approach can contribute to augment our knowledge about the environment and improve the quality of the elicitation task. More experiments are needed to confirm these results.

Table 5. Summary of the qualitative evaluation of the reports

	Evaluation Criteria	G1 –Interviews	G2 – Observations
General	**Completeness**		
	Description of the domain*	N/A	N/A
	Description of the environment	No	Yes
	Description of the working posts	No	Yes
	Glossary of Terms	Yes – 2 terms	Yes – 10 terms
	Description of the existing systems	Yes	No
	Description of the system functionalities and user interfaces	No	Yes
	Questionnaires	Yes	No
Activities	**Details**		
	Indication about who is responsible for activities	Yes	Yes
	Indication about their duration	No	Implicit in the video records
	Indication about the artifacts (documents, lists, forms, etc.)	Yes, but implicit	Yes, very explicit
	Information about Input / Output	No	No
	Completeness		
	G2 presented a report with higher level of detail, specifying the users' activities and the system response, besides those not supported by the system.		
	Quality of the reports produced by the groups		
	G1 presented data flow diagrams without specifying their performers. It was necessary to refer to the descriptions in order to identify the performers. On the other hand, G2 presented too many data flows. No consolidation was carried out in these data flows.		
Problems and Diffi-culties	**Detail and quality of the supporting documents and material**		
	Indication of activities that could be supported by the new system	Yes, but very few	Yes, twice of G1
	Indication of failures and inefficient actions	Yes, but very few	Yes, higher than G1
	Information about the causes	Little information	Yes
	Information about the consequences	No	Yes
	Indication about the context	No	Yes
	Presented supported material	No	Yes, video records
Requirements	**Detail**		
	Justification for them	No, requirements separated from the problems	Yes, requirements associated to each problem
	Discussion about the benefits	No	Yes
	Point to other users who could benefit of this functionality?	No	No
	Related with the organization goals	No	No

Obs.* each group received a previous document with the activity domain description

6 Conclusions

This paper has presented a review of the most traditional and the emerging approaches to system requirements elicitation. Then a proposal for integrating some approaches is discussed through the Observation Conceptual Model and Collaborative Observation method, including CSCW elements to improve teamwork productivity. Some technical aspects of the groupware tool are also outlined.

It is evident that work analysis through ethnographic field studies is coming into prominence. The more we learn and understand about the work domain, the better the requirements elicitation activity should be. Therefore, the Observation Conceptual Model and the Collaborative Observation Method presented here are complementary approaches extending the traditional requirements elicitation techniques. We believe it is able to increase and improve shared understanding about the working activities and context, so as to guide the development of more complete system requirements that better support stakeholders' organizational goals, minimizing system projects failure due to inadequate requirements. Moreover, it can help identify factors that might increase resistance to adopting new systems, if requirements engineers take into consideration the development of systems that expand users' human potential and knowledge.

The teamwork aspects are relevant both at the observer side and at the workers´ side. In the first case, the work under observation is so complex that a group of observers is necessary to capture the different perspectives of this work. This situation will require a groupware tool that helps observers to combine their observations into an integrated report, in our case a set of integrated requirements. In the second case, the work under observation is carried out by a team. The observation should focus both on the individual work and the interaction between team members. In this case the requirements of the new system should cover both individual tasks and communication and interaction tasks. A study about both observation approaches is under development.

As a result of the initial observations generated by the experiment, a groupware tool has been specified and developed. The tool was not used in the experiment reported, but a second ongoing experiment is using this tool. The tool is not part of the scope of this paper and it is not described here. Details about the *EyeonAction* groupware can be found in [12].

Acknowledgments

Renata Guanaes was partially sponsored by a grant from CENPES. Marcos R.S. Borges and Jose Orlando Gomes were partially supported by grants No. 305900/2005-6, 479374/2007-4 and 484981/2006-4 from Conselho Nacional de Desenvolvimento Científico e Tecnológico CNPq (Brazil).

References

1. Anderson, B.: Work, ethnography and systems design. In: Kent, A., Williams, J.G. (eds.) The EncyclopeDay of Microcomputers, vol. 20, pp. 159–183. Marcel Dekker, New York (1997)
2. Araujo, R.M., Santoro, F.M., Borges, M.R.S.: A conceptual framework for designing and conducting groupware evaluations. International Journal of Computer Applications in Technology 19(3-4), 139–150 (2004)

3. Crandall, B., Klein, G., Hoffman, R.R.: Working Minds: A practitioner's guide to Cognitive Task Analysis. MIT Press, Cambridge (2006)
4. Davis, A., Dieste, O., Hickey, A., Juristo, N., Moreno, A.: Effectiveness of requirements elicitation techniques: Empirical results derived from a systematic review. In: Proc. of the 14th IEEE International Requirements Engineering Conference (RE 2006), pp. 176–185. IEEE Press, New York (2006)
5. Dourish, P.: Implications for design. In: Proceedings of CHI, Design: Creative and Historical Perspectives, Montreal, Canada, pp. 541–550. ACM Press, New York (2006)
6. Guerlain, S., Shin, T., Guo, H., Calland, J.F.: A team performance data collection and analysis system. In: Proc. of the Human Factors and Ergonomics Society 46th Annual Meeting, Baltimore, MD (2002)
7. Herskovic, V., Pino, J.A., Ochoa, S.F., Antunes, P.: Evaluation Methods for Groupware Systems. In: Haake, J.M., Ochoa, S.F., Cechich, A. (eds.) CRIWG 2007. LNCS, vol. 4715, pp. 328–336. Springer, Heidelberg (2007)
8. Hollnagel, E.: Handbook for Cognitive Task Design. Lawrence Erlbaum Associates, London (2003)
9. Iqbal, R., James, A., Gatward, R.: Design with ethnography: an integrative approach to CSCW design. Advanced Engineering Informatics 19(2), 81–92 (2005)
10. Jirotka, M., Luff, P.: Supporting Requirements with Video-based analysis. IEEE Software 23(3), 42–44 (2006)
11. Kotonya, G., Somerville, I.: Requirements Engineering: Processes and Techniques. John Wiley & Sons Ltd., New York (1998)
12. Machado, R.G.: An Ethnography and Collaborative Method for Requirements Elicitation. Master Science Dissertation. Graduate Program in Informatics, Federal University of Rio de Janeiro (2008) (in Portuguese)
13. Mackay, W.: Is paper safe? The role of flight strips in air traffic control. ACM Transactions on Computer Human Interaction 6, 311–340 (1999)
14. Macleod, I.S.: Real-world effectiveness of Ergonomic methods. Applied Ergonomics 34(5), 465–477 (2003)
15. Millen, D.R.: Rapid ethnography: time deepening strategies for HCI field research. In: Proc. of the Conference on Designing Interactive Systems: Processes, Practices, Methods, and Techniques, pp. 280–286. ACM Press, New York (2000)
16. Myers, M.D.: Investigating information systems with ethnographic research. Communications of the Association for Information Systems 2, Article 23 (1999)
17. Nardi, B.: The use of ethnographic methods in design and evaluation. In: Helander, M.G., Landauer, T.K., Prabhu, P. (eds.) Handbook of Human-Computer Interaction, pp. 361–366. Elsevier Science, Holland (1997)
18. Nuseibeh, B., Easterbrook, S.: Requirements engineering: a roadmap. In: Proc. of International Conference on Software Engineering, pp. 4–11. ACM Press, New York (2000)
19. Pinheiro, M.K., Lima, J.V., Borges, M.R.S.: A framework for awareness support in groupware systems. Computers in Industry 52(1), 47–57 (2003)
20. Rubinstein, D.: Standish Group Report: There's Less Development Chaos Today. Software Development Times,
http://www.sdtimes.com/article/story-20070301-01.html
21. Sommerville, I.: Software Engineering, 7th edn. Addison-Wesley, Harlow (2004)
22. Sommerville, I., Rodden, T., Sawyer, P., Bentley, R., Twidale, M.: Integrating ethnography into the requirements engineering process. In: Proc. of IEEE International Symposium on Requirements Engineering, pp. 165–173. IEEE Press, New York (1993)

23. Standish Group International: The CHAOS Report,
    http://www.standishgroup.com
24. Twidale, M., Rodden, T., Sommerville, I.: The Designers Notepad: supporting and under-standing cooperative design. In: Proc. 3rd European Conference on Computer-Supported Cooperative Work (ECSCW), pp. 93–108. Kluwer Academic Publishers, Norwell (1993)
25. Valenti, S., Panti, M., Cucchiarelli, A.: Overcoming communication obstacles in user-analyst interaction for functional requirements elicitation. ACM SIGSOFT Software Engi-neering Notes 23(1), 50–55 (1998)
26. Viller, S., Sommerville, I.: Social analysis in the requirements engineering process: from ethnography to method. In: Proc. IEEE International Symposium on Requirements Engi-neering, pp. 6–13. IEEE Press, New York (1999)
27. Viller, S., Sommerville, I.: Ethnographically informed analysis for software engineers. In-ternational Journal of Human Computer Studies 53(1), 169–196 (2000)

Author Index